# SMART HOME HACKS™

*Gordon Meyer*

O'REILLY®

Beijing • Cambridge • Farnham • Köln • Paris • Sebastopol • Taipei • Tokyo

**Smart Home Hacks™**
by Gordon Meyer


Printed in the United States of America.

Published by O'Reilly Media, Inc., 1005 Gravenstein Highway North, Sebastopol, CA 95472.

O'Reilly books may be purchased for educational, business, or sales promotional use. Online editions are also available for most titles (*safari.oreilly.com*). For more information, contact our corporate/institutional sales department: (800) 998-9938 or *corporate@oreilly.com*.

**Editors:** Brian Sawyer, Rael Dornfest
**Series Editor:** Rael Dornfest
**Executive Editor:** Dale Dougherty
**Production Editor:** Marlowe Shaeffer
**Cover Designer:** Hanna Dyer
**Interior Designer:** Melanie Wang

**Printing History:**

October 2004: First Edition.

This book uses RepKover™, a durable and flexible lay-flat binding.

ISBN: 0-596-00722-1
[C]

# Contents

# Credits

## About the Author

Gordon Meyer (*http://www.gordonmeyer.com*) is a writer and software engineer who lives in a fully automated home located in the heart of Silicon Valley. His house, nicknamed *CoasterHaus* for its proximity to an amusement park, is better wired than most businesses and is smarter than your average house cat. It makes sure he gets up on time in the morning, watches over his dog while the family is away, and contacts him about missed visitors and telephone calls during the day. It even tucks the family in with a weather report and story each night.

When not living like George Jetson, Gordon is a designer who specializes in technology for writing and delivering onscreen instructional material. For more than a dozen years, his uncredited manuals and help systems have been used by millions of people worldwide. This book is his foray beyond the anonymous world of corporate technical publications.

Gordon holds degrees in sociology and broadcasting. He is the author of a seminal study of computer culture, *The Social Organization of the Computer Underground*. Prior to devoting himself to the study of technology and deviance, Gordon worked as a professional magician, an avocation he continues to this day, which perhaps explains his interest in making the difficult seem easy, throughout all of his careers.

## Contributors

The following people contributed their hacks, writing, and inspiration to this book:

- Ido Bar-Tana is a microprocessor design engineer by day and a home-automation hacker by night. His web site (*http://www.geocities.com/idobartana*) is the Google-ranked #1 site for X10 home automation. He can be reached via email at *ido_bartana@yahoo.com* and welcomes any home-automation-related questions and comments.

- Matt Bendiksen's mission is to bring intuitive home automation to geeks everywhere. Over the last 20 years, Matt has been using computers to monitor and control home environments. In 2002, Matt founded Perceptive Automation and has received rave reviews on Indigo, its Mac OS X home control server. Visit *http://www.perceptiveautomation.com* to download a free trial version.
- Edward Cheung was born and raised in Aruba, where he started his hobby of electronics as a young boy. After high school, he continued his education in the United States. After completing his Ph.D. in electrical engineering from Yale University, he joined NASA at the Goddard Space Flight Center, where he currently works on the Hubble Space Telescope Project. Edward enjoys developing hardware for space flight and his hobby of home automation.
- Dean Davis operates AfterTen Software (*http://www.afterten.com*) and is the author of WeatherManX, CIDTrackerX, and other Macintosh programs.
- Arthur Dustman has been doing home automation for about three years. He has expanded from the X10 standard package to a 5×5 room to hold all his equipment. He uses HAL2000, Adicon Ocelot with C-max, nine Secu-16s, one Secu16-IR, 94 Relays, one scm-810 mixer, nine pzm-10 microphones, nine Xantech keypads, nine custom wall panels, and much more. He has created many custom circuits and devices to automate devices that have no "factory" interfaces. His motto: "There's always a way to automate anything."
- Michael Ferguson (*http://www.shed.com*) is the principal author and designer of XTension for Mac OS. Michael began experimenting with electronics in 1955, began writing code and designing computer systems in 1964, and has done nothing more useful since then.
- Bill Fernandez is an avid electronics hobbyist, home automation aficionado, do-it-yourselfer, home handyman, fine-art photographer, martial artist, husband, father, and all-around nice guy. Bill is known for having introduced Steve Jobs and Steve Wozniak to each other, and for having been Apple Computer's first employee. Professionally, Bill is a consulting user interface architect.
- Dan Fink is the technical director of Otherpower.com (*http://www.otherpower.com*), which is a web site dedicated to alternative-energy enthusiasts and their experiments. Dan has lived in the remote, mountainous area west of Fort Collins, Colorado, for 13 years, with no grid

electricity or phones, and generates his power with solar, wind, and hamster. He administers the company's servers remotely from home over VSAT satellite Internet, which leaves extra time for experimenting with homebrewed wind turbines, weird science projects, and silly ideas such as hamster-powered night lights. Dan's publication credits include articles in *Home Power*, *Back Home*, and *Zymurgy* magazines.

- Richard Gensley is a founder of HomeTech Solutions (*http://www.hometech.com*), a home automation products company in Cupertino, California. He currently teaches seminars on lighting control and automation including the use of X10 and related technologies. He authored a number of "Mr. Module" feature articles in *Electronic House Magazine*, which addressed subtle and clever applications of the X10 technology. His automation experience includes work as an automated flight controls engineer and serving in key management positions with computer and industrial automation firms.
- Mark Kelly
- David A. Kindred (*david@retroscape.com*) is a richly blessed home automation enthusiast who resides in Kemptown, Maryland, with his sons Andrew, Kit, and Josh, and his loving and understanding wife, Pam.
- Robert Ladle is a retired electrical engineer living in the Seattle area. He is a self-taught computer nut who has been using Mac computers since Apple's Mac SE model first came out in the late 1980s. Bob has been interested in automation of all types, beginning with radio control models in the early days to his current home automation activities. He first started using XTension in 1991 and continues to develop and experiment with new uses and applications to enhance the field.
- Guy Lavoie currently works for a telephone company doing software support. He has been involved in home automation since 1998 and enjoys working with Applied Digital's Ocelot and Leopard controllers. He holds a degree in Industrial Electrotechnology and enjoys programming microcontrollers in assembler.
- Reynold Leong (*http://www.kjsdoghouse.com/autohouse*) is currently in year 10 of his never-ending home automation project. He spends a lot of time automating stuff so that he can have more time to be lazy. "Laziness is our inspiration." Donations are encouraged and welcomed.
- Rob Lewis remembers when X10 equipment was sold under the BSR brand, and still can't understand why most people have never heard of it. He's currently vice president of business development at Shared Media Licensing, Inc. (*http://www.weedshare.com*), a new company that pays music fans to share files legally on the Internet.

- Warren G. Lohoff founded WGL & Associates (*http://www.wgldesigns.com*) in 1992. He designs, manufactures, and markets leading-edge home automation and irrigation products.
- Don Marquardt (*kyham@k9soa.net*) has been playing around with home automation for almost 30 years. Each move to a new home just brought more toys to play with. His current home has just about everything he could think of. In total, it has almost eight miles of various types of wire to handle just about anything. Now, even his car talks to the Internet. Don's web site (*http://www.k9soa.net*) is open to visitors to take a little tour of Jeannie, The House That Listens. When you visit, be sure to sign his guestbook.
- Roger C. Meyer was an electronics technician, auto mechanic, carpenter, and gardener. His ability to design and build anything, including a fully custom basement with built-in intercoms, a workshop, ingenious storage areas, and a model train table that descended from the ceiling, continues to inspire his son: the author of this book. Roger was truly a smart home hacker before his time.
- Frank Perricone's first gadgets were imaginary, cardboard mockups of TRS-80s. At age 14 he delivered newspapers to buy a Sinclair ZX-81. Since then, the toys have improved, but not much else has changed. He's currently building his dream house with his wife and cat in rural Vermont, wishing for DSL.
- Jerry Prsha owns a Macintosh consulting business in St. Louis, Missouri, and is a member of the Apple Consultants Network. His home is automated with the application XTension. His hobbies are entertaining the kids (and parents) with elaborate Halloween displays, creating garden ponds, and of course, scaring the wits out of the unsuspecting repairman.
- Steve Simon is a software engineer who has been using home automation for personal convenience since 1997. He maintains a modest personal web site that can be found at *http://homepage.mac.com/snsimon*.
- Smarthome, Inc. (*http://www.smarthome.com*) is the world's largest retailer of home automation products. Today, the company sells more than 5,000 products, including X10 remote control, lighting, wireless security, and home entertainment systems. Visit Smarthome's web site or call (800) 762-7846.
- Doug Smith (*http://www.smithsrus.com*) works in information technology by day and enjoys tinkering with home automation, computers, electronics, robotics, and music in his spare time. He occasionally has worked in the home automation industry and has contributed many articles to home-automation-related magazines and web sites.

- Greg Smith (*http://homepage.mac.com/gregjsmith*) is a longtime Mac geek living in New Mexico. Although he has to be a Windows "expert" for his day job, he would prefer to spend the day working on his garden, taking photos, working on his home automation system, or just doing about anything else. His family would appreciate it if they didn't have to be beta testers for his smart home.
- Richard (Rick) Tinker has been actively involved in home automation since 1992, working professionally in the field since 1998. His interests include his family, home automation, home improvements, movies, and music. Currently, he is the CTO of HomeSeer Technologies, LLC of Bedford, New Hampshire (*http://www.homeseer.com*). His favorite home automation project is his PC-based MP3 jukebox. The automation items that have the highest WAF (wife approval factor) in his home are the laundry washer and dryer reminders.
- Henk van Eeden immigrated to the USA in 1962 from the Netherlands. He earned a bachelor's degree in electrical engineering at Virginia Tech. His career started at a large northeastern utility. While there, he worked on a master's degree in electrical engineering at Bucknell University. His last position before retirement was superintendent of a hydro and coal-fired electrical generating plant. In 1980 his interest in computers and home automation started to blossom. He purchased his first Macintosh in 1986 and has owned at least one ever since. In addition to other duties, the Mac has been running and monitoring his house for the past 14 years.
- Jon Welfringer is currently an IT director and has been working in the technology field for more than 15 years. His love for technology follows him home from work, as evidenced by his never-ending desire to integrate new technology into his family's house. The number-one goal of his automation endeavors is to always make sure his family and guests feel at home with technology. As he says, "Automation in the home should be smart and intuitive without the need to perform training sessions for your family."

## Acknowledgments

I would like to thank all those who contributed their ideas, techniques, tricks, hacks, observations, and code to this book. Your willingness to share your experiences and advice with others is very much appreciated.

Thanks to the following communities for their inspiration, camaraderie, and like-minded spirit: XTension Discussion List (*http://www.shed.com/maillist.html*), Indigo Forum (*http://www.perceptiveautomation.com/phpBB2/index.php*), HomeSeer Message Board (*http://ubb.HomeDeer.com/*), HomeToys (*http://www.hometoys.com*), and CocoonTech (*http://www.cocoontech.com*).

Thanks to Rael Dornfest, who saw the seeds of this book in my presentation at the O'Reilly Mac OS X Conference. I am forever grateful for this opportunity and for your encouragement, patience, and persistence that got this project off the ground.

Jackie Streeter, vice president of hardware engineering at Apple Computer, not only granted me the freedom to tackle this *extra-curricular* project, but also she has supported my professional growth in many new directions. Thank you, Jackie.

Brian Sawyer, who came on board at just the right time to help bring this book to fruition. Thanks for providing the perfect mix of editor, taskmaster, and friend. I'm looking forward to the day we meet in person.

Matt Bendiksen, Michael Ferguson, Richard Gensley, and Rick Tinker covered my back, technically speaking, and each provided immeasurable ideas, criticism, and clarifications throughout the book. The combined home automation knowledge of these four technical editors is truly encyclopedic. Any remaining deficiencies in this work are truly of my own doing. Thank you, gentlemen.

Even as I write these final words, the staff at O'Reilly—production, graphic design, marketing, and others—are busily working to take this book across the finish line and beyond. Without your expertise and contributions, all of this would be for naught. Thank you.

Finally, for my wife, Gale, who truly makes our house a home. I love that you not only tolerate my endless home automation experiments and whacky projects, but also that you'll (occasionally) admit you enjoy them, too. You're my best friend and biggest supporter, and I can't wait to spend the rest of my tomorrows with you.

# Preface

For generations, scientists and marketers have been promising flying cars, personal robots, and the automated home. Of these, only the automated home is within reach today. Thanks to the combination of inexpensive yet powerful computers, open source and scriptable software, and a 20-year-old method of controlling everyday lights and appliances, you can live in a home that would make any futurist proud.

One of the best things about home automation is that it's easy to get started and doesn't require any commitment or major changes to your home or lifestyle. This book shows you how others have applied simple techniques to enhance their families' enjoyment of their homes. You can adopt just the ideas that match your vision of what a smart home should be, or you can become inspired by what's possible and undertake great changes to create your very own home of the future.

## Why Smart Home Hacks?

The term *hacking* has a bad reputation in the press. Journalists use it to refer to someone who breaks into systems or wreaks havoc with computers as his weapon. Among people who write code, though, the term *hack* refers to a "quick and dirty" solution to a problem, or a clever way to get something done. And the term *hacker* is taken very much as a compliment, referring to someone as being *creative* and having the technical chops to get things done. O'Reilly's Hacks series is an attempt to reclaim the word, document the good ways people are hacking, and pass the hacker ethic of creative participation on to the uninitiated. Seeing how others approach systems and problems is often the quickest way to learn about a new technology.

Using the term *hacking* to refer to your home might seem odd at first, but there's a long-standing tradition of shaping a personal habitat to meet individual needs. As the saying goes, "There's no place like home," and for many, there's no better place to apply your creative energy and technical prowess, especially when used for the betterment of your family's life. Most technical innovations, and energy, are devoted to cleverly solving problems that apply to the outside world. Instead, join the community of home automation enthusiasts who do their best and most rewarding work, quite literally, at home.

## What Is Home Automation?

The essence of home automation is using specialized equipment that can control your lamps, appliances, heater, and air conditioning, and perhaps sense where in the house people are located. Every automation system is built upon these building blocks. Once these methods are in place, it's really not much of a stretch (as you'll see throughout this book) to have some actions performed for you automatically. For example, once you have the ability to detect when a person has left a room, turning off any lights that she might have left on isn't much of a technical challenge.

Indeed, it's so small a challenge that the method of how it's done is almost inconsequential. The hard part is making sure it's done in a way that is convenient, nonintrusive, and truly a service to your family. In other words, turning off the light is easy. Making sure it's *appropriate* to turn off the light is what puts the *smart* in *smart home.*

That's not to say that you won't need some equipment. You will, and this book focuses almost exclusively on using X10-based home automation technology. Technically speaking, X10 is a *power-line carrier* (PLC) communications method. This means it sends signals using the electrical wiring in your home, not unlike Morse code. Your appliances, lights, and everything else you have plugged into your home's electrical system continue to work; X10 sneakily adds its signals onto the wires in a way that standard AC equipment doesn't notice. However, when you have equipment that is tuned to listen for X10 commands, that equipment can be controlled as if by magic, reacting to signals sent by you, your computer, or even other X10 devices.

Now, some old-timers will tell you that X10 is too unreliable or "old" to be useful. Many of them tried X10 when it first arrived on the market, circa 1978, and X10's quirks and the clunky software of that era turned them off. Like most new technologies, when X10 was first introduced, it didn't quite live up to its introductory hype.

But time and technology march on, and X10 has gotten a lot better over the years—thanks to the ingenuity of smart hackers, many of whom have contributed to this book, and to companies that continue to push the envelope by adding new capabilities, and increased reliability, to the many products that work with X10.

And because of the ubiquity of X10, it's today's best choice for automating your home. It is much less expensive than other technologies, it is more widely supported and available, and you can easily grow and expand your system as you become enamored with your new hobby. Later, if you really get into home automation, you might want to try out some of the technologies that provide additional capabilities (see the "Beyond X10" sidebar for examples). But in my experience, X10 will serve you well for years, and even people who have moved on to other methods of automation still maintain several X10 devices in their homes.

This underlines a key point to remember about this book: although the hacks are described in X10 terms, you easily can apply them to any home automation system. The concept of sending commands to specific devices, telling them to turn on or off, is the same, regardless of the underlying protocol that transmits the commands. Aside from some troubleshooting information that's specific to X10's implementation, these hacks aren't inherently reliant on X10.

## Crossing Platforms

Just as the fundamental principals of X10-based automation can be used with other protocols, the same philosophy applies to home automation software. Most of the hacks in this book are implemented using one of three different home automation programs: HomeSeer **[Hack #19]** for Windows, Indigo **[Hack #18]** for Macintosh, or XTension **[Hack #17]** for Macintosh. Other software packages are available (see "Add a Brain to Your Smart Home" **[Hack #16]** for a list of some of the others), but just as X10 concepts can be applied to other technologies, so, too, can programming techniques.

If you come across a hack you want to use, and it's described using software you don't have, in most cases, you can apply its technique to another software package. For example, a hack that uses Indigo might use *trigger actions* to react to signals from motion detectors. In HomeSeer, you'd use *events* instead. Techniques that rely on platform-specific capabilities, such as Macintosh's AppleScript, won't be translated easily, but, in many cases, taking an idea from one world to another will be possible.

If you need help translating from one system to another, refer to the appropriate *Get to Know* hack, such as "Get to Know XTension" [Hack #17], for an overview of how that application approaches automation. This probably will give you enough information to map the technique to software with which you're more familiar.

### Beyond X10

The home automation landscape is littered with failed attempts to supplant X10's dominance as the protocol of choice. At the time of this writing, three contenders might have a chance at finally dethroning the king:

*Z-Wave (http://www.zen-sys.com)*
: A wireless radio frequency (RF) protocol for home automation. It avoids the most common problem with X10—interference on the power line—by sending its signals over the air. It's much faster and more secure than X10 because devices will accept commands only from registered controllers. Because it uses a completely different method of communication, a Z-Wave can't communicate with an X10 device. However, thanks to applications such as HomeSeer [Hack #19], you can bridge the two systems using a computer.

*UPB (Universal Powerline Bus; http://www.pcslighting.com)*
: Transmits over the power line as X10 does, but with a much stronger and faster signal. It also supports having more devices, and it will retransmit commands if the recipient does not acknowledge them.

*INSTEON (http://www.insteon.net)*
: A combination RF and power line-based protocol. It's also said to support communicating over USB and Ethernet connections, and it can be intermixed with X10 devices for backward compatibility. The low projected prices might make it quite appealing when it reaches the market in 2005.

## How X10 Works

X10 is a protocol for sending data over standard, household electrical wiring. Not many people think of their home electrical system as a network, and it certainly has an unusual topology, but it surely is a network. Everything on a circuit is wired to other endpoints on the circuit—switches and outlets—and each circuit (most homes have several) is connected at the circuit breaker panel. X10 simply exploits this existing network of wire by using it to send data signals, carefully mixed in with the electrical current it already carries, to all points in your home.

The technical details of how X10 signals are interjected on the power line are really quite impressive. They involve a lot of critical timing and exact frequencies to encode data so that the signals can be slipped into the power line without getting mangled or lost. Also, the data that has to be sent can be somewhat complex. Each X10-capable device is constantly listening to the power line for a signal telling it that an X10 command is about to be sent. Then, the device listens to see if the command is intended for it, based on its delivery address **[Hack #1]**. If it is, the device keeps listening to learn what it should do, such as turn on or off.

Transmitting all this information takes a little bit of time—about 1.5 seconds—so X10 isn't an *instant-on* technology. You do have to get used to the delay between sending a command and seeing its result.

To be more specific, X10 commands are sent in a 1 ms burst of 120 kHz, during the zero crossing point of the 60 Hz AC cycle. It takes 11 cycles of the power line to send a full X10 command, and each code block is sent twice, once in complement. For all the nitty-gritty details, visit *http://www.x10.com/technology1.htm* to read about X10 Technology Transmission Theory.

It's certainly true that X10 is not a perfect technology. But like any communications system, you can improve its reliability by creating an environment in which it has a better opportunity to succeed. Just as microwave ovens and cordless phones can befuddle WiFi connections, electrical noise and poor connections can stymie X10 systems. Several hacks in this book will help you over the rough spots, starting with "Groom Your Home for X10" **[Hack #12]**.

## X-What, You Say?

Unfortunately, the name *X10* causes a bit of confusion. X10 is the name of three things: the protocol used to communicate over the power line, the eponymous name of a company that manufactures home automation equipment, and the brand under which the equipment is sold. Let's sort it all out.

The X10 communications protocol debuted in 1978 in the form of the *X10 Home Automation System* sold and manufactured by X10 Corporation. In addition to selling modules under the X10 brand name, X10 Corporation is an Original Equipment Manufacturer (OEM) for devices sold by others under several different names. See "Shop for Secret X10 Devices" **[Hack #23]** for a listing of some of the names under which you'll find X10-compatible devices sold.

Because of the many meanings of X10, and due to the propensity of the home automation community to be technically inclined, when you see a reference to X10, it's usually about the protocol, not the company. That's how it's used in this book, except where otherwise noted.

Let's clear up another potential point of concern. A company related to X10 Corporation, called X10 Wireless, filed for Chapter 11 bankruptcy protection in October 2003. Its main business is selling Xcam-brand wireless cameras, and it's infamous for its annoying, yet pioneering, use of pop-up and pop-under ads on the Web. If you hear that X10 is "out of business" and that you shouldn't use it to automate your home, rest assured that whoever is telling you this probably is thinking of X10 Wireless and is understandably confused about the differences between X10 technology and its corporate namesakes.

## How This Book Is Organized

You can read this book from cover to cover if you want, but each hack stands on its own, so feel free to browse and jump to the different sections that interest you most. If there's a prerequisite you need to know about, a cross-reference will guide you to the right hack.

Chapter 1, *A Foot in the Front Door*
: The hacks in this chapter form the cornerstones of your home automation system. It's a good place to learn the basic techniques, get started on the right foot, and decide which software you want to use to automate your home.

Chapter 2, *Office*
: If you work at home, or bring work home, hacks in this chapter will help you make more efficient use of your telephone, calendar, and other office equipment.

Chapter 3, *Kitchen and Bath*
: These hacks focus on monitoring important conditions you need to know in the two places where we spend the most time and where we want technology to intrude the least.

Chapter 4, *Bedroom*
: Learn how your computer can gently wake you up in the morning, help you start your day, and make your bedroom a more relaxed and comfortable place.

Chapter 5, *Garage and Yard*
: The workplace of the home is your garage and yard. The hacks in this chapter will help you save money, work smarter, and get more done in these areas.

Chapter 6, *Security*

Home security is one of the biggest reasons people get into home automation, and the hacks in this chapter reflect the diversity and interest in techniques for keeping your family safe.

Chapter 7, *Advanced Techniques*

This grab bag of hacks represents some of the most cherished secrets of experienced home automation enthusiasts. When you're ready to take your home to the next level, here's where to turn.

## Conventions Used in This Book

The following is a list of the typographical conventions used in this book.

*Italic*

Used to indicate device names, URLs, filenames, filename extensions, and directory/folder names. For example, a path in the filesystem will appear as */Developer/Applications*.

`Constant width`

Used to show X10 commands, code examples, the contents of files, and console output, as well as the names of variables, commands, and other code excerpts.

**`Constant width bold`**

Used to highlight portions of code, typically new additions to old code.

*`Constant width italic`*

Used in code examples and tables to show sample text to be replaced with your own values.

Color

The second color is used to indicate a cross-reference within the text.

You should pay special attention to notes set apart from the text with the following icons:

This is a tip, suggestion, or general note. It contains useful supplementary information about the topic at hand.

This is a warning or note of caution, often indicating that your money or your privacy might be at risk.

The thermometer icons, found next to each hack, indicate the relative complexity of the hack:

## Using Code Examples

This book is here to help you get your job done. In general, you may use the code in this book in your programs and documentation. You do not need to contact us for permission unless you're reproducing a significant portion of the code. For example, writing a program that uses several chunks of code from this book does not require permission. Selling or distributing a CD-ROM of examples from O'Reilly books *does* require permission. Answering a question by citing this book and quoting example code does not require permission. Incorporating a significant amount of example code from this book into your product's documentation *does* require permission.

We appreciate, but do not require, attribution. An attribution usually includes the title, author, publisher, and ISBN. For example: "*Smart Home Hacks* by Gordon Meyer. Copyright 2005 O'Reilly Media, Inc., 0-596-00722-1."

If you feel your use of code examples falls outside fair use or the permission given here, feel free to contact us at *permissions@oreilly.com*.

## How to Contact Us

We have tested and verified the information in this book to the best of our ability, but you might find that features have changed (or even that we have made mistakes!). As a reader of this book, you can help us to improve future editions by sending us your feedback. Please let us know about any errors, inaccuracies, bugs, misleading or confusing statements, and typos that you find anywhere in this book.

Please also let us know what we can do to make this book more useful to you. We take your comments seriously and will try to incorporate reasonable suggestions into future editions. You can write to us at:

O'Reilly Media, Inc.
1005 Gravenstein Highway North
Sebastopol, CA 95472
(800) 998-9938 (in the U.S. or Canada)
(707) 829-0515 (international/local)
(707) 829-0104 (fax)

To ask technical questions or to comment on the book, send email to:

*bookquestions@oreilly.com*

The web site for *Smart Home Hacks* lists examples, errata, and plans for future editions. You can find this page at:

*http://www.oreilly.com/catalog/smarthomehks*

For more information about this book and others, see the O'Reilly web site:

*http://www.oreilly.com*

## Got a Hack?

To explore Hacks books online or to contribute a hack for future titles, visit:

*http://hacks.oreilly.com*

CHAPTER ONE

# A Foot in the Front Door

## Hacks 1–24

Getting started with home automation can feel like entering a strange, new world. Familiar things such as light switches and electrical outlets take on new roles and capabilities. You hear about house codes, controllers, and sensors that can tell when someone has entered a dark room. And what in the world is a Powerflash? Before you can create your own smart home, you'll need to educate yourself.

The hacks in this chapter form the foundation upon which you'll build your smart home. Start at the beginning and get the basics of what does what, or dive into the middle and discover something surprising. You'll learn how to turn on lights [Hack #2], take control of your appliances [Hack #3], and find automation equipment that's masquerading at your local hardware store [Hack #23].

But before you get sucked in too far, it's a good idea to prepare both your house [Hack #12] and your housemates [Hack #14] for the adventure upon which you're about to embark. With these hacks in hand, you're sure to get off on the right foot.

### Know the X10 Address

**To send commands to X10 devices, you need to know their addresses. Here are the basics of deciding which addresses to use and how to set them.**

The X10 protocol works by sending commands—using the power line in your home—to modules that know how to listen for and respond to requests to turn on, turn off, or brighten the lamp or appliance which they're controlling. A lot of careful timing and engineering are involved to send and receive the commands, but all you really need to know is that each module understands which commands to react to and which ones to ignore.

Here's how it works. Unlike the Postal Service, which (hopefully) delivers mail directly to your address, X10 crudely broadcasts its commands far and wide, throughout your entire electrical system. It's up to each module to continuously listen for commands and discern whether each one is meant for it. To facilitate this, every X10 command is prefaced by an *address*. If the command's address matches that of the module, the module acts on the command; otherwise, it ignores it and waits to see if the next command is meant for it instead.

An X10 address is made up of two parts: a *house code*, whose value is `A` through `P`; and a *unit code*, whose value is `1` through `16`. Together, these form a complete address. For example, the appliance module that you use to control the lava lamp in the den might have an X10 address of `B5`. Therefore, to turn on the lava lamp, you send the command `B5 On`. This module will ignore commands prefaced with any other address.

## Setting an Address

Each X10 module needs to be configured to listen for its address. You set the house and unit codes separately, usually by turning dials on the front of the module, as shown on the lamp module in Figure 1-1. In this example, the module is set to address `A1`.

You program some modules, such as Smarthome LampLinc (*http://www.smarthome.com/2000sc.html*; $13), by sending a series of commands over the power line **[Hack #13]**. Still others, such as X10 motion detectors, require you to push and hold buttons inside the unit to set their addresses. Regardless of how you do it, all X10 modules require you to configure their addresses; just be sure to keep the instructions **[Hack #22]** that come with the modules in case you decide to change their settings later.

## Planning an Addressing Scheme

The ability to choose from 16 house and 16 unit codes allows for up to 256 individual X10 addresses. That's probably a lot more units than you'll need, even in the largest homes, because each module doesn't necessarily need its own unique address. For example, you might want to set all your hallway lights to `C12` because it's unlikely you'll need to control them individually. Because each module is listening for its commands, all of them will respond in unison to a single command, such as `C12 On`. Having several lights responding at once can be a nice effect and can simplify your system greatly.

It's a good idea to give some thought to how you will allocate the house codes you use in your system. It's often convenient to set modules that are in proximity to each other to the same house code. For example, all the

*Figure 1-1. Using the house and unit dials to set a module's address*

lamp modules on the second floor of my home might be assigned to house code B. This makes it possible to turn on every light upstairs in an emergency by sending an `All Lights On B` command, and it means I need only a single transceiver for wireless control of my lights **[Hack #5]**.

If you will be using wireless controllers, such as the Palm Pad, it's important to give some thought to the unit codes, too. The Palm Pad, and its wired sibling the minicontroller, have only eight buttons. That is, they can send commands to unit codes `1` through `8`, or unit codes `9` through `16`, depending on how you have them configured. It can be annoying to have to change the setting on a Palm Pad just to send commands to different modules, so try to stick to unit codes within the same ranges to save you some hassles.

Another good reason to plan which unit codes you'll be utilizing is that every X10 motion detector takes up two addresses **[Hack #6]**. You need to account for this when programming your motion detectors; otherwise, you'll end up quite confused. A good way to keep track of all this is to maintain a spreadsheet of your units and their addresses. Or, if you're more visually inclined, you can use a map, as shown in Figure 1-2.

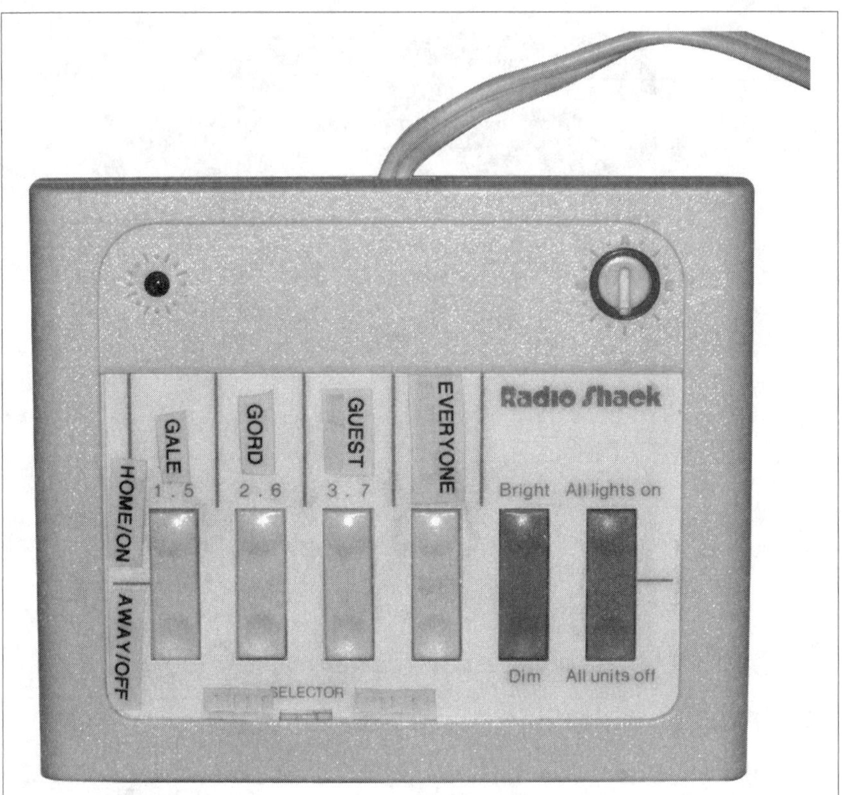


*Figure 1-2. Using a map to plan your system*

## Addresses to Avoid

During the last several years of X10 usage, the culture of home automation enthusiasts has discovered some X10 addressing folklore and techniques that might be helpful. Due to the fragility of transmitting data on electrical power lines, and the limited bandwidth available to do so, some addresses seem more susceptible to problems than others. It's said, for example, that electrical noise introduced by motors turning on and off (such as your furnace blower) can mimic an `M1 On` command. In fact, if you dig deeply enough into the collective wisdom of the community, you'll find some that contend the entire `M` house code is unreliable, due to the technical details and timing necessary to send that particular X10 code. I've never experienced it myself, but then, because so many other house codes are available, I don't use `M` at all.

Another, more pragmatic tip is to avoid using the address `A1`. It's simply too common because it's both the default address for many modules and the address that a non-savvy neighbor, who unknowingly has purchased an X10-based wireless doorbell, is likely to use. Better still, if you don't assign any modules to use `A1` and suddenly you start seeing that address in your system's log, you know either a module has forgotten its address due to a bad battery or failing electronics, or your neighbor has discovered X10 and it might be time to consider a signal blocker **[Hack #86]**.

### Final Thoughts

Although you definitely should consider which addresses you'll use before diving in, rest assured that eventually you'll change them around anyway. As your system grows you'll think of new ways to organize your units, or you might buy an X10 thermostat that takes up an entire house code just for itself **[Hack #41]**, so never consider your addressing scheme a permanent arrangement.

## HACK #2 Turn On a Light

**If you're new to home automation, the best way to start is with a few lamp modules.**

Lamp modules are the basic building block of home automation, and it's not unusual for a smart home to have one connected to nearly every lamp in the house. Lamp modules can vary in appearance, but, essentially, all of them look like the one shown in Figure 1-1 **[Hack #1]**.

Setting up a lamp module is very simple. Set the address you want this module to respond to by turning the house and unit code dials **[Hack #1]** on the front of the unit. Then plug a lamp into the front of the module and plug the module into a wall outlet. You'll want to put it into the bottom plug so the module doesn't block the rest of the outlet. If you need to use the top plug for some reason, plug the module into a short extension cord, instead of directly into the outlet, so the bottom outlet remains clear.

Next, turn on the lamp's switch. If it doesn't turn on, don't be alarmed; just turn the switch until it does. We'll talk about why this is necessary in a moment. Now, send an `Off` command to the module's address using a Palm Pad **[Hack #5]** or minicontroller **[Hack #4]**. If this is the first time you've seen X10 in action, don't feel self-conscious as you stand there and make it go on and off repeatedly; it's nearly impossible to resist.

Next, try changing the light's brightness. The lamp module is unique in its ability to brighten and dim whatever is connected to it. To do this, press the On button on the Palm Pad or minicontroller, and then press the Dim/Bright switch. In a second or two, you'll see the lamp react. Whee!

This ability to brighten and dim is why you should use the lamp module only to control lights. Never use it with an appliance, a fan, halogen or fluorescent lights, or a lamp that has a built-in dimmer. For these items, use an appliance module [Hack #3] instead.

Most lamp modules have a feature called *local sense* or *local control*. This enables you to use the lamp's switch to turn it on and off, but you'll notice that you might have to turn the switch twice before it takes effect. The extra turn is sometimes necessary—depending on the type of module you're using—for it to detect that you're trying to control the lamp locally. Note, however, that if you turn the lamp off at the switch, the lamp module will never be able to turn it on. So, making sure the lamp's switch is on is the very first step in troubleshooting why a light won't turn on when you tell it to. No worries, though: as you get used to living in an automated home, you'll soon get out of the habit of turning on lights manually, and you'll eventually think of doing so as quaint and old-fashioned.

The only drawbacks to lamp modules are that they're not very attractive, and you can't use them to control lights that don't plug into the wall, such as overhead lighting. If either of these limitations becomes bothersome, you might want to replace your electrical outlets and switches with built-in modules [Hack #14].

### Hacking the Hack

You don't have to plug the lamp module directly into a wall outlet; you can use it with an extension cord. This makes it easier to use a lamp module in hard-to-reach areas, and it is a great technique for controlling Christmas lights. For outdoor use, you'll want to be sure to keep the lamp module dry [Hack #59].

## HACK #3 Master Your Appliances

**If you want to automate a fan, coffeepot, or radio, use an appliance module. Thanks to their versatility, appliance modules are an integral component of a smart home.**

In many ways, the X10 appliance module fulfills the promise of a *Jetsonian* future. You can turn fans, radios, coffeepots, garden lights, and popcorn makers into your obedient servants with the addition of this little module. Although nearly identical in appearance to the lamp module [Hack #2], the appliance module is a lot more versatile. You can use it to control nearly anything that turns on or off.

Although you can use an appliance module to control a lamp, albeit without the ability to control the brightness, never ever use a lamp module to control an appliance. If the lamp module is dimmed accidentally, the appliance might overheat and cause a fire.

The appliance module doesn't understand many X10 commands; it responds only to `On` or `Off`. To set the module's address, use the house and unit dials **[Hack #1]** on the front of the module. For example, the module shown in Figure 1-3 is set to the address `B12`.

*Figure 1-3. An X10 appliance module*

After you've set the address, simply plug the module into a wall outlet and plug the appliance that you want to control into the module's plug receptacle. Then, make sure the appliance's switch is turned to the on position.

Send an X10 `On` command [Hack #4], and a moment later you'll hear a click, followed by your appliance springing to life.

Unlike lamp modules, some appliance modules don't have a local control feature that enables you to turn on the appliance by using its switch. If an appliance module does have this feature, you can turn it on by turning it on, off, and then on again. If not, you'll need to either disconnect the appliance module or send an X10 command to the module.

You can't control some appliances using this module because they don't have a switch that stays in the on position when the power to the appliance is turned off. That's how the appliance module works: it controls the flow of electricity to the connected appliance. When the module is off, it's as if you've unplugged the appliance from the wall. If the appliance you want to control has a *soft power switch*—one that is controlled electronically instead of physically—it won't work with an appliance module. If you're not sure you can tell the difference, here's how to test it. Turn on the appliance, and then unplug it from the wall. Wait a few seconds, and then plug it in again. If the appliance comes back on when you reconnect the power, it will work with an appliance module. If it remains off, you're out of luck and will need to figure out some other way [Hack #11] to control the appliance.

You'll also want to make sure the appliance you're connecting doesn't exceed the rated range of the module. Appliance modules typically support up to 15 amps or 300 watts, but check the label on your module to make sure. Modules are also available with two- or three-pin electrical plugs, so make sure you get one that matches the type of plug used by the appliance you're controlling. You also can get modules that work with 220-volt appliances, and sometimes these heavy-duty modules are necessary for 100-volt appliances that draw a lot of current, such as vacuum cleaners.

The most common question that arises about the appliance module is how to make it quieter. The loud *click* (some say it's a *clack*) is the sound of its heavy-duty relay switch turning on or off. You'll just have to learn to live with it, although if it bothers you too much, you might consider using a Smarthome ApplianceLinc (*http://www.smarthome.com/2002s.html*) instead of the standard X10 module; the ApplianceLinc makes less noise. Or, adjust your attitude and recognize that the click provides confirmation that the signal was received and acted upon. See, now the click is a feature, not a problem!

For more information about putting an appliance module to use, see "Brew Your Morning Coffee" [Hack #37].

# Send X10 Commands

**The computer you use for home automation often will be sending commands to your X10 devices. But when you want to send a command yourself, to turn on a light or change a setting, for instance, you'll need an X10 controller.**

The most common way to send an X10 command is to use a *minicontroller*, which is a small box (shown in Figure 1-4) that plugs into a wall outlet and has an array of switches that send `On` and `Off` signals to X10 addresses.

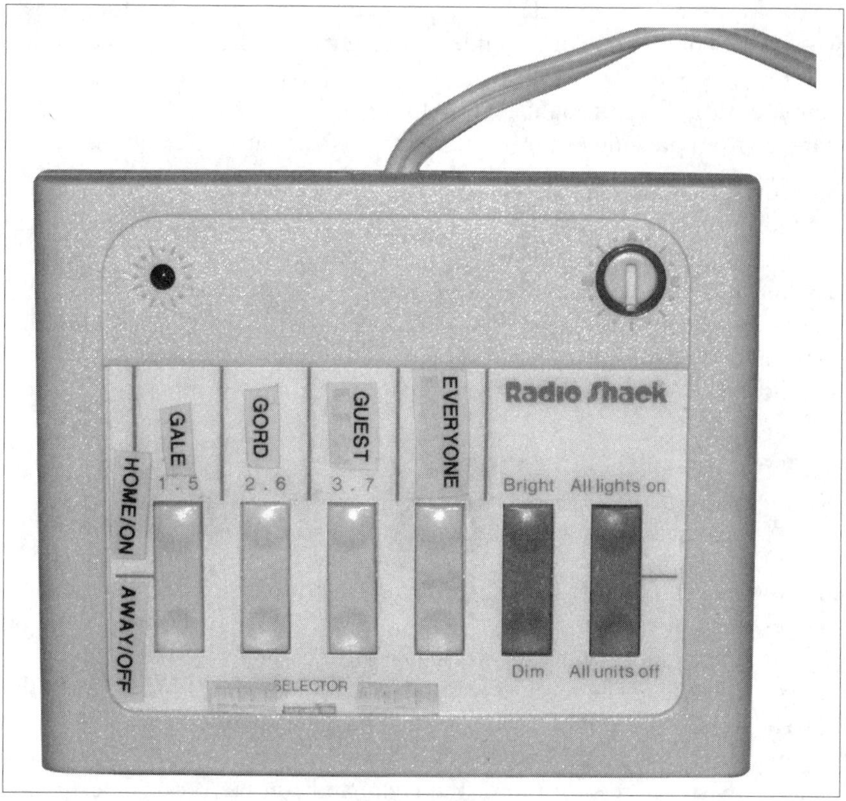

*Figure 1-4. A minicontroller with labeled buttons*

It's called a *minicontroller* because it has four buttons that, when combined with the selector switch at the bottom of the unit, can address only 8 units—half of the 16 possible units in a given house code. This is something to keep in mind when you're selecting the module addresses you want to use; if you spread the addresses out too far, you'll need two controllers to easily send them all commands. Or, buy a maxicontroller instead.

Set the dial on the minicontroller to a house code, and then press one of the numbered buttons to send a command. For example, with the house code set to `H`, press in the top of the first button to send `H1 On`, press the second button to send `H2 On`, and so on down the line. To send a `Dim` command to a lamp module, first press the On button for the lamp you want to control, and then press Dim or Bright.

Minicontrollers are useful for sending commands directly to a module, such as a lamp, but they're better used to communicate with your home automation software, where a single button press can kick off a script that does a whole series of events. For example, you might use a bedside minicontroller to signal that you've gone to bed for the night **[Hack #48]**, which causes the house to turn off all the lights, check for open doors, and lower the thermostat—all from pressing just one little button. Now that's automation!

Minicontrollers are very handy and it's a good idea to keep one or two around the house. Occasionally, you'll want to send an X10 command to adjust something in your home. Perhaps a light is too bright, or it's a cloudy day and you want a lamp on before sunset. Or, you might have a houseguest to whom you want to give an easy way to control some key areas, such as the hallway lights near the guest bath. All of these are good reasons to use a minicontroller or its wireless brethren, the Palm Pad **[Hack #5]**.

## Send X10 Commands Wirelessly

**Wireless remote controls are handy power tools for home automators, but you need to understand their quirks to get the most out of them.**

In addition to modules that plug into your electrical system, several wireless devices work with X10-based systems. All wireless X10 devices work by transmitting a radio frequency (RF) signal to a nearby receiver that's plugged into an AC outlet. The receiver translates the command and issues it to the power line so that other devices, including your computer and its home automation software, can see it. These devices are called *transceivers*. Ultimately, it's still necessary to put the commands on the power line—that's the core modus operandi of X10, after all—but the ability to initiate a command without a connecting wire is a great freedom.

The most common wireless X10 devices are motion detectors **[Hack #6]**, key-chain-size remote controls, and Palm Pad remote controls.

### Key-Chain Remotes

Tiny key-chain remotes, such as the one shown in Figure 1-5, enable you to send `On` or `Off` commands for one or two addresses. A watch battery powers them, so the range is limited, but it's still pretty good given their tiny power requirements. In most styles of key-chain remotes, you set the address that the first set of buttons control, and the second set of buttons are programmed automatically for the next incremental unit. That is, if you set the first buttons to transmit `A4 On` and `A4 Off`, the other buttons will send `A5 On` and `A5 Off`. You'll want to keep this unalterable relationship in mind when assigning addresses to your devices.

*Figure 1-5. A typical key-chain remote control*

Key-chain remotes are handy for scattering about the house, such as at your bedside, for a convenient way to send a command. They're also lightweight and easy to mount on the wall or under a table edge with double-stick tape.

Keep the programming instructions **[Hack #22]** that come with your remote. To set the remote's address, you'll have to press different combinations of the buttons, and the instructions vary between different models.

## Palm Pad Remotes

The Palm Pad remote is very versatile, but not very attractive, as you can see in Figure 1-6. It enables you to send commands to all 16 addresses in a single house code **[Hack #1]**. You set the house code for the remote and then press the On or Off buttons for each address you want to control. For example, set the house code dial to `N`, and then press the third On button to send `N3 On`.

You also can use the Palm Pad to set the brightness level for lamp modules. First, press an On button; then press the Dim or Bright button at the bottom of the remote. The commands will be sent automatically to the last device to which you sent the `On` command.

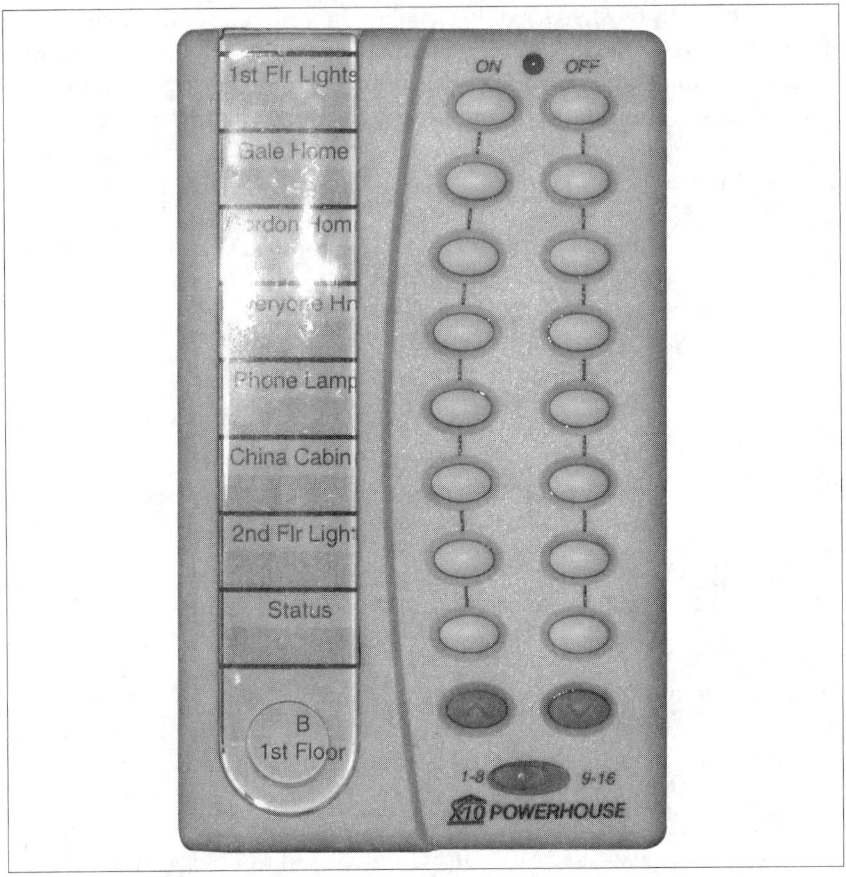

*Figure 1-6. The ugly but versatile Palm Pad remote control*

If you have FileMaker Pro, download a Palm Pad label template from the XTension web site (*http://www.shed.com/xtension/pplabels.sit.hqx*).

You'll notice that the Palm Pad has only eight buttons. So, how can it send to 16 addresses? The secret is the slider switch at the bottom of the unit. Slide it to one side and the buttons send to addresses `1` through `8`; slide it the other way and the buttons send to addresses `9` through `16`. This is a handy way to cram a lot of functionality into a single Palm Pad, but in order for it to work correctly, you'll need a transceiver capable of receiving all the unit codes, as discussed later in this hack.

*Multimedia* remote controls that combine X10 functionality with infrared (IR) transmission are also available. You can use these to control your home stereo and your lights, too.

## Transceivers

Transceivers (such as the one shown in Figure 1-7) are an essential link in wireless X10. In fact, they do all the real work, and without one, the remote control is useless. The transceiver plugs into a wall outlet and silently waits for a signal from a remote control, motion detector, or other wireless device. When the signal comes in, the transceiver translates it into an X10 command and transmits it on the power line so that other X10 devices, such as lamp modules, can respond to it. Additionally, most transceivers feature *collision avoidance*. That is, before they send the command to the power line, they'll listen to make sure another device isn't already transmitting an X10 command. This makes your wireless devices more reliable. But when combined with the inherent wireless delay, as detailed in the following section, this can slow down the transmission of wireless commands.

An important thing to know about transceivers is that you need one, and only one, for every house code that your wireless devices will be using. For example, if you have two Palm Pads, one set for house code `A` and the other for house code `L`, you'll need two transceivers: one to listen for `A` commands and the other set for house code `L`. Additionally, some transceiver models can listen for commands to units `1` through `8`, or they can listen for `9` through `16`, but not the entire unit range at the same time. You determine which range they listen for by setting a slide switch on the transceiver.

Some models also include a built-in appliance module **[Hack #3]** with a push-button switch on the front of the transceiver that enables you to turn the appliance plug on or off manually. This can be handy, but modules that have this will dedicate the first unit code to the appliance connection—either unit `1` or unit `9`, depending on the range of addresses for which they're set to listen. You can control the appliance plug wirelessly, but with some models, the transceiver does not send the command to the power line; it simply toggles its built-in unit and swallows the command. This means you cannot use that unit code for another purpose, nor can you track its state using your home automation software. Again, consult the specifications for your transceiver.

*Figure 1-7. A transceiver, the workhorse of wireless X10*

Transceivers have good reception range, but if you're having trouble receiving wireless commands, avoid the temptation of using two transceivers that are set for the same house code. If both transceivers occasionally receive the same wireless signals, both will try to handle it, and that can cause problems that are tricky to diagnose. Instead, add a transceiver set to a different house code and reprogram the nearby wireless devices to match. Or, if you want to get fancy, remap commands received on one house code so that they appear to have originated on another **[Hack #94]**.

## The Wireless Delay

One thing you'll notice about wireless commands is that they seem slower than commands sent using wired controllers. That's because of the time it takes for the transceiver to receive the command and then send it to the power line. This is in addition to the inherent delay [Hack #1] in sending and receiving X10 commands. The extra delay can be exacerbated if there's a lot of wireless traffic at the same time (from multiple motion detectors, for example). In this case, the transceiver has to wait for the power line to clear to avoid signal collisions. In the end, you'll notice at least an additional 1.5-second delay—sometimes a little longer. That doesn't sound like much, but when you're standing in a dark room waiting for the lights to come on, it seems like an eternity.

To mitigate this delay, you can replace your transceivers with a direct-to-computer wireless receiver [Hack #83]. Or, position your motion detectors so that the signal is received before you reach the location where you need the lights [Hack #85].

## Keep Watch with Motion Detectors

**A key element in any smart home, motion detectors enable your system to react to you and your visitors as you move about your house.**

The X10 wireless motion detector is a beautiful thing in the world of home automation. It's small, inexpensive, and incredibly useful when it comes to creating a smart home. Nothing does a better job of impressing your friends than the lights coming on automatically when you enter a room and turning off after you've left. And that's just the tip of the iceberg for what you can do with these beauties.

A typical motion detector, the X10 MS-14A Eagle Eye, is shown in Figure 1-8. About 2.5 inches square and powered by two AAA batteries, the MS-14A (*http://www.smarthome.com/4086.html*; $20) is mounted easily on the wall or ceiling with double-stick tape. When the detector senses motion, it sends a signal wirelessly, so you'll need an X10 transceiver [Hack #5] nearby to receive and convert the signal to an X10 command for your computer and other modules to see.

The detector's transmitting range is about 25 feet, but that varies depending on the strength of the batteries and the construction of your home. If you have a stucco home, for example, the detector's signal won't penetrate walls easily, so you might need to put the transceiver in the same room as the detector. Also, keep in mind that the transmitting range will drop off as the detector's batteries weaken. This can lead to troubleshooting confusion when you can see that the LED on the detector indicates it's seeing motion, yet no signal is being sent—a prime indicator that it's time to change the batteries.

*Figure 1-8. An X10 motion detector*

Motion detectors use *passive infrared* (PIR) to sense when something around them is moving. They see the world in terms of temperature, and they detect motion by watching for moving patterns of warmth. When you walk into a room, you might feel like a living, breathing being; but to a PIR detector, you're just a moving blob of body heat. Don't take it personally; it treats everyone the same way.

When the detector notices the blob (you), it sends a signal that is picked up by the transceiver, which then passes an On command to your other X10 modules and your computer **[Hack #16]**. There are two things to keep in mind about this. If your body temperature isn't very different from the ambient room temperature, the detector will have a harder time detecting your presence. For example, if it is 98 degrees in your garage on a summer day, and you walk into your house with your approximately 98-degree body temperature, the motion detector might miss your entrance but will probably pick you up after you've walked a few steps. This brings up the second key point. They're called *motion* detectors for good reason; if you walk into a room and stop, waiting for the light to come on, don't expect them to trigger. You'll get better results if you take a few steps into the room, so the detector can better see you for the hot mass you really are.

Once the detector has figured out you're in the room, it sends an On command about every 10 seconds until you leave or stand still. This can be handy for identifying when a room has become empty (though more reliable methods [Hack #26] do exist). But the constant stream of `On` signals can be problematic if you have multiple detectors and a home with a lot of active people. If several motion detectors are sending signals at the same time, their signals will collide and get missed, so it's best to limit the number of motion detectors you have in a home or use a direct-to-computer receiver [Hack #83] instead of a transceiver.

## Selecting Addresses

Each Eagle Eye motion detector uses up two X10 addresses [Hack #1]. The first address is used to indicate when motion has started and ceased; the unit's dusk detector uses the second address.

For example, if you set the address to `C1`, the motion detector sends `C1 On` when it senses motion. When it no longer sees movement, it sends `C1 Off`. When a lit room becomes dark, it sends `C2 On`. When the light returns, it sends `C2 Off`.

If that seems backward, remember that it is a *dusk* detector. That is, it's turned on when the room darkens and turned off when it brightens. Now, in practice, the dusk `On` and `Off` signals aren't all that useful, although they can come in handy for confirming that you haven't left the light on in the garage or basement when going to bed at night. The most important thing to remember about the dusk sensor is that although you can ignore it, you can't turn it off. After you program a motion detector with an address, the next sequential address always will be taken up by the dusk signal. Keep this in mind when you're planning for the addresses you'll use in your home.

One thing you can change about the dusk sensor is whether the Eagle Eye watches for motion when it is turned off. In other words, you essentially can turn off the motion detector so that it no longer sends signals if the room is lit when it senses movement. If you're using the motion detector primarily to turn on lights for a dark room, this is a decent option to consider. Once the lights are turned on, you won't get another signal from the detector until motion ceases and it sends an `Off` command. But if you're using the motion detector to generally know when someone might be present, such as when a visitor comes to your front door [Hack #74], you probably want to turn off the dusk-only mode.

You can also set how soon after motion ceases that the detector sends the `Off` command. You can set the off delay to 1, 2, 4, 8, 16, 32, 64, 128, or 256 minutes. It's nice to have this flexibility if you're using the signal to turn off lights in the room [Hack #7].

The process for setting an Eagle Eye's address, dusk mode, and off delay is more than a little convoluted. It involves repeatedly pressing two tiny buttons located in the battery compartment while counting little blips of light as the detector acknowledges your actions. I'm not making this up. Luckily, every motion detector has a piece of paper glued to the inside of the battery compartment door that explains the procedure. Get used to consulting it often; you'll be reprogramming the unit every time you change its batteries.

## Positioning Motion Detectors

Now that you've configured the motion detector, you'll need to mount it on the wall. For best results, position it so that movement occurs across its field of view. It's looking for a moving object, and a person walking toward the motion detector won't set it off as quickly as someone walking past it will.

The detector's field of view is cone-shaped and extends out about 30 feet. If you expect people will be passing close to the detector, mount it at about chest or waist height to put more mass in the field of view. You also might want to turn it sideways, as shown in Figure 1-8, so that the widest part of the field is oriented vertically rather than horizontally.

## Other Motion Detectors

The Eagle Eye motion detector is also sold as the Hawk Eye motion detector. They're functionally the same, except that the Hawk Eye is black rather than white, and the cover has a rubber gasket that seals the unit for outdoor use.

Another wireless motion detector, the X10 DM10A (*http://www.smarthome.com/4086.html*; $30), has better transmitting range than the Eagle Eye, but it is so unsightly due to its large size (about the size of a softball) that your spouse is unlikely to let it anywhere inside your home. However, if you can find a suitable place to use it, the DM10A does have another advantage: it is a lot less chatty than the Eagle Eye and sends a signal only every minute or so. Another key difference is that it never sends an `Off` signal when motion ceases. It simply stops sending the `On` signal.

When you're shopping for motion detectors, be careful about buying ones that are for use with the X10 security consoles. Their signals are different, and X10 transceivers cannot receive them. You'll need to use a direct-to-computer interface to use security motion detectors **[Hack #78]**.

# HACK #7 Turn On the Lights When You Enter a Room

Here's an easy way to turn on the lights when you enter a room and have them turn off automatically after you leave. You don't even need a computer!

Although many of the hacks in this book use a computer to automate your home **[Hack #16]**, computers are not always strictly necessary. X10 modules work just fine without a computer, and in the case of motion-activated lights, you can improve response time and keep your system simple by setting them up to work without a computer-based controller. Even if you generally want to have a computerized smart home, setting up a few lights so that they work independently is a good idea to ensure some basic functionality even when the computer is turned off.

I use that approach for the overhead light in my garage. It's a single light, so it's a good candidate for having it controlled strictly with X10. Also, there aren't any windows, so it's very dark, and I want the light to come on quickly and reliably, even if I happen to be tinkering with my home automation system and it's offline or terribly confused.

I use a motion detector **[Hack #6]** to turn on the overhead light whenever someone enters the garage using the door from the house. Here's how it's set up:

- An Eagle Eye motion detector, mounted near the door, is set to address `C2`.
- A built-in X10 light switch **[Hack #14]** is set to address `C2`.
- An X10 transceiver, set to house code `C`, is plugged into a wall outlet in the garage.

Here's how it works. When someone enters the garage, the motion detector sends `C2 On`. The transceiver picks up this signal and relays it to the power line. The X10 light switch sees the `C2 On` command, which matches its address and turns on the garage light. This whole process takes just less than two seconds, due to the timing necessary for the wireless and X10 command transmissions.

After the motion detector hasn't sensed any motion for five minutes, it sends `C2 Off`, which causes the garage light to turn off. The five-minute delay is programmed into the motion detector; you might want to set a longer delay **[Hack #6]** if you're using this technique in a larger room or in an area where you might not be moving around too much. This will ensure that the light won't be turned off when the room is still occupied.

The home automation software running on my computer will see this activity take place because it is listening to the power line and keeping track of all the X10 commands it receives, but it does not send any commands.

Involving the computer in this process would allow for a more sophisticated response [Hack #8], but would introduce an additional delay while it responded to the signals and made decisions about how to react.

## Turn On the Lights When They're Needed

**Combining a computer-controlled home, motion detectors, and lamp modules can ensure that your lights come on only when you really need them.**

The simple approach [Hack #7] of setting a motion detector to the same address as the light you want to control is often useful, but there's a more flexible approach. Instead of having the motion detector control the lights directly, use your home automation software [Hack #16] to add some intelligence to the equation.

Here's the situation:

- An Eagle Eye motion detector [Hack #6] set to address `B2`
- An X10 transceiver to receive the signal from the motion detector
- A lamp module [Hack #2] set to address `B4`
- A computer with home automation software, such as XTension, with units defined for both the motion detector and the lamp module

When a person enters the room and triggers the motion detector, it sends `B2 On`, which the transceiver receives and transmits to the power line. XTension sees the command and turns on the unit that represents the `B2` motion detector in the software.

### The Code

The unit has the following On script, which XTension executes [Hack #17] when it turns on the unit:

```
if (status of "daylight") = false then
    turnon "bedroom light"
endif
```

In this script, the lamp module named `bedroom light` will be turned on only if it's nighttime. Compare this with the technique in "Turn On the Lights When You Enter a Room" [Hack #7], which turns on the light every time the motion detector is activated, day or night. It's a basic idea, but it illustrates how simple logic and just a little scripting can elevate your automation to smart behavior.

## Hacking the Hack

In the preceding example, how is the light turned off? There are at least two good approaches. First, you could modify the On script to automatically add a scheduled event that turns off the light five minutes later:

```
if (status of "daylight") = false then
    turnon "bedroom light" for 5 * minutes
endif
```

This approach works, but it will result in a new scheduled Off event every time the motion detector sends an On command, so it's not very efficient. It does work well, however, for hallways, garages, or other areas where people tend not to linger.

A different approach is to use an Off script that executes when the motion detector sends an Off command when it detects that motion has stopped **[Hack #6]**. The Off script doesn't have to be complicated; it simply tells the lamp module to turn off:

```
Turnoff "bedroom light"
```

Like so often in computing, the simplest approach is often the best. Once you've gotten the hang of using your computer to react to motion detectors, you'll likely want to do even more. The actions you take aren't limited to controlling lights; you also can make voice announcements **[Hack #74]** or decide if you should sound an alarm **[Hack #71]**.

HACK #9

## Ring a Bell to Alert the House

**The chime module provides an easy way to add audible alerts to your home automation system. Simply plug it in, and then tell it to sound off with a single command.**

The chime module, shown in Figure 1-9, is quite simplistic. You need only to set its house and unit codes, and then to plug it in. When it receives an On command, it makes a doorbell-like "ding-dong" sound that repeats three times. There's no way to make it ring in any other pattern, and there's no volume control, so this module is a take-it-or-leave-it affair.

The chime module most commonly is used as a remote doorbell and is helpful particularly if your home's bell is too quiet to hear throughout the house. You can set up a Powerflash to trigger the module with your existing doorbell button, or set its house and unit code to the same as a motion detector and have it chime automatically every time the motion detector signals that someone is nearby. This probably would get annoying quite quickly because it will chime every time the motion detector is triggered, so it's better to control it with a script that limits how often it goes off **[Hack #85]**.

*Figure 1-9. The chime module*

In a less highly trafficked area, such as a back gate, the chime module is useful for alerting you of comings and goings where there usually aren't any. Another useful application is to associate the chime module with the arming or disarming of your X10-enabled burglar alarm system so that everyone knows when the system is being armed. Many alarm consoles can send an X10 command when they're armed or disarmed, so simply set the chime module to the address sent by the console and you'll get a nice audible beep when the alarm changes state. Another way to use a chime module is to have it alert you to when the mail is delivered **[Hack #62]**.

## Hacking the Hack

If there's one thing to be said about the chime module, it's that it's loud and it gets your attention. You can quiet it down a bit by placing duct tape over the speaker grille. Or, consider using a universal module instead **[Hack #11]**. Its "beep, beep, beep" is quieter than the chime module's "ding-dong, ding-dong, ding-dong."

# Sense What's Happening

**Combine the Powerflash module with switches and sensors to monitor conditions in and around your home.**

The oddly named Powerflash module (*http://www.smarthome.com/4060.HTML*; $24), shown in Figure 1-10, often is overlooked by beginning home automation enthusiasts because at first glance it doesn't seem to do much of anything. It doesn't have the ability to control lamps or anything else, and, in fact, it doesn't respond to any power-line commands at all. So, you might be wondering, what's it good for?

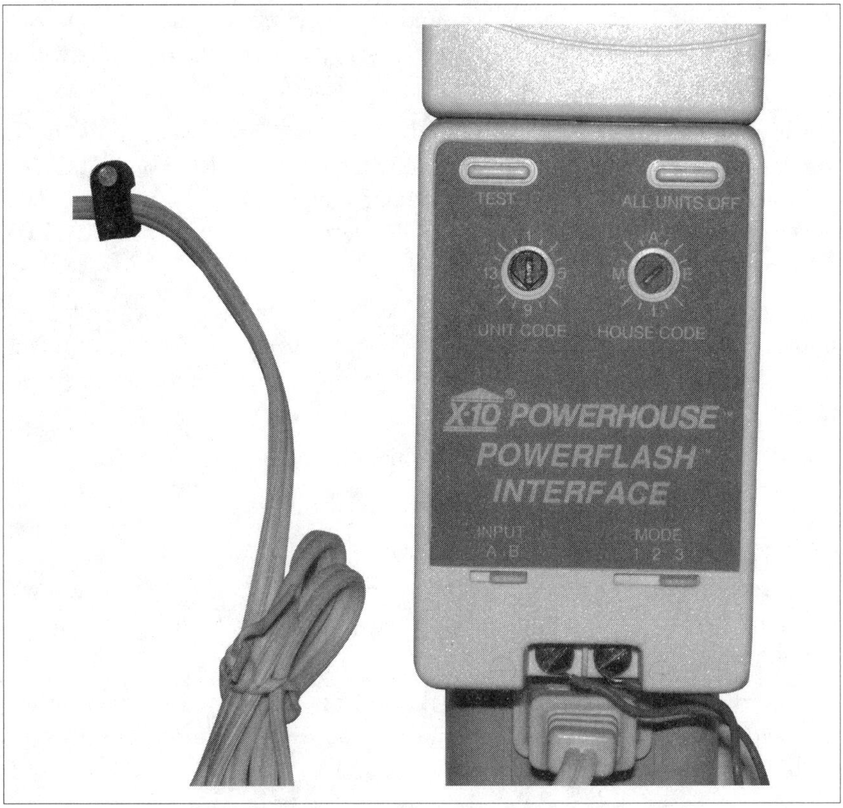

*Figure 1-10. The misunderstood Powerflash module*

The Powerflash module is an essential tool for knowing what's happening in your home because you can use it to turn virtually any switch or sensor into an X10-savvy signaling device and have your home react automatically to changes in conditions and states. Thanks to the Powerflash and to the wide variety of sensors that are commonly available, you can integrate nearly any condition you care to monitor into your smart home—water, temperature points, current detectors, pressure mats, breaking-glass sensors, and more—nearly any of which you can bring into the reach of your home automation controller.

Simply connect a switch to the two screw-terminals on the front of the Powerflash. The switch doesn't need to know anything about home automation; it just sends its usual signal (`On` or `Off`), which the Powerflash detects and translates into an X10 signal to which other modules or your home automation controller can react.

If at any time you find yourself thinking, "I wish I could tell when [something] happens," think Powerflash and go Googling for sensors that complete the puzzle. With a little creative thinking, a solution is probably within reach.

For example, connect a magnetic reed switch to a Powerflash and you can know when the garage door has been opened or closed [Hack #55]. If you have a problem with rainwater seeping into your garage during storms, hook up a water sensor to a Powerflash and you'll get a signal that alerts you to the encroaching puddle [Hack #44].

You'll need one Powerflash for every sensor you want to use, and many home automators find they have at least one or two in their setups. In fact, the artful use of a Powerflash or two is often what separates the newbies from the old hands in the ranks of home automation.

There are three steps to hooking up a Powerflash module:

1. Configure the Powerflash for the type of switch you're using.
2. Connect the switch to the Powerflash.
3. Select a signaling mode and set the house and unit codes, which will determine the X10 commands the Powerflash sends when the sensor is triggered.

A Powerflash is only as useful as the sensor or switch to which you've attached it. Thankfully, the Powerflash is quite versatile and works with both dry-contact and low-voltage devices.

## Using Dry-Contact Switches

A dry-contact switch is a common type of switch; it works by opening or closing a circuit. An everyday light switch is a dry-contact switch; when the switch is closed, the circuit is complete, and electricity can reach the light bulb, turning it on. When the switch is open, no electricity can reach the bulb, so it's off.

Magnetic reed switches, mercury tilt switches, pushbuttons, and pressure-mat sensors are examples of dry-contact switches used in home automation. In essence, a dry-contact switch is little more than a safe and reliable way of touching the ends of two wires together and pulling them apart again.

To use a dry-contact switch, set the Input selector on the front of the Powerflash to position `B`.

### Using Low-Voltage Switches

A low-voltage switch, sometimes known as a digital sensor, works by changing the amount of power on a circuit. It's up to some other component, in this case the Powerflash, to interpret these fluctuations.

If you're connecting the Powerflash to a spare relay on your home alarm system, or using a digital I/0 controller connected to a computer, you're using a low-voltage switch.

To send its signal, a low-voltage switch has to have some sort of power source, typically in the form of a battery or power supply. The Powerflash will work with devices that supply between 6 and 18 volts (AC or DC), so be sure the power output of the switch you're connecting falls within that range. In addition, some sensors are finicky about polarity, so pay attention to the positive (+) and negative (–) indicators on the Powerflash (near the screw-terminals) when you connect the sensor to the Powerflash.

To use a low-voltage switch or sensor, you need to set the Powerflash Input selector to position `A`.

### Selecting a Switch

Whether a switch is dry-contact or low-voltage is determined by its design, so there's no decision about that on your part; you simply set the Powerflash to the correct setting for your switch. But before you buy a switch, consider whether you want a momentary or toggle switch.

A momentary switch is like a doorbell button; it automatically reverts to its previous state when pressure is removed. A toggle switch is like a light switch—it stays in one state or another until it is reset manually.

Be sure to choose the appropriate switch for the application you have in mind. If it is important to know the exact state of something, choose a toggle switch. For example, if you want to know whether the garage door is currently open or closed, use a reed or mercury-based toggle switch. You'll be able to determine the state of the door by watching the state of the switch, as indicated by the On or Off received when the switch moves into each distinct position. If you only want to know that the door has been moved, and you don't care whether it's open or closed, use a momentary switch. When connected to a momentary switch, the Powerflash sends an On command followed quickly by an Off command when the switch resets itself.

Regardless of the type of switch or sensor you buy, it's most important that you get one that's classified as normally open (N/O). The Powerflash will not work correctly with a normally closed (N/C) switch. The reason lies in its roots as a module for security applications. Alarm system components are manufactured to be normally open so that a burglar cutting a connecting wire will trigger the alarm.

## Connecting to the Powerflash

The next step is to connect your switch to the screw-terminals on the front of the Powerflash. Use good quality solid-core wire, as well as some wire mounts or wraps to keep the wires securely fastened for the length of the run.

Don't worry too much about the length of the wire; either your low-voltage device or the Powerflash itself (in input mode B) will provide sufficient voltage for powering the circuit. But it's always good practice to test the setup so that you know it will work before spending a lot of time carefully running, hiding, and securing the wires.

## Setting the Powerflash House and Unit Code

You set the house and unit codes by turning the dials on the front of the Powerflash module. But unlike almost all other X10 modules, this selects the codes the module *sends*, not the code to which it responds. When the attached switch is triggered, the Powerflash will send its commands using the codes you've set. The exact sequence of commands it sends is determined by its mode setting.

## The Powerflash Modes

Aside from the terminals where you connect a sensor, the most important part of the Powerflash is its mode selector switch. Selecting the proper mode setting is truly the key to getting the most out of this module.

The Powerflash reacts in one of three ways when the sensor is triggered. You indicate how you want the Powerflash to react by using the Mode selector on the front of the unit.

In the following examples, assume that the Powerflash's House Code dial is set to `B` and the Unit Code dial is set to `9`.

*Mode 1*

When the attached sensor closes, the Powerflash sends an `All Lights On` command to `House Code B` and then sends `B9 On`. When the sensor's contacts open, the Powerflash sends `B9 Off`.

This mode of operation is probably useful only as part of a security setup to activate multiple lights and set off a siren. After the `All Lights On` command, you'll need to manually push the All Lights Off button on the front of the module.

*Mode 2*

When the attached sensor closes, the Powerflash repeatedly sends `All Lights On` and `All Lights Off` to `House Code B`, causing all the lights to blink on and off. (Now you know where this module gets its name.)

When the sensor opens again, all the lights are left on. That is, the last command sent always will be `All Lights On`. This mode also is useful for security applications, unless you need to be alerted to a condition using a silent, but obvious, signal.

*Mode 3*

Thankfully, this Powerflash mode is quite useful for general home automation. In this mode, `B9 On` is sent when the sensor's contacts close, and `B9 Off` is sent when the contacts open.

A Powerflash set to Mode 3 is easy to incorporate into your home automation system. If you have a switch attached to a door, for example, `B9 On` tells you the door has been opened. When the command `B9 Off` is received, the door has been closed. In the case of a water sensor, `B9 On` means your garage is getting wet and `B9 Off` means the sensor is once again high and dry.

## Robustness Considerations

A downside to the Powerflash is that it sends each command, when in Mode 3, just once. It does not signal repeatedly, like a motion detector might do, so you'll want to be prepared for the occasional lost signal if knowing the exact state of the sensor is critical to your implementation.

For example, if you're counting on a low-temperature sensor to alert you to when it's too cold for your prize-winning koi to survive in their outdoor pond, build in some redundancy by using two sensors or another different method to ensure you get a chance to rescue the little carps from Mother Nature.

# HACK #11 Control the Uncontrollable

The universal module is an important part of an advanced or intermediate home automation system. With it you can control things that don't have built-in X10 capability.

Let's cut right to the bottom line: the *universal module* is a relay switch that you control with X10 commands. If there's something you want to automate and it has a switch you can replace or bypass, the universal module, shown in Figure 1-11, is just what you're looking for. Plug this module into the wall and connect wires from its terminals to the device you want to control, and you're pretty much all set.

*Figure 1-11. A universal module*

A common use of the universal module is to control a garage door **[Hack #56]**. In this case, the module becomes a second switch, acting in parallel to the push button you use to manually open and close the door.

## Setting Up

You set the module's address [Hack #1] by turning the House and Unit dials on the front of the unit. The universal module can act as either a momentary-contact switch—like a doorbell button—or a continuous switch. You determine in which mode the universal module operates by moving the slider on the front of the unit to one of these settings.

*Momentary mode*
: The module briefly closes the relay switch when it receives an `On` command.

*Continuous mode*
: An `On` command closes the switch until an `Off` command is received.

The universal module also has a built-in chime. It's louder than the standard chime module [Hack #9] and sounds off in a more pleasing tone.

The universal module's chime has three different settings:

*Relay Only*
: The module's chime is turned off. The relay switch functions silently.

*Sounder & Relay*
: When the relay switch is turned on or off, the bell will ring four times. This is useful as a warning tone that alerts bystanders that something is about to happen, such as a garage door opening or closing.

*Sounder Only*
: Puts the module into chime-only mode. The relay-switch portion of the unit is disabled and its switch will never be opened or closed. An `On` command causes the bell to ring four times. `Off` commands are ignored in this mode. This mode is useful as a remote doorbell or as part of a simple alarm or visitor alert system.

## Final Thoughts

The universal module is easily confused with the Powerflash module. They appear to be similar, and both have terminals you use to wire them into another device. Here's an easy way to remember how they differ: the universal module enables you to control (turn on and off) the universe. The Powerflash is like a flashbulb on a camera: it provides a snapshot of the state of something (on or off).

# Groom Your Home for X10

The X10 method of sending signals over your home's power system is quite clever, but it's subject to interference from other electrical devices and any anomalies you have in your power system. With a little bit of effort and equipment, you can greatly improve the reliability of X10 in your home.

Most homes in North America have two 120-volt power lines from the utility company coming into the home. These two lines meet at the home's breaker box, where the circuits that feed light switches, plug-in outlets, and appliances are supplied with electricity. Half of the circuits are fed by one of the 120-volt lines, and the second 120-volt line feeds the other half. The intermittent operation of X10 modules usually happens when the transmitter is sending signals on one line and the receiver module is plugged into an outlet on the other line. For the signals to get to the receiver, they actually leave the home, travel to the utility company's transformer, and then come back into the home on the other power line. Not surprisingly, by the time the X10 signal completes this circuitous journey, it might be too weak for the module to detect, particularly in large homes.

The first order of business, then, is to install an X10 *coupler-repeater*, also known as an *amplifier*. A coupler-repeater will detect the incoming X10 signal, regenerate it, and then blast it out over both 120-volt lines. If your home is larger than 3,000 square feet, consider installing a coupler-repeater. In smaller homes, a device known as a *signal bridge* might be enough to get good results. A signal bridge performs the same function but does not amplify the signal when it passes it on to the other power line. See "Improve X10 Reliability" **[Hack #86]** for more information about options for choosing between when to use an amplifier and when to use a repeater.

Once the signal has been amplified, it's time to preserve it. X10 signals are like water under pressure in a pipe: they go everywhere they can, not just to the receiving module. This means they reach every electrical device in your home, and some of them will affect the strength of the X10 signal. Computers, video gear, and high-end electronics are likely culprits of interfering with X10 signals. The more complicated the electrical power supply in a device, the more likely it is to disrupt X10 signals because the engineers who design power supplies build in traps to filter out and kill electrical noise. Unfortunately, the X10 signals look like electrical noise to these devices, and when the signals are filtered, they get weaker and harder for the intended recipients to receive.

These are the most common sources of signal loss:

- Televisions
- Audio and video equipment

- Computers
- Uninterruptible Power Supplies (UPSs) and power strips
- Power supplies for laptops and cell phones

If you suspect a device is absorbing signals, unplug it and retransmit the X10 signal. It is important that the device is *unplugged* and not just turned off because some devices can cause problems simply by being plugged in. If the X10-controlled product begins working after the device is unplugged, you'll need a plug-in X10 filter to prevent the offending device from interfering with your home automation. A typical house requires four or five filters to ensure X10 signals can be sent reliably. For more on using X10 filters, see "Improve X10 Reliability" **[Hack #86]**.

Another factor that can degrade X10 signal quality is the number of X10 transmitters installed in your home. Each X10 transmitter contains a tuned circuit that, when it's not sending X10 signals, is absorbing them! Generally, the closer your devices are to each other, electrically speaking, the more signals are weakened by absorption, particularly if you have *two-way* X10 devices that have circuitry to send and receive X10 commands. If you find your system getting less reliable as you add more devices, here's a good rule of thumb: if you have more than 20 devices, consider that you might have reached the point where you need an amplifier or repeater.

*—Smarthome, Inc.*

## HACK #13 Set Addresses for Modules Without Dials

**Some X10 modules don't have mechanical dials to set their addresses or other options. To configure these types of devices, you need to send commands over the power line.**

Some of the X10 modules from Smarthome, Inc. have shed the old-style dial method of setting the module's house and unit code **[Hack #1]**. For example, the LampLinc and ApplianceLinc modules, as shown in Figure 1-12, have smooth fronts with no code wheels.

Eliminating the code wheels has some advantages because they can be prone to failure and have a certain air of cheesiness about them, but it's more complicated than that. By teaching the module to listen for configuration commands over the power line, more options can be supported. That's exactly what Smarthome has done. With the SmartLinc lamp module, for example, you can set how quickly it dims, its initial brightness level when it's on, and whether using the switch on the lamp will turn on the module.

Whew. That's a lot of choices to make when setting up a single lamp module, and, correspondingly, some special steps are required to program it.

*Figure 1-12. The LampLinc and ApplianceLinc, which require commands to set their addresses*

Because the X10 protocol supports only a handful of commands, setting the options involves sending addresses in a specific sequence, within a short window of time. For example, to set the address of a LampLinc 2000 model, you plug it into the wall, hold down its Set button for five seconds, then send the X10 address you want to set the module to respond to (such as `B11`) within 30 seconds. Then, send an `On` command if you want the module to turn on when the lamp's switch is used, or send `Off` if you want to disable the local control feature.

To set other features, such as the initial brightness level, you might need to send several addresses in sequence. For example, to set the LampLinc 2000SLS's default brightness you send `O16`, `N16`, `M16`, `P16`, and `M16`. Then send one or more `Dim` commands to adjust the lamp to the desired level. Then, send `P16`, `N16`, `M16`, `O16`, and `M16` to lock in the setting.

The sequence of commands you use varies depending on the module, so be sure to check the documentation that came with your unit. Because the sequence is so lengthy and any X10 commands sent on the power line from another source such as a motion detector can interrupt the progression, these units can be easier to program if you set up a closed X10 system. Plug an inexpensive power strip into an X10 filter **[Hack #86]**, and then plug the filter into the wall. Plug the module you want to program into one of the power strip's outlets, and plug an X10 minicontroller into another of the power strip's outlets. The filter will prevent the minicontroller's commands from being sent onto your home's power line, but the module you're programming still will see them because it's plugged into the power strip, too.

More importantly, the filter also will prevent commands from sources reaching the module and interfering with your programming steps.

Alternatively, do as Smarthome suggests in its documentation and disable all your transceivers, home automation computer, or any other device that might decide to send an X10 command while you're in the process of configuring the module. Either way, it's a bit of a hassle.

Instead of using a minicontroller to program the modules, you probably will have more success, and less setup, by using your home automation software instead. It still will be possible for the programming sequences to be interrupted accidentally by another command, but the speed at which your computer can send commands makes it less likely to happen—or, at least, less of a hassle to start all over again.

## HomeSeer

For HomeSeer, you use the X10 control panel. Choose Control Panel from the View menu. Enter an X10 address in the Device Address field, and then click the appropriate command button. To send just the address, without an `On` or `Off` command, click the Address Only button. To send to multiple addresses at the same time, enter each address separated by commas, as shown in Figure 1-13.

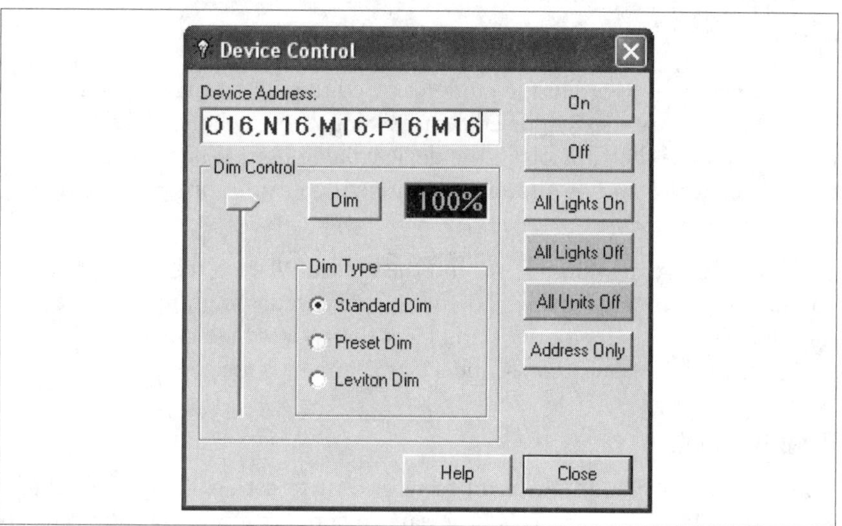


*Figure 1-13. Sending multiple X10 addresses with HomeSeer*

### XTension

If you're an XTension user, the Command Line window provides the simplest way to send a command to a single address. Choose Command Line from the Window menu. Type in an AppleScript command to execute, and then press Return. To send just an X10 address, one that does not include an `On` or `Off` command, use the `send only address` command as shown in Figure 1-14.

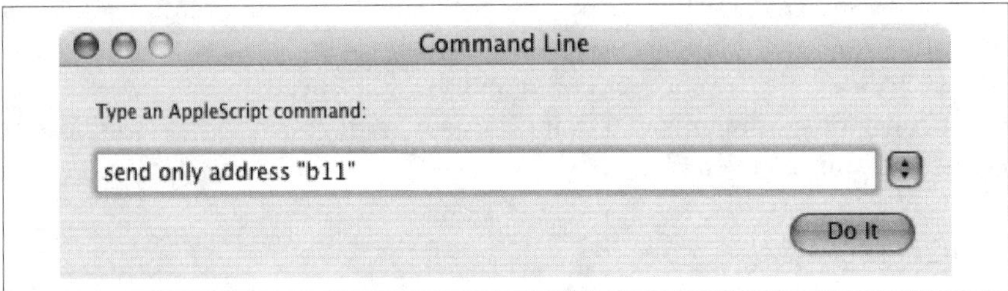


*Figure 1-14. Sending an X10 address with XTension*

If you have to send multiple addresses, you can write a script that uses this same approach, or simply enter and send each command in the Command Line window. Click the button with the arrows at the end of the text field to select commands you entered previously if you need to resend any.

### Indigo

With Indigo, you program the modules by editing its settings in the application. Create a new device, or edit an existing one to change settings made earlier, then use the controls in the device dialog box. When the Type is set to a device which Indigo supports, additional controls appear in the dialog, as shown in Figure 1-15.

Set the address you want to use, and then click OK to send the appropriate commands to the power line. To change the default brightness level or the dimming speed, set the options and click the appropriate download button. Indigo sends the appropriate sequence of commands for you.

### Final Thoughts

If all this seems like an awful lot of work just to set an X10 address, it is. Eliminating the code wheels is necessary to support more sophisticated settings, and it cleans up the appearance of the modules, but progress doesn't come without a price. Luckily, it's not something you have to do very often.

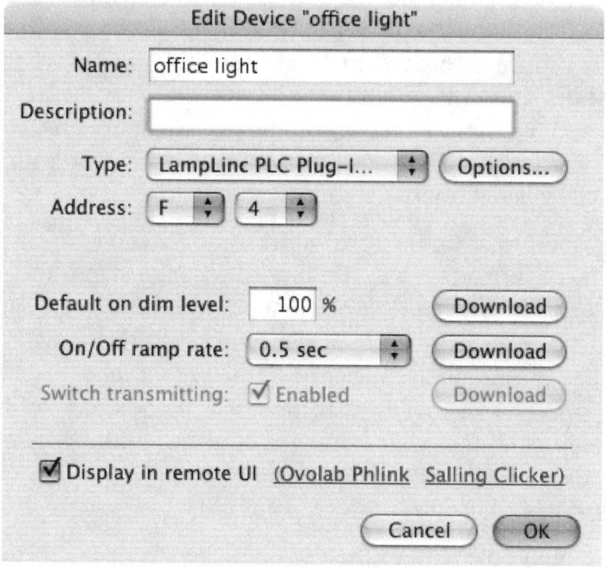

*Figure 1-15. Setting device options with Indigo*

## HACK #14 Increase the Spousal Approval Factor

**Living in a smart home requires, at the very least, tolerance from your family or housemates. If they're indulging your desire to create a home of the future, here are some tips for returning the favor by smoothing out some of the rough spots.**

Not everyone in your household will be as enthusiastic about home automation as you are. That's to be expected—everyone has different interests—but unlike some hobbies, automating your home has a profound impact on others. If it's not done in a careful and considerate fashion, it can disrupt and bring frustration to a family's ultimate retreat: their home. For this reason, and just for common courtesy, it's a good idea to discuss your plans *before* you implement them. The results of some automation projects can be surprising, such as a talking house [Hack #28], so it's best to make sure you aren't the only one who will enjoy them.

### Use Better Modules

Something else to keep in mind is that installing X10 modules changes the behavior of everyday objects, sometimes in ways that make them virtually unusable by normal methods. For example, using an appliance module to control the Vornado air fan [Hack #3] is a neat idea, but it means you might not be able to turn on the fan by using its switch; some appliance module might prevent that from happening. In addition, if the fan is operating and you turn it off by using its switch, the home automation system won't be able to turn it back on later. You should make sure that everyone in the house is aware of this to reduce frustration with the new approach, and that the benefits of automation—such as energy savings or being able to have the fan turned on only when it's needed—outweigh the drawbacks.

Other X10 modules have their own quirks, too. If you're using lamp modules to control lighting **[Hack #2]**, you'll need to understand how local sense works. This enables you to turn on the light by turning its built-in switch, but it often takes an extra turn of the switch before the light comes on. Moreover, like the air fan, if the light is turned off at the switch, it's beyond the control of X10. For motion detectors, it is a good idea to discuss how they work **[Hack #6]** so that your housemates understand that taking a single step into a dark room and then stopping to wait for the light to come on won't work very well; it's better to keep moving.

Another common objection to the use of X10 concerns aesthetics. Most of the devices are far from attractive, as you can see in Figure 1-16, and their size means they not only draw attention to themselves, but also they're often in the way, such as when you want to use them behind a couch that you'd like to abut against the wall. Fortunately, if you're willing to spend a little more money, you can use built-in X10 modules that are virtually identical to standard equipment.

Replacing external X10 modules with built-ins is one of the surest ways to boost your home's spousal approval factor (SAF) rating. Not only do built-in modules look better, but also they often add features that you can use to make your home automation system even less intrusive. For example, some built-in light switches are *two-way*. That is, when the light switch is used, it sends a signal back to the home automation system reporting that it has been turned on or off. These types of switches, such as the SwitchLinc (*http://www.smarthome.com/2380.html*; $70) shown in Figure 1-17, also respond to X10 status requests, which can enable you to confirm that a light has turned on or off as the result of a command, or to see its current level of brightness.

In addition to the higher cost of these units, you'll want to make sure your home automation software works with their extra features, which are special-purpose extensions to the X10 protocol and require additional support.

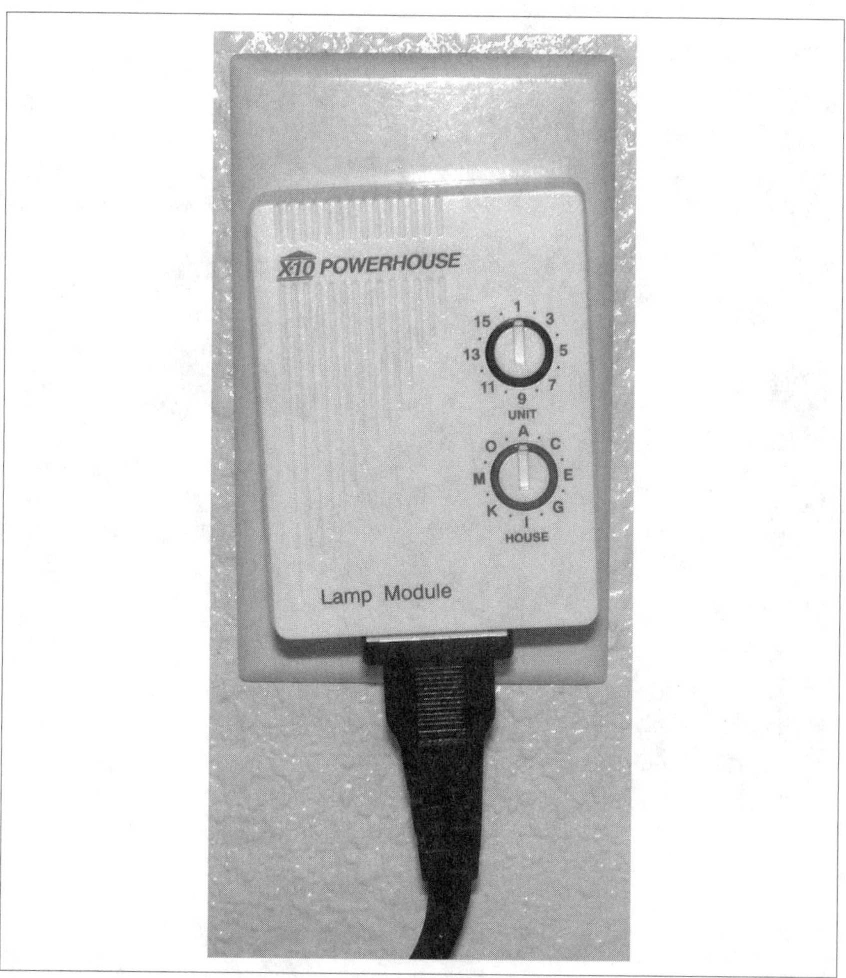


*Figure 1-16. A sight only an enthusiast could love*

Other types of built-in modules include power outlets in which one of the plugs is X10-controllable and the other is not, and multiple-function controllers that mount in the wall, such as the KeypadLinc (*http://www.smarthome.com/12073w.html*; $90) shown in Figure 1-18. A device such as this can replace not just a light switch, but also a plugged-in minicontroller **[Hack #4]** that you might use for triggering scripts.

It certainly costs more to use these better-looking modules, in terms of both the price of the equipment and the work required to install them, but the results can be worth it. To temper the costs, consider using them only where you'll get the biggest benefit. Put some nice switches in the kitchen, for example, but not in the garage or little-used spare bedroom. Also, as with all

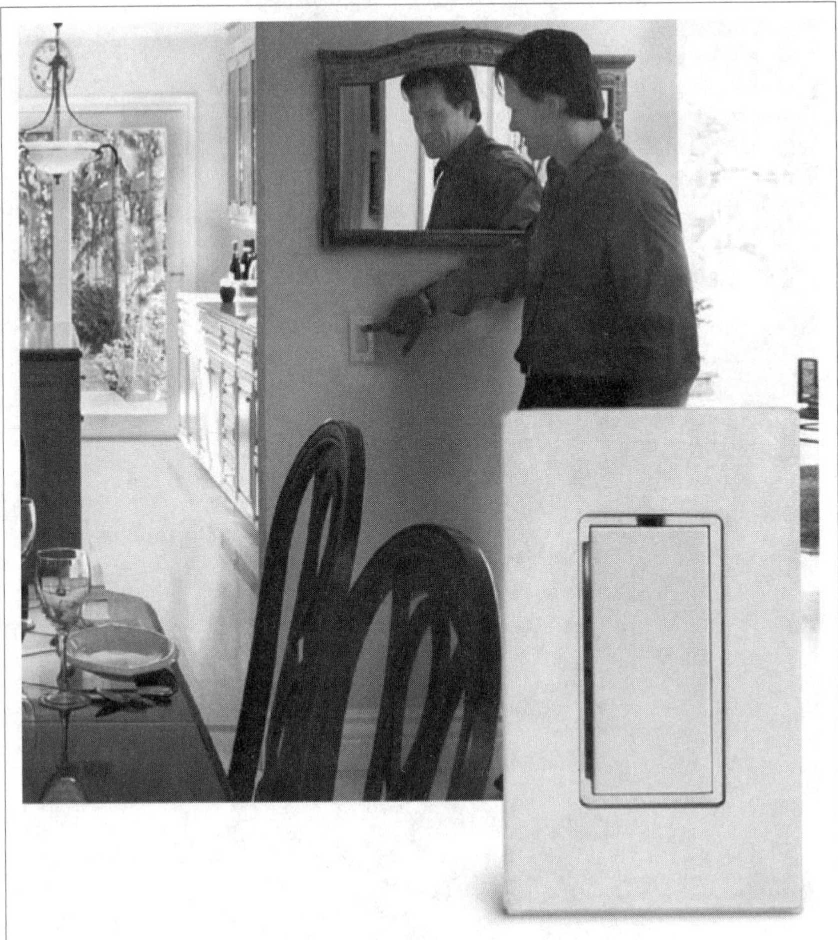

*Figure 1-17. A built-in light switch, an attractive alternative to a lamp module*

things related to home automation, it's a good idea to expand your system little by little to avoid introducing too much change all at once, which is often the cause of hard-to-solve problems.

## Program for Politeness

In addition to the X10 equipment you choose to install, keep your family in mind when you program your home automation system. Consider how and when they might want to override the behavior of your design. For example, although it can be very nice to have lights come on automatically at sunset, occasionally you might want to turn them on earlier, such as when it's dark and stormy outside. In such cases, provide a Palm Pad or another controller that triggers your sunset routine [Hack #20] on demand.

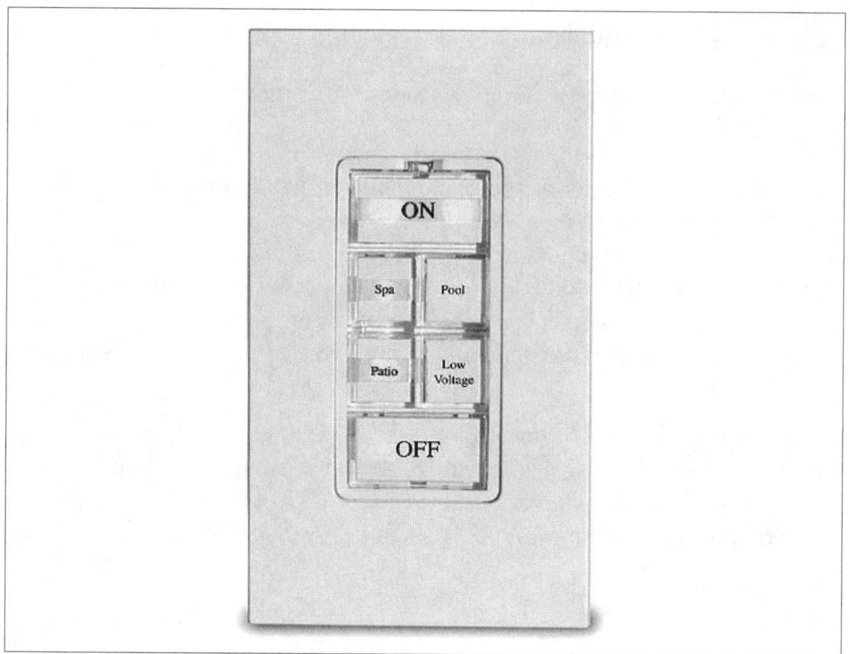


*Figure 1-18. A built-in multifunction controller*

If you frequently have houseguests, consider altering your system so that it works well for them, too. Many visitors will get a kick out of your smart home, but expecting them to adapt to lights that come on by themselves might be asking too much. If you've set up your system so that it knows who is at home [Hack #70], it's easy to create a mode that simplifies, or bypasses, the automations that guests are most likely to encounter. In fact, this can be a good idea for use with anyone in your home who might be less enthusiastic about home automation. For example, you could have the computer play spoken announcements only when you are at home, remaining silent if you're gone.

## Final Thoughts

Like any technology, home automation ultimately is useful only if you use it in a way that enhances or makes your life easier. For some alpha geeks, the fact that you *can* put your computer in charge of your home automatically means that you *should*, but usually that's not a recipe for success. For better results, select a few key things to automate that benefit everyone in the home and that are easy to live with, and grow the system from those starting points. The best measure of cool technology is that you genuinely miss it when it's not functioning, and your home automation system will fit that definition in spades, if you build it with human interest at its core.

# HACK #15 Unplug Your Computer

**Using your computer to control your home makes your home smart. But don't overlook the benefits of a slightly less sophisticated approach.**

You can create an automated home by using different approaches. These approaches vary in terms of the equipment needed and the degree of control and automation they provide. If you simply want to turn on lights or appliances without leaving your easy chair, all you need are a few X10 modules and a wireless Palm Pad **[Hack #5]** or a minicontroller **[Hack #4]**. This approach gives you complete *control* (you push buttons to make things happen), but it provides very little *automation* (if you don't push those buttons, nothing happens).

If you want the lights to come on by themselves at certain times during the day or night, you can use a standalone X10-capable timer, such as the Mini Timer (*http://www.smarthome.com/1100x.html*; $30) shown in Figure 1-19. This box plugs into your wall and you can program it to turn lights on or off, up to twice a day.

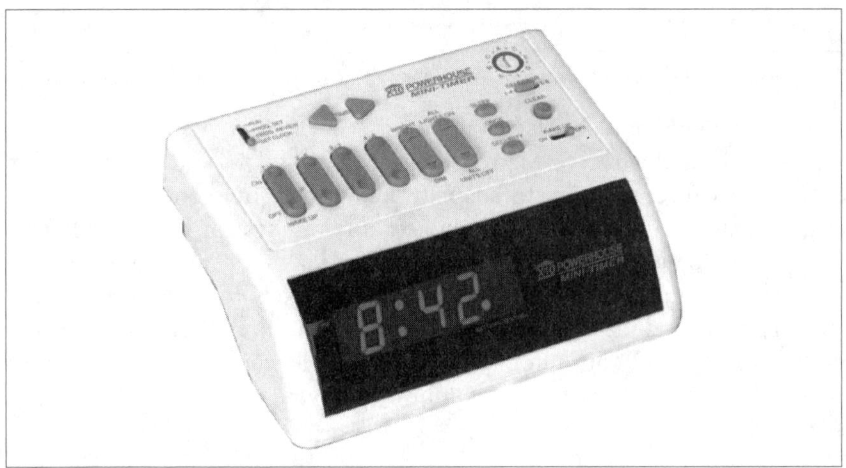


*Figure 1-19. The X10 Mini Timer*

The Mini Timer also has a *security mode* that varies the on-and-off schedule you've entered to provide the appearance that someone might be home. You also can use it for an alarm clock and, optionally, have it turn on your coffee maker **[Hack #37]** when the alarm goes off. With a timer-based system, you're doing more than remotely controlling your home; you're beginning to move toward automation.

Moving up in sophistication are X10 controllers that can use simple logic to make automation decisions, execute a series of X10 commands defined as a *macro*, and execute macros at scheduled times or at sunrise and sunset. You need a personal computer to program these controllers, but once you have done so, they operate by themselves. See "Choose the Right Controller" [Hack #21] for a discussion of some controllers that fall into this category.

Next up, providing the most sophistication and flexibility are computer-based home automation systems. These systems enable you to use sophisticated logic in your automation, such as reacting appropriately based on which house members are at home and whether it's a holiday [Hack #24]. You also can use the other capabilities of your computer, such as voice synthesis [Hack #28] and its Internet connection [Hack #64], to make your home seem smart. See "Add a Brain to Your Smart Home" [Hack #16] for more on this approach.

Each approach has its advantages, and, thankfully, it's not necessary to choose just one. In fact, a mix-and-match approach can result in a very robust and reliable home automation system. Use the technique that's appropriate for the problem you're trying to solve.

For example, you might use a Mini Timer to control your landscape lighting; it needs scheduled events only, not sophisticated logic. If you travel frequently, you might decide to use computer-based home automation when you're at home so that you can benefit from its capabilities. But when you go on a trip, you might put a standalone controller in charge so that you don't have to leave a computer turned on while you're away.

Most of the other hacks in this book focus on using a computer-based system. Because it's simple to program a Mini Timer, let's focus on using a standalone controller here.

## Advantages of a Standalone Controller

In general, standalone controllers are very reliable. They rarely crash or lock up, and they don't have mechanical components that might fail, such as a hard drive. For mission-critical home automation tasks, using a standalone controller can make a lot of sense.

Most standalone controllers can function independently, or in a mode where they pass commands only to the computer to which they're connected. When disconnected from a host computer, they use the logic and scheduled events you've stored in their memory. An example is the USB PowerLinc Controller (1132CU) from Smarthome, Inc. (*http://www.smarthome.com/1132CU.html*; $70). With Indigo [Hack #18], the process of updating the controller for standalone operation is done automatically when you quit the application.

## Programming the Controller

To prepare for standalone use, you need to specify which portions of your home control logic and schedules for which you want the PowerLinc to assume responsibility when Indigo quits. Let's say you have several X10-controlled exterior and interior lights, and you have motion detectors on the front porch, in the backyard, and over the driveway. You have defined trigger actions to turn on the exterior lights for 30 minutes when motion is detected, and you have defined time/date actions to turn on the interior lights one hour after sunset. Additionally, this time/date action is randomized by 25 minutes to give the house a lived-in look and to discourage would-be burglars. These are the actions you want your PowerLinc to continue to perform when Indigo is not in charge.

To flag these actions for standalone operation, choose Upload from the Interface menu. The Upload Settings for Interface dialog appears, as shown in Figure 1-20.

Upload Settings for Interface

| Upload | Compat. | Action Name | Trigger Description |
|---|---|---|---|
| ☑ | good | back yard motion | x10 "backyard motion detector" on received |
| ☑ | good | front porch motion | x10 "front porch motion detector" on received |
| ☑ | good | garage motion (driveway) | x10 "garage motion detector" on received |
| ☑ | okay | dining room lamp with randomize | sunset every day |
| ☑ | [illegible] | [illegible] | [illegible] |
| ☑ | okay | track lighting with randomize | sunset every day |
| ☑ | poor | front porch motion email | x10 "front porch motion detector" on received |

Interface incompatibility: time randomize changed to 15 minutes;

Cancel | Save Changes | Upload Actions

*Figure 1-20. Uploading events to a PowerLinc*

Indigo provides an *upload compatibility rating* for all your home-control actions. This rating tells you how compatible a particular action is with the PowerLinc's standalone mode. Some actions translate perfectly to the controller (*good* compatibility), some partially translate (*okay* compatibility), and some will not translate to work at all (*poor* compatibility).

If an action gets a less-than-perfect rating, select it in the list and Indigo displays more information about it at the bottom of the window, as shown in Figure 1-20. In this case, the action *living room lamp with randomize* has been adjusted automatically to a 15-minute period of randomization. That's the maximum value PowerLinc supports.

The action retains the 25-minute randomization setting when used by Indigo. It's adjusted for the PowerLinc only when it's uploaded to the controller for standalone use.

A standalone PowerLinc can't execute some actions. For example, *front porch motion email*, shown in Figure 1-20, cannot be selected for uploading to the PowerLinc because the action includes steps that send email, which the PowerLinc is incapable of doing.

Even if an action can be used with a standalone PowerLinc, you might not want to upload it to the controller. The controller can store about 1,000 commands (32 KB), but you should consider omitting actions that don't need to take place when you're away from home, or at other times when you're using the standalone controller.

Once you've set up Indigo to upload to the PowerLinc, switching between controlled and standalone mode is as simple as quitting the Indigo application. When you're ready to put the computer in charge again, just launch Indigo and it will resume control.

## Hacking the Hack

If you're a dedicated computer-in-control type of automator, consider adding a standalone PowerLinc to your tool belt. You can simplify your computer-based system by offloading basic automation, such as sunrise and sunset events **[Hack #20]**. However, keep in mind that if both your computer-based system and the standalone controller are working at the same time, you'll need to ensure they don't respond to the same events. This would result in a too-many-cooks-in-the-kitchen situation and would increase the likelihood of X10 signal collisions. It also could make the system confusing to debug.

—*Matt Bendiksen*

# HACK #16 Add a Brain to Your Smart Home

**You don't need a computer to use X10, but you'll be missing out on many great techniques if you don't have one. Take your automation to the next level by letting your computer pilot your home.**

At the most fundamental level, the only thing you need for an automated home is a few X10 modules and a way to send X10 commands, such as a Palm Pad **[Hack #5]** or a minicontroller **[Hack #4]**. For some people, a simple setup that enables you to push a button on your nightstand to turn off the lights in the next room, or have the lights come on automatically when you enter a room [Hack #7], is more than enough. And it's even fun, at least until you discover that it's pretty limited.

But that's so 20th century. You still have to push a button to control the lamp, so all you've really done is relocate the lamp switch from the room next door to your nightstand. That's not automation: that's remote control. Sure, it's a bit more sophisticated than using The Clapper (*http://www.chia.com*), but only because X10 doesn't make you clap your hands together to trigger it.

To move up from remote control to automation, you need to have something making decisions—something that's dedicated, tireless, and works for little pay. Something exactly like a personal computer! Computers excel at doing repetitive, boring tasks such as waiting for an X10 command to which they can respond. It will sit happily, waiting for the sun to set, and then turn on your outdoor lights the instant the sun dips below the horizon. Or, if you're not home, it will make your house look occupied [Hack #74], convincing nearby prowlers that they should look elsewhere for an easy mark.

## What You Need

Of course, it's not quite as simple as just adding a computer into the mix. You'll need a few bits of hardware and some software that's ready to run your house. Let's take a look at each piece individually.

**Computer.** To get started, you'll need a computer that you're willing to leave switched on all the time. You can turn off the monitor, of course, but the CPU needs to be powered on so that it can keep track of what's going on in your house. You can use the computer for other server-like things, such as plyaing iTunes music. But you'll need to keep the home automation software running all the time, so a system that is often occupied playing CPU-hogging games, or switched off at night for some peace and quiet, isn't the best candidate.

The good news is that you don't need the latest, fastest computer to run your home. If you've got an older system sitting around, or if you're looking for an excuse to upgrade to a newer main machine, you've just found a great way to put "ol' pokey" to use. If you're lucky enough to have an old laptop, all the better because they're low-energy and power-failure-resistant (for a couple of hours, at least), as well as quiet and easy to stash out of the way.

**Computer-to-X10 translator.** You'll also need a *power-line interface*, also called a *controller*, which is a box that enables your computer to send and receive X10 commands. It plugs into an electrical outlet, just like a lamp module [Hack #2], and has a serial port for connecting to your computer. The power-line interface acts like a *babelfish* that speaks both X10 and computer code. It listens to the power line for X10 commands, which it translates and sends to your computer over the serial connection. Your computer, in turn, can send commands to the interface; the commands are converted into X10-speak and sent down the power line to all your modules.

You have several different interfaces to choose from, but the most common are X10 Corporation's CM11A, Marrick Limited's LynX-10 or LynX-10 PLC, and Smarthome's PowerLinc USB, shown in Figure 1-21.

*Figure 1-21. The PowerLinc USB power-line interface*

*The CM11A (http://www.x10.com/automation/x10_ck11a_1.htm)*

The CM11A is the most used, least expensive, and most-often-derided power-line interface available. Its reliability and speed aren't the best, but its ubiquity means you won't have a problem finding software that can work with it. The CM11A often is bundled with home automation starter kits that include several modules, which enables you to keep a spare on hand and stay within a reasonable budget. If you just want to try out home automation and aren't sure it's something you'll really dig, buy a CM11A; it will be quite a while before you outgrow its capabilities.

Unique to the CM11A is its ability to function when the computer it's attached to is turned off. It has a tiny bit of built-in RAM to store a series of X10 commands as a macro. This is a fine approach for very simple automation projects, but most serious automators generally avoid it because the macros don't enable you to utilize programming logic for making decisions.

*The LynX family (http://www.marrickltd.com/lynx105.htm)*

The LynX-10 and LynX-10 PLC interfaces, in contrast to the CM11A, are reliable and much loved. They're also more expensive, so you might want to save your pennies until you are seriously hooked on this new hobby. The LynX interfaces have several specialized functions that your home automation software will need to support to be useful. But most software packages can use them in their basic modes, and even if you don't take advantage of the fancy stuff, you'll appreciate how quickly the LynX sends X10 commands.

*The PowerLinc USB (http://www.smarthome.com/1132u.html)*

The PowerLinc USB is one of the newest interfaces on the market, so it hasn't yet established a reputation for itself. Because it's a newcomer, it doesn't yet enjoy the widespread software support of the CM11A or LynX. But if the software you want to use will work with it, it's well worth considering. For one thing, it's the only interface that has a built-in USB port. That makes it much easier to connect to any new-ish computer. The PowerLinc USB also has a pass-through electrical plug on the front, so it doesn't completely block access to the outlet you plug it into, a feature that increases its acceptance around the home [Hack #14].

*Other interfaces*

Experienced home automators might be sputtering over the omission of the Adicon Ocelot (*http://www.appdig.com*) and the JDS Stargate (*http://www.jdstechnologies.com*). Both are capable products, but more commonly are installed and supported by home automation consultants. Check out the companies' web sites if you want to learn more. Both offer unique functions, such as infrared control, which you might find useful.

The X10 FireCracker, the TW523 Two-Way, and the X10 CP290 interfaces, however, are best avoided. The CP290 and TW523 interfaces are quite limited in their abilities, and the FireCracker is a send-only device. That is, you can use it to send X10 commands from your computer, but not to react to commands from other devices. It's fun to play with, but it's not capable of running a smart home.

For more about the features of each controller, and how they compare, see "Choose the Right Controller" [Hack #21].

## Hooking It Up

This brings us to the boring, yet important, topic of adapters and cables. Both the CM11A and LynX have an old-school, nine-pin serial port for connecting to the computer, so you're probably going to need a serial-to-USB converter. The Keyspan High Speed USB Serial Adapter (Part #USA-19HS, *http://www.keyspan.com/products/usb/USA19HS/*) is a good choice, though similar adapters should work just as well.

If you're using a desktop computer with a spare expansion slot, consider adding a standard serial card and foregoing the USB-serial conversion. If you're using an older Macintosh that has a round DIN8 serial port, a palmOne Mac Serial Adapter (*http://www.amazon.com/exec/obidos/tg/detail/-/B00000JHVS/qid=1090876873*) will do the trick.

## Choosing Your Software

The last piece of the puzzle is home automation software. As luck would have it, several good software packages are up to the task of running your home. All of these programs can send and receive X10 commands and run scheduled events, and each enables you to script your own sequence of actions so that you can fine-tune your system to your every desire. Indeed, it's in the scripting where all the fun and personalization become possible, as evidenced by so many hacks in this book.

Read this list to get a brief overview of each application, and then refer to the hacks that walk you through the basics of using and configuring each package. Naturally, you'll find the applications' web sites to be really useful.

**Applications for Macintosh.** XTension, by Sand Hill Engineering (*http://www.shed.com*), is the granddaddy of Macintosh home automation software. It has a multiwindow interface that you can use to control your X10 devices directly, and it has extensive support for AppleScript so that you can tailor the system to your heart's content. For more information about XTension, see "Get to Know XTension" [Hack #17].

Indigo, by Perceptive Automation (*http://www.perceptiveautomation.com*), is a Cocoa-based application that works with AppleScript but also has a unique interface that enables you to define home automation tasks without writing any code. It also works with the PowerLinc USB controller, which eliminates the need to buy a USB-to-serial adapter. For more information about Indigo, see "Get to Know Indigo" [Hack #18].

**Applications for Windows.** ActiveHome, by X10 Corporation (*http://www.x10.com/support/support_soft1.htm*), is bundled with home automation starter kits and is free to download and use, so it's the first introduction to home automation for many people. You can use it to create macros that are stored in the CM11A and to program simple timed events. It's not sophisticated enough to implement most of the hacks in this book, but it's OK for getting started.

HomeSeer, by HomeSeer Technologies LLC (*http://www.homeseer.com*), is a very popular home automation package that supports a wide variety of add-on devices and includes a built-in web server for controlling your home via the Internet. You can script HomeSeer using VBScript, JScript, or Perl. For more information about HomeSeer, see "Get to Know HomeSeer" [Hack #19].

Home Automated Living (HAL), from Home Automated Living, Inc. (*http://www.automatedliving.com*), is also popular, and it features some very interesting options for using voice recognition to control your home. For example, with the right kind of modem, you can pick any extension phone in your house and tell HAL what you want it to do.

**Applications for Linux.** As is typical with Linux, a lot of interesting home applications are available, but to get a full-featured solution, you'll probably have to use two or three separate applications in conjunction with each other. For example, the powerful Heyu (*http://heyu.tanj.com/heyu/*) is a command-line application that can send X10 commands. Many automators use it with Xtend (*http://www.jabberwocky.com/software/xtend*), which can execute commands when the computer receives an X10 signal. By using both, you get two-way control. But when you use them separately, each is only half of the solution.

If you want to access your Linux-based home automation system over the Web, you'll want a copy of BlueLava (*http://www.sgtwilko.f9.co.uk/bluelava/*), which is a CGI that works with Xtend and Heyu to provide enough tinkering room to keep you busy for a long time.

Most of the Linux programs are available for Mac OS X, too, if you want to get down-and-dirty with the command line instead of using Indigo or XTension. Chuck Shotton's port of Heyu and Xtend to Mac OS X is available at *http://www.shotton.com/muscle/*.

Another popular choice for Linux is MisterHouse (*http://misterhouse.sourceforge.net/*). Written entirely in Perl so that it works with Mac OS X as well as Windows, it has an impressive array of features including support for voice, web access, and integration with Outlook scheduling.

### Final Thoughts

A lot of people who have dabbled with home automation and then lost interest because it seemed so limited never discovered the almost limitless possibilities that are added when you use a powerful home automation software package. Sure, it takes some effort to decide what you want to do and to learn how to do it, but it's something that's easy to grow into as your skills develop. And don't be shy about asking for help. Most of the products mentioned here have a thriving community of users who are friendly and helpful to newbies, so be sure to explore these resources as you decide which package will work best for you.

## HACK #17 Get to Know XTension

**To implement many of the hacks in this book, you'll need to know some basic techniques for programming the XTension home automation application for Mac OS X.**

XTension is the granddaddy of X10-based home automation software for the Macintosh, having originated several years ago for the Classic Mac OS.

In fact, if you have an older Macintosh, even a lowly Mac Plus, XTension (*http://www.shed.com/*; $150) is a great way to put your doorstop-Mac back in service.

But don't let its modest origins fool you. Beneath the old-school user interface lies an automation powerhouse. Its designer and chief programmer, Michael Ferguson, helped NASA automate its space shuttle operations, and he makes sure XTension is every bit as capable for those of us who aren't rocket scientists.

For more detail and the latest skinny, be sure to consult the XTension user guide. Choose Download Manual from the XTension menu in the application to get the latest version, or visit *http://www.shed.com/download.html*.

## Getting Connected

XTension can handle all the home automation programming you can throw at it, but to send commands to X10 modules, it needs a power-line interface **[Hack #16]**. XTension currently works with the CM11A, LynX-10, and LynX-10 PLC. These are standard serial devices, so if you're using a newer Macintosh, you'll also need a USB-to-serial interface, such as the Twin USB Port Serial Adapter from Keyspan (*http://www.keyspan.com*). If you're using a Macintosh that has a round serial port, you'll need an adapter to convert the connection to a standard nine-pin configuration. A palmOne Mac Serial Adapter (*http://www.palmone.com*) will do the trick.

After you've connected the power-line interface to your Mac, you'll need to configure XTension. To do this, choose Preferences from the XTension menu, and then click Communications. Check the Enable X-10 checkbox and select the interface you're using, such as a CM11A, from the pop-up menu. See Figure 1-22 for an example.

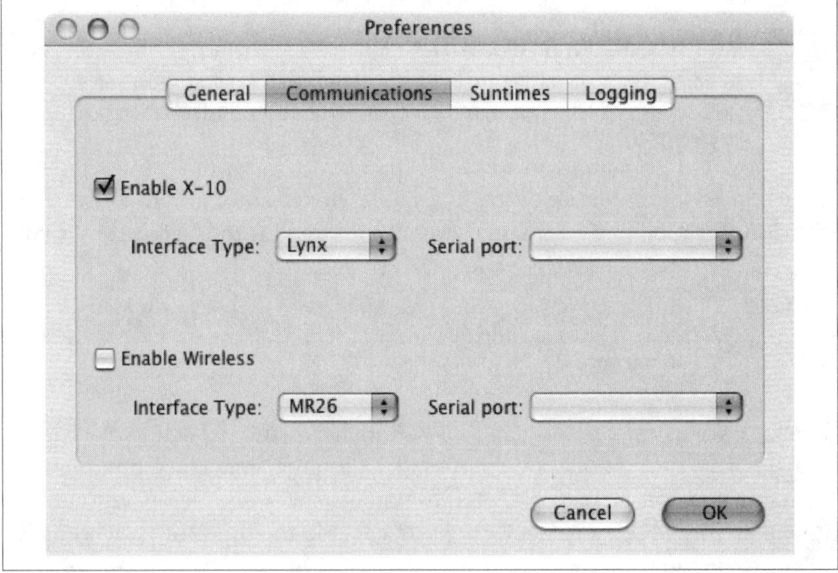


*Figure 1-22. Selecting a home automation interface*

Why in the world do you have to enable X10 before you can choose an interface? Well, you can use XTension as a wireless-only controller or simply as a way to execute scripts and automated tasks automatically—such as sending email or anything else you can do with AppleScript. If you leave both wireless and X10 turned off, you still can use XTension to kick off anything you want to happen on a regular basis.

XTension also works with two wireless X10 receivers, the MR26A and the W800 **[Hack #83]**. These are excellent ways to improve your wireless motion detectors and the like, but wireless X10 is not a replacement for the powerline interfaces discussed earlier. If you're just getting started, don't worry about wireless X10 yet. Just know that it will work with XTension when you're ready to go there.

Now that you've got XTension set up, let's start adding the details it will need to run your home.

## Adding a Unit

XTension needs to know about the X10 devices you have installed in your home so that it can send commands to them in response to events, your actions, or scripts that execute on a scheduled basis. XTension refers to X10 devices—lamp modules, motion detectors, appliance modules, and so on—as *units*. To add a new one, choose New Unit from the Edit menu. The New Unit dialog appears, as shown in Figure 1-23.

Every unit needs a name, so fill in the Name field with something meaningful. Because you might be referring to this name often in your scripts, choose something that is easy to type and reminds you of the device it represents. *Bedroom Lamp*, *LampMBR*, or *MBRL1* are common approaches to unit naming.

The next important field is the X10 Address **[Hack #1]**. Here, type in the address that you assigned to this X10 module. In this example, the lamp in my bedroom is plugged into a lamp module whose address is set to `House Code A`, `Unit Code 10`—a.k.a. `A10`.

The Description field is optional, but you might find it handy. You'll be able to see it in the Master List of units (see Figure 1-24), but if you're planning on using a hack that dynamically sets the description **[Hack #27]**, don't bother filling it in, as whatever you put there now will just be changing later.

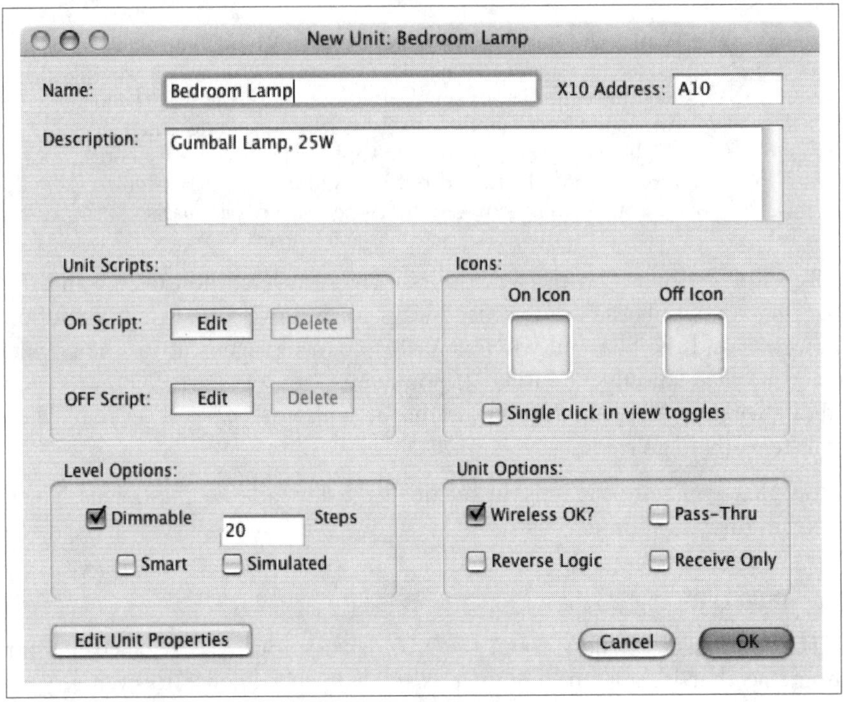

Figure 1-23. Adding a new unit

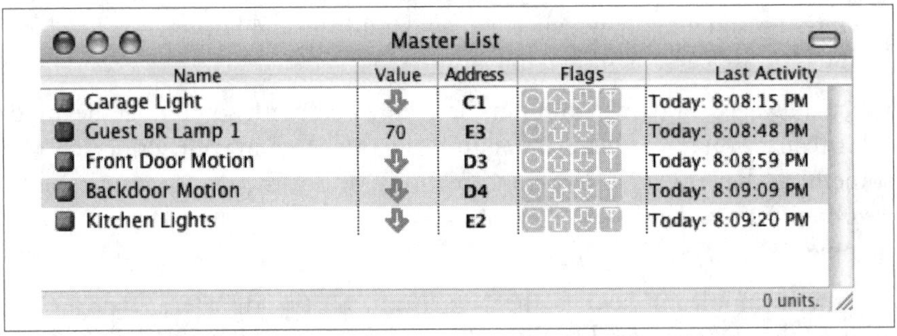

Figure 1-24. The Master List of units

If you're adding a lamp module and you want to be able to set the brightness level, be sure to check the Dimmable checkbox. When you do this, the Master List will show the current brightness level in the Value column. In Figure 1-24, the unit `Guest BR Lamp 1` is dimmed to 70%. If you don't select this, `Dim` commands will be ignored, which can be quite perplexing to troubleshoot unless you notice the entry that gets put in the log when this occurs.

If the unit you're adding sends its signals wirelessly (primarily this will be a motion detector) and you have a Wireless-to-X10 interface [Hack #5], check the "Wireless OK?" checkbox. This tells XTension to accept wireless commands for this unit. If you don't check this, incoming wireless commands for this address will be noted in the log, but otherwise ignored.

## Adding Pseudo Units

Just as you're getting used to the idea of units, let's throw in a measure of possible confusion for the fun of it. It turns out that a particular unit doesn't necessarily have to correspond to a particular X10 module. You can, in fact, add units that don't specify an X10 address. These are called *pseudo units* and you'll use them for tracking the states of things in your house, such as the temperature, who is at home, or the last time it rained.

To create a pseudo unit, you choose New Unit from the Edit menu, as you do for adding an actual unit, but leave the X10 Address field blank. Figure 1-25 shows a unit for storing the temperature and sky conditions.

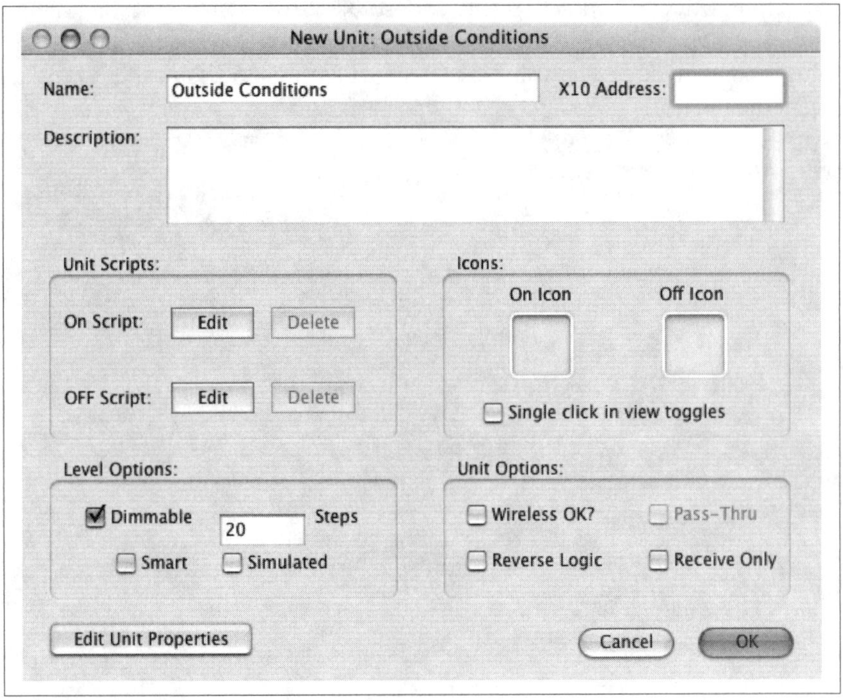

*Figure 1-25. Creating a pseudo unit*

Notice that the Dimmable checkbox has been checked. This is so that we can use the *value* of the unit to store the outside temperature. Usually, the value indicates the brightness level of a lamp module, but in reality it holds any negative or positive whole number, so even if you live in chilly Chicago, or steamy Key West, XTension can more than adequately track the current conditions **[Hack #64]**.

You also can use the Description field of a unit to store bits of information, as mentioned previously, or a unit's Properties list if you need to keep track of even more metainformation **[Hack #61]**.

## Scheduling Events

If you want something to happen at a specific time in the future, either once or repeatedly, set up a scheduled event. For example, you might want the air cleaner in the office, which is connected to an appliance module, to come on for three hours every other day. Or perhaps you need to get up early tomorrow, and you want the coffeepot to be turned on at 4:30 a.m.

To create a new scheduled event, choose New Event from the File menu. The New Event dialog is where you specify what you want to happen and when and how often you want it to occur. Figure 1-26 shows the XTension Event dialog.

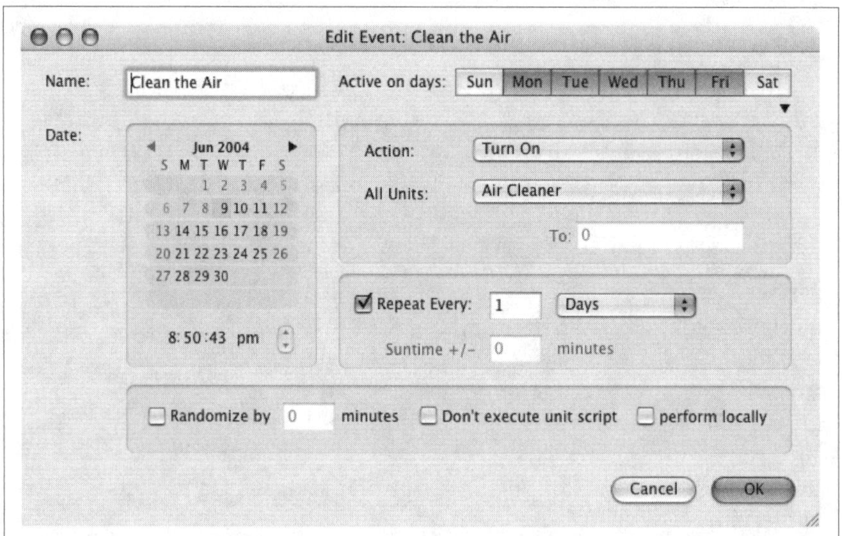


*Figure 1-26. Creating a scheduled event*

Enter a name for the event so that you can identify it easily in the Scheduled Events window (see Figure 1-26), and then select the days on which you want this event to be active and the time of day when it should happen. If you want it to occur more than once, check the Repeats checkbox and enter a value. Choose an interval, such as Daily, from the pop-up menu.

Select what you want this event to do by selecting an action from the pop-up menu, and then a target from the All Units menu. For example, to turn on a unit or group, choose Turn On. Then choose the unit name. To run a global script you've written, choose Execute Script and select the script from the second pop-up menu.

Figure 1-26 shows an event called `Clean the Air` that turns on the `Air Cleaner` unit every weekday at 8:50 a.m.

To view all of your system's scheduled events, choose Scheduled Events from the View window, as shown in Figure 1-27. In the window, click the "Sort by Date" column to show the events in the order in which they'll occur. The first column shows some rather cryptic but useful symbols. The `R` next to an event indicates it is set to repeat, and the upward arrow indicates that the unit shown below the event name, such as `Kitchen Lights`, will be turned on. The notation "as" indicates that the event runs a scripted action.

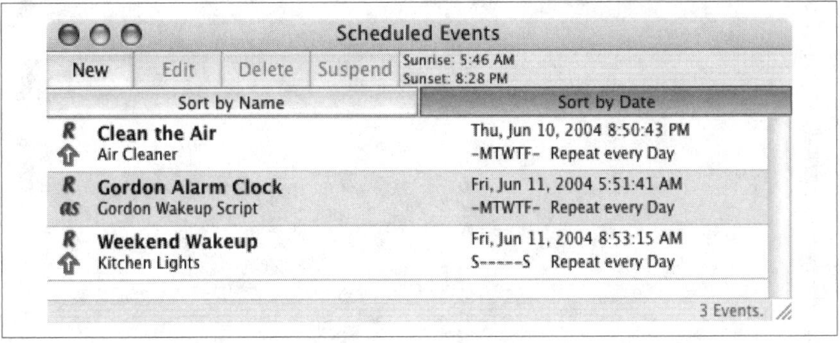


*Figure 1-27. Scheduled events*

## Responding to Events with Unit Scripts

Although scheduled events are handy, having your home automation system respond to stimuli puts the "smart" in your *smart home*. Like the man behind the curtain in *The Wizard of Oz*, it's XTension responding to events—motion detectors triggering, garage doors opening—which makes it all happen.

For example, when you press a button that tells your home you're departing **[Hack #70]**, XTension runs a script when it receives the On command sent by the button. Or, after you've gone downstairs in the morning, the Off command sent by the motion detector in the bedroom triggers a script that turns off all the upstairs lights.

To define what happens when a unit turns on or off, select the unit in the Master List, then choose Edit Unit from the Edit menu. In the Edit Unit dialog box, click the Edit button next to On Script or Off Script. A little window opens where you can edit, or import, an AppleScript. When an On command is received for this unit, XTension executes the script you enter in the On Script window. You've probably guessed that it executes the Off script when it receives an Off command, right?

In Figure 1-28, we're entering the On script for the Air Cleaner unit at address K2. This unit is turned on every weekday with a repeating event, as described earlier, but instead of setting up another repeating event to turn it off a few hours later, the unit's On script schedules it to turn off automatically. Whenever the air cleaner is turned on, the script creates a one-time scheduled event that automatically turns off the unit three hours later.

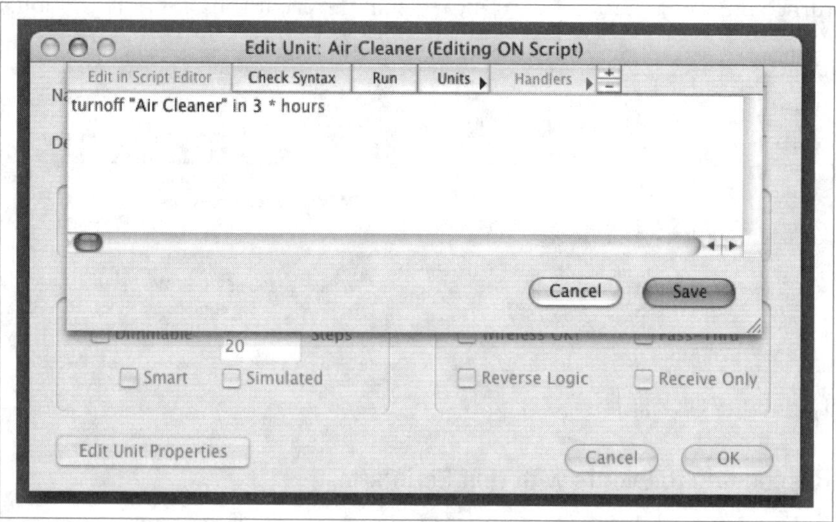


*Figure 1-28. Editing an On script*

You can test this by using a Palm Pad **[Hack #5]** to send an On command to the unit's address (K2). When you do, the air cleaner turns on, and then the On script is executed and a new event appears in the Scheduled Events window, as shown in Figure 1-29. The 1 in the first column indicates that this event will happen only once (that is, it's not a repeating event) and the down arrow shows that it will turn off the `Air Cleaner` unit. XTension generates the name of the scheduled event, `Air Cleaner #1`, automatically by adding a number to the unit name. This ensures that each event has a unique name.

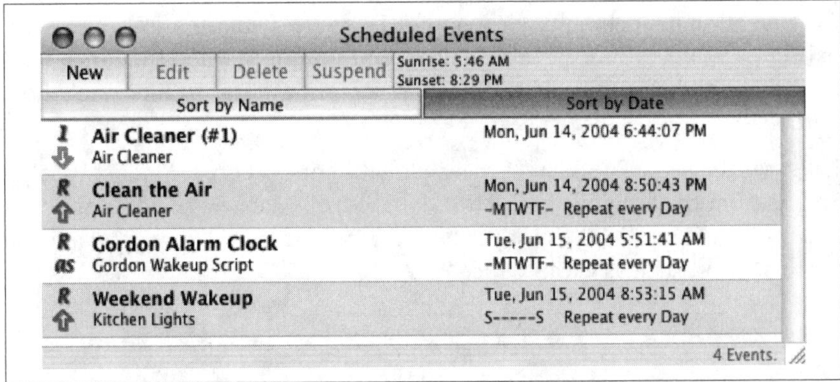


*Figure 1-29. Scheduled event added by the On script*

You should remember two key things about unit scripts. First, the On or Off scripts are executed whenever that unit receives the indicated command. It doesn't matter how that command was initiated—by a scheduled event, by pressing a button on a Palm Pad, by clicking the unit in XTension, or via another script—the On and Off scripts will run. In our example here, this permits us to ensure that the air cleaner is always turned off after three hours, without regard to how it is turned on.

The second thing to remember is that each unit script is available only to the unit for which it is defined. That is, you can't share scripts between multiple units, nor can you execute a script without also turning the unit on or off. If you have a script that you want multiple units to be able to use, or you want to use it directly without having to trigger a unit, you should use a global script instead.

A common mistake for new users is to create an Off script that includes a command that turns the unit off. If you do this, you'll create an infinite loop when an incoming `Off` command runs an Off script that sends an `Off` command that runs a... Well, you get the picture. Don't do this.

## Using Global Scripts

A *global script* is an AppleScript that a scheduled event can execute or another script can call. In XTension, global scripts are the glue that ties your whole automation system together.

Earlier in this discussion, we created a scheduled event that turned on an air cleaner every weekday, and then we used a unit script to create a second scheduled event that turned the air cleaner off after three hours. That's pretty neat, but it's not very smart; it doesn't take into account anything else that might be happening in your home. To make it smarter, let's use a global script to make sure the air cleaner comes on only when the office window is closed because it doesn't make much sense to try to filter the whole outdoors.

Let's get started. To create a new global script, choose Manage Global Scripts from the Scripts menu. The Global Script Manager window appears, as shown in Figure 1-30. Click the New Script button and then enter a name for this script, such as Clean Office Air.

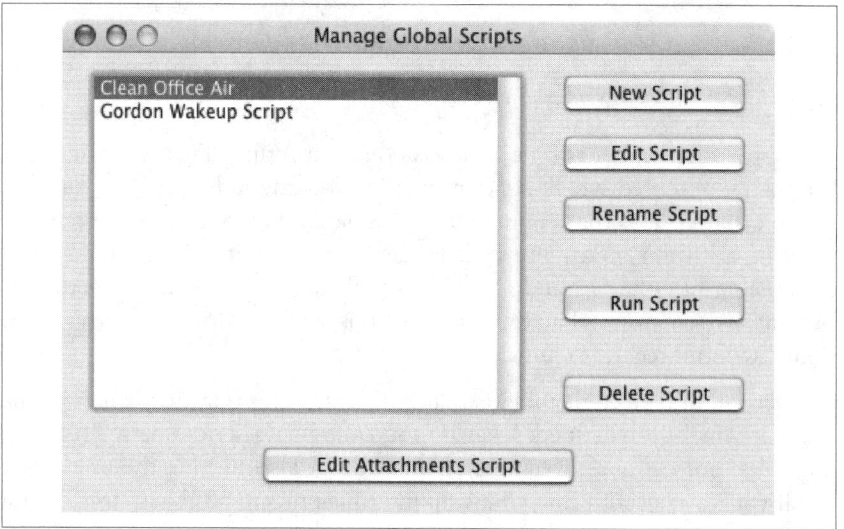


*Figure 1-30. The Global Script Manager window*

The Script Editor window (Figure 1-31) is where you enter your script. After you've typed it in (or imported one you've written in Apple's Script Editor), click the Check Syntax button to make sure it compiles without any errors.

When you've got it perfected, click Save. The new script now appears in the Global Scripts Manager window, and XTension adds it to the Scripts menu so that you can execute it easily by selecting it. But in this example, we want to run this script automatically, so let's modify the scheduled event we created earlier.

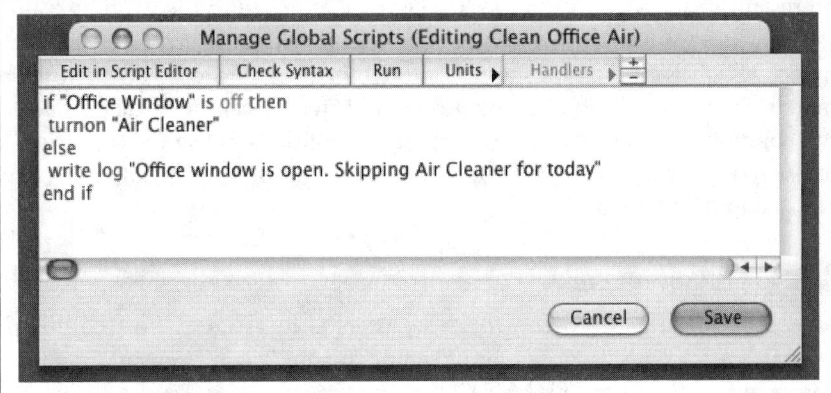


*Figure 1-31. A smarter Air Cleaner script*

Choose Scheduled Events from the Window menu, and then double-click Clean the Air in the Scheduled Events window (Figure 1-29) to edit the event. In the Action pop-up menu, choose Execute Script. Then, in the All Units pop-up menu, choose the global Clean Office Air script we just created. Click OK to save the changes. The `Clean the Air` scheduled event now appears as shown in Figure 1-32.

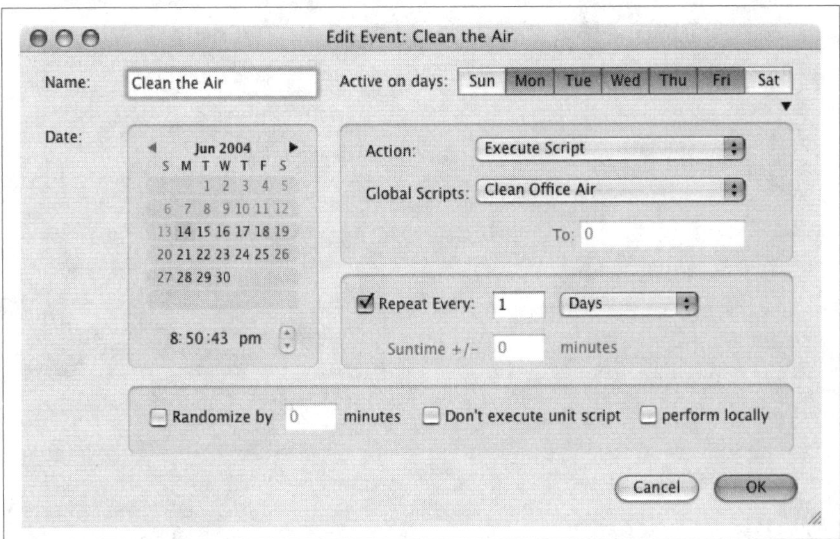


*Figure 1-32. A scheduled event to execute a global script*

Let's take a closer look at this script, which you can see in the Global Script Editor window in Figure 1-31. It's simple, but it illustrates some key points. First, it checks the status of something called `Office Window`. That's a unit representing an X10 security module [Hack #78] attached to the window frame in the office. When the window is opened, the module sends an `On` command, which XTension sees and dutifully changes the status of the unit to match. So, the script, in checking to see if `Office Window` is off, determines if the window is currently open. If not, it turns on the air cleaner; otherwise, a message is written to the XTension log indicating that the cleaning cycle is being skipped this time.

## Creating Groups of Units

One of the best ways to control many units is to assign them to a *group*. Using a group can save you a lot of typing in your script; you simply send a command to the group and XTension automatically repeats the command for each unit individually. To create a group, choose New Group from the File menu. Then drag units from the Master List window to the "Units in this Group" list in the New Group dialog box, as shown in Figure 1-33. The order of the units in the list determines the order in which they'll be commanded. You can change the lineup at any time by dragging the units into a new order.

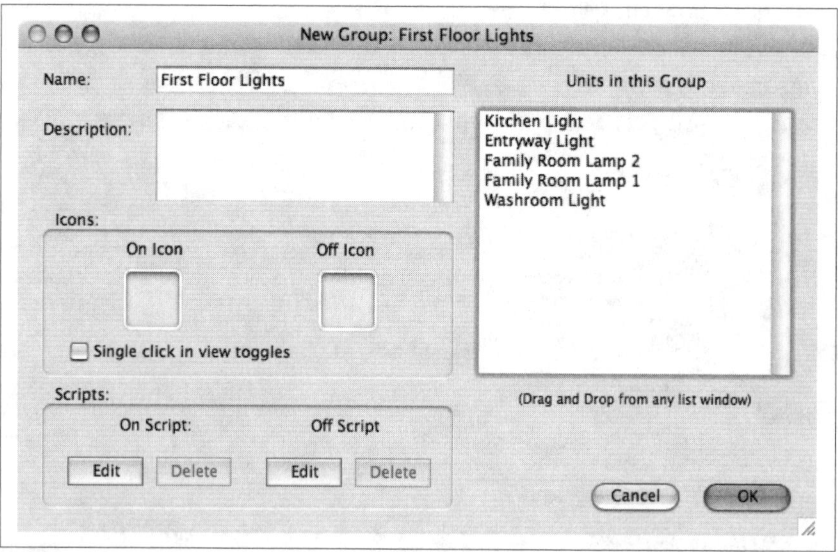


*Figure 1-33. Creating a group*

Be sure to give the group a distinctive name, and if you're an advanced user, you might even want to create On and Off scripts for the group. As with the unit scripts discussed earlier, XTension will execute these when the group receives an `On` or `Off` command.

Note that groups don't have an X10 address, so you can't send them a command using a Power Pad or minicontroller; they're accessible only from within the XTension user interface or scripts. To command a group in a script, you address it as if it were an individual unit:

```
Turnoff "First Floor Lights"
```

For more information about using groups effectively, see "Control Lights in a Group" **[Hack #95]**.

## Using the XTension Log

Speaking of the log file, XTension keeps an extensive record that enables you to see exactly what's going on with your home automation system. Not only does the log keep track of everything XTension does, but also, because the application is listening to the power line, you'll see X10 commands that are sent from other sources as well, such as minicontrollers **[Hack #4]**. Figure 1-34 shows just a few minutes of activity from a typical day. To show the log, choose Log Window from the Window menu.

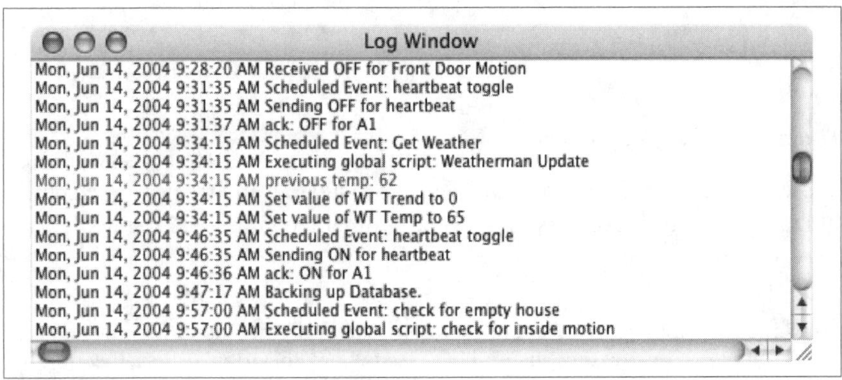


*Figure 1-34. A typical XTension log file*

All of this detail makes the XTension log a very valuable debugging tool, but it also means it's overwhelming to view if everything is running smoothly. To reduce some of the clutter, you can set XTension preferences so that only critical information is displayed in the Log Window, as shown in Figure 1-35.

If you select the "Database items only" option, the Log Window shows X10 commands for the units that you defined in XTension. In other words, if XTension sees a command on the power line and it doesn't have a unit that corresponds to that address, that command will not be shown in the Log Window. If you select the "Exceptions only" option, the Log Window will

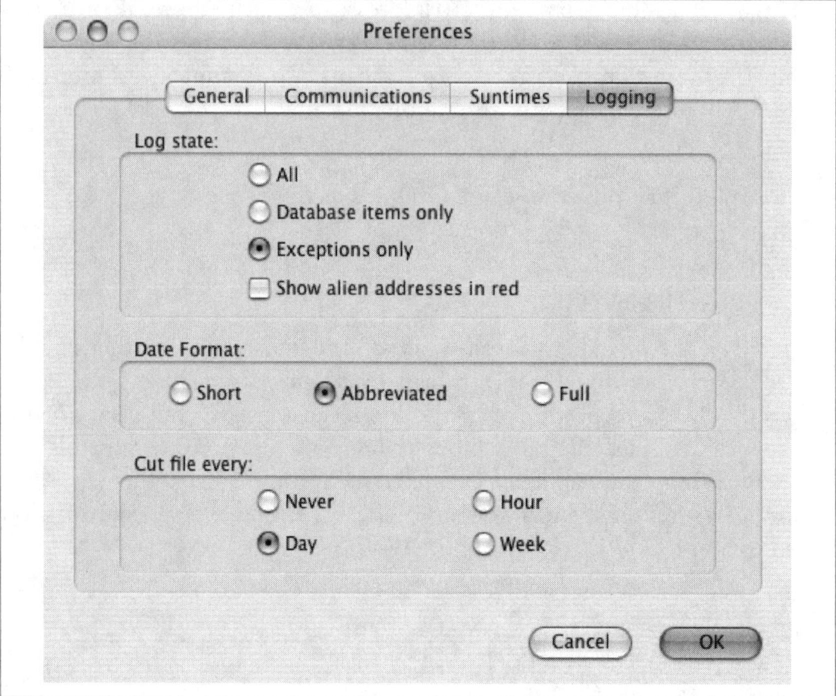


*Figure 1-35. Setting log preferences*

show a message only when there's an error, such as an attempt to execute a global script that doesn't exist, or a problem communicating with your power-line interface.

Either of the logging options can really cut down on the visual clutter in the Log *window*, but they don't affect the information written to the XTension log *file*. XTension always writes all activity to the log file, so even if you've filtered the log display by changing XTension's preferences, you always can get the full story by opening the log file with TextEdit or another text editor.

## Learning More

Now you know enough about XTension to begin building your system, but be sure to visit Sand Hill's extensive web site (*http://www.shed.com*) and peruse the extensive technical notes and tutorials, too. You'll also want to check out the product's active, and friendly, email discussion list. Even if you're not a Mac user, it offers a daily dose of practical home automation techniques, news, and chit-chat.

# HACK #18 Get to Know Indigo

**Using the Indigo home automation application for Mac OS X requires some basic techniques that you'll need to know in order to implement many of the hacks in this book.**

Indigo, from Perceptive Automation, is the newest home automation software for the Mac, but it already has earned an enthusiastic following for its clean, Cocoa-based implementation and its support of the latest equipment. Its designer and chief programmer, Matt Bendiksen, has a hand in programming some of the best mainstream Macintosh applications, and he's applied that experience to making sure Indigo is approachable for beginners and experts alike.

For more detail and the latest skinny, be sure to consult the Indigo user guide—available from Indigo's Help menu—and the online tech support area (*www.perceptiveautomation.com/indigo/support.html*).

## Getting Connected

For Indigo to send commands to X10 modules it needs a power-line interface **[Hack #16]**. Indigo currently works with the CM11A, the LynX-10 PLC, and the PowerLinc Serial, PowerLinc USB (1132U), and PowerLinc Controller (1132CU). With the exception of the PowerLinc USB, the others are standard serial devices, so if you're using a newer Macintosh you'll need a USB-to-serial interface, such as the High-Speed USB Serial Adapter from Keyspan (*http://www.keyspan.com/products/usb/USA19HS/*), to make the connection to your computer. If you're using a Macintosh that has a round serial port, you'll need an adapter to convert the connection to a standard nine-pin configuration. A palmOne Mac Serial Adapter (*http://www.palmone.com*) will do the trick.

If you don't already have a power-line interface, you should consider getting the PowerLinc USB (*http://www.smarthome.com/1132u.html*). It's inexpensive, and because it has a USB port rather than a standard serial port, you won't have the added expense of buying an adapter. At the time of this writing, Indigo is the only Macintosh home automation software that works with the PowerLinc USB, and it also supports the capability of storing macros in the PowerLinc Controller (1132CU), so you can have some automation even when your computer is turned off **[Hack #15]**.

After you've connected a power-line interface to your Mac, you'll need to configure Indigo. To do this, choose Preferences from the Indigo menu and then click Interface. Check the "Interface communication online" checkbox, and then check the "Enable X-10 interface" checkbox. Select the type of interface you're using from the pop-up menu. See Figure 1-36 for an example.

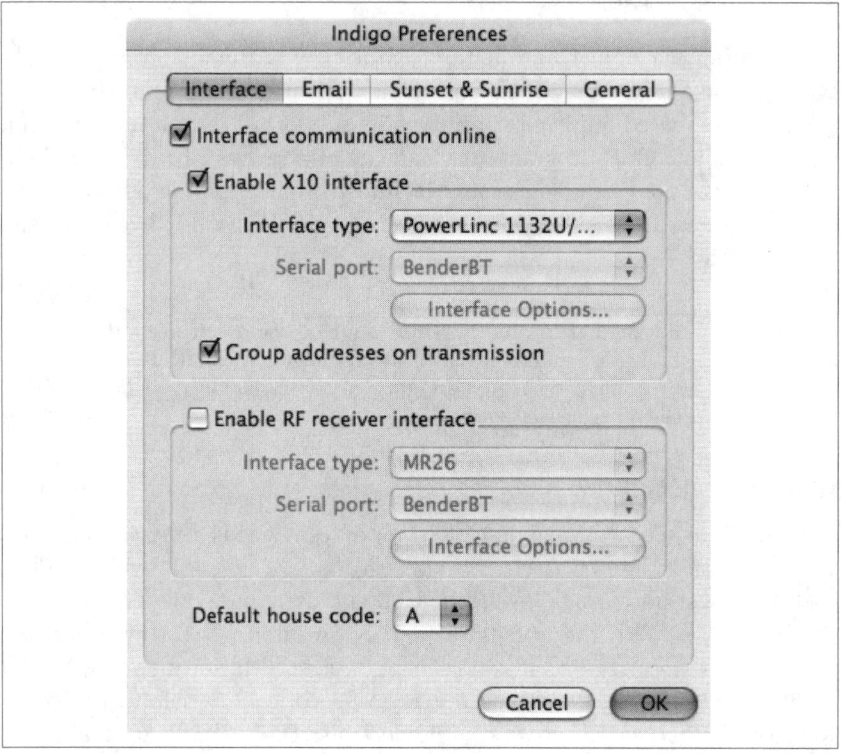


*Figure 1-36. Selecting a home automation interface*

Why do you have to turn on "Interface communication online" *and* "Enable X10 interface" before you can choose an interface? Well, Indigo can be used as a wireless-only controller or simply as a way to automatically execute scripts and automated tasks, such as sending email or anything else you can do with AppleScript. If you leave both wireless and X10 turned off, you still can use Indigo to schedule things you want to happen on a regular basis.

Indigo also works with two wireless X10 receivers, the MR26A and the W800RF32 **[Hack #83]**. These are excellent add-ons for improving your system, but wireless X10 is not a replacement for the power-line interfaces discussed earlier. If you're just getting started, don't worry about wireless X10 yet. Just know that it will work with Indigo when you're ready to go there.

Now that you have Indigo set up, let's start adding the details it will need to run your home.

## Adding a Device

Indigo needs to know about the X10 devices you have installed in your home so that it can send commands to them in response to events, your actions, or scripts that execute on a scheduled basis. To add a device to Indigo, choose New Device from the Edit menu. The Edit Device dialog appears, as shown in Figure 1-37.

*Figure 1-37. Adding a new X10 device*

Every device needs a name, so fill in the Name field with something meaningful. Because you might be referring to this name often in your scripts, choose something that is meaningful and easy to type. *Bedroom Lamp*, *LampMBR*, or *MBRL1* are common approaches to unit naming.

Select the type of module you're adding, using the Type pop-up menu. It's important that you choose the correct type of device, as this tells Indigo what commands it can accept. For example, when you choose Lamp Module, Indigo knows this module can accept `Dim` commands to set the lamp's brightness level.

The next important field is the X10 Address **[Hack #1]**. Here, select the address that you assigned to this X10 module. In this example, the gumball lamp in my bedroom is plugged into a lamp module whose address is set to `House Code A`, `Unit Code 10`—a.k.a. `A10`.

The Description field is optional, but you might find it handy. You'll be able to see it in the list of units Indigo displays when you click the Devices button on the left side of its window, as shown in Figure 1-38.

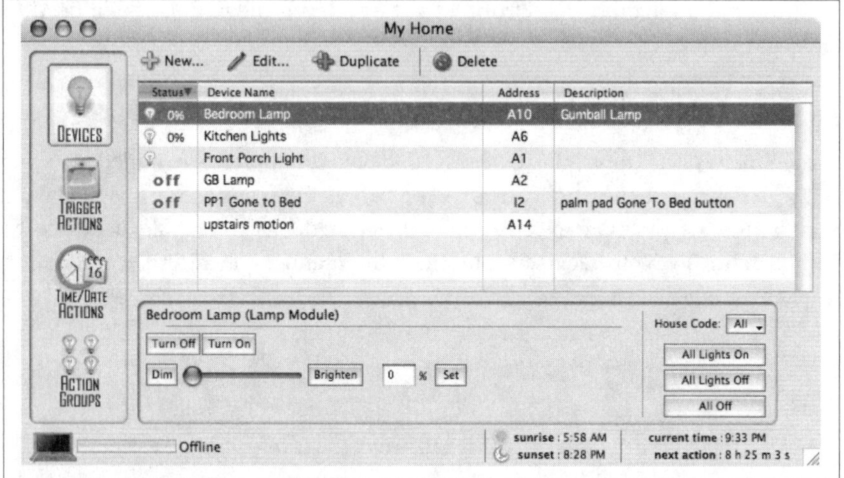


*Figure 1-38. Indigo's list of devices*

## Adding Variables

Sometimes you need to keep track of the states of things in your house, such as the temperature, who is at home, or the last time it rained. To do this in Indigo, you use a *variable*. A variable is a place where you can store some text, a number, or a true/false Boolean value.

Some hacks in this book, which are described as implemented in XTension **[Hack #17]**, will refer to pseudo units. Indigo users should use variables instead. Keep this in mind if you're adapting a hack that's described using XTension for use in Indigo.

To create a variable, choose Variable List from the Window menu, then click the New button in the window's toolbar, shown in Figure 1-39. In the dialog that appears, enter a name for the variable and an initial value, such as `false`.

Indigo variables are useful for tracking things such as the current outside temperature **[Hack #64]** or knowing which house members are currently at home **[Hack #70]**.

## Scheduling Events

If you want something to happen at a specific time in the future, either once or repeatedly, set up a scheduled event. For example, you might want the air cleaner in the office, which is connected to an appliance module, to come on

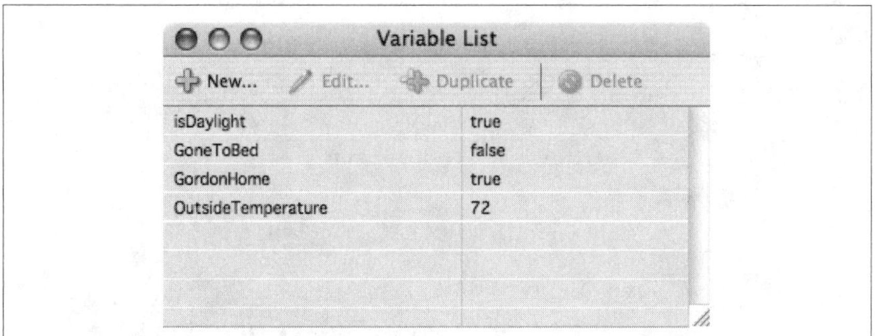


*Figure 1-39. Indigo's Variable List window*

for three hours every weekday. Or, perhaps you need to get up early tomorrow, and you want the coffeepot to be turned on at 4:30 a.m.

To create a new scheduled event, choose New Time/Date Action from the File menu. The Create New Time/Date Action dialog is where you specify what you want to happen, when, and how often it occurs. Figure 1-40 shows an event called `Clean The Air` that turns on the `Air Cleaner` unit every weekday at 8:50 a.m.

Enter a name for the event so that you easily can identify it later. Then, click the Time/Date Trigger and select the days on which you want this event to be active, and the time of day when it should happen. If you want it to occur more than once, make sure the "Delete this action after first trigger time" checkbox is unchecked.

Select what you want this event to do by clicking the Action tab and then make a selection from the Type pop-up menu. For example, to turn on a unit, choose Send Device Action and choose the device name. To run a script you've written, choose Execute AppleScript and then either select the script file on your hard disk, or enter the script into the dialog box's editing field.

To see all of your system's scheduled events, click the Time/Date Actions button on the left side of the main window, as shown in Figure 1-41. To view details about the scheduled action, select it in the list and Indigo displays a summary near the bottom of the window.

## Responding to Events with Trigger Actions

Although scheduled events are handy, having your home automation system respond to stimuli puts the "smart" in your *smart home*. Like the man behind the curtain in the *The Wizard of Oz*, it's Indigo responding to events—motion detectors triggering, garage doors opening—which makes all the magic happen.

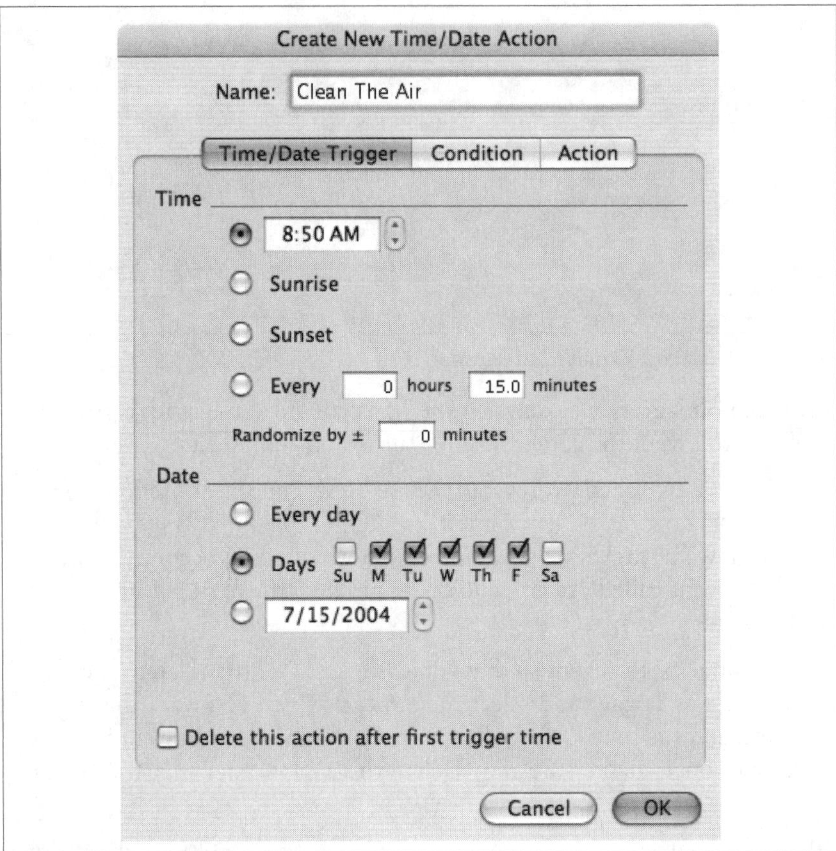


*Figure 1-40. Creating a scheduled event*

For example, when you push a button that tells your home you're departing [Hack #70], Indigo runs a script when it receives the On command sent by the button. Or, after you've gone downstairs in the morning, the Off command sent by the motion detector in the bedroom triggers actions that turn off all the upstairs lights.

To define what happens when an event occurs, choose New Trigger Action from the File menu. The Create New Trigger Action dialog box appears, as shown in Figure 1-42. Enter a meaningful name for this action, and then select an action from the Type pop-up menu. This defines what action will cause this trigger to execute, such as receiving an X10 command. The rest of the dialog changes depending on the action you select. In this example, we define that this trigger is active when the Air Cleaner device receives an On command.

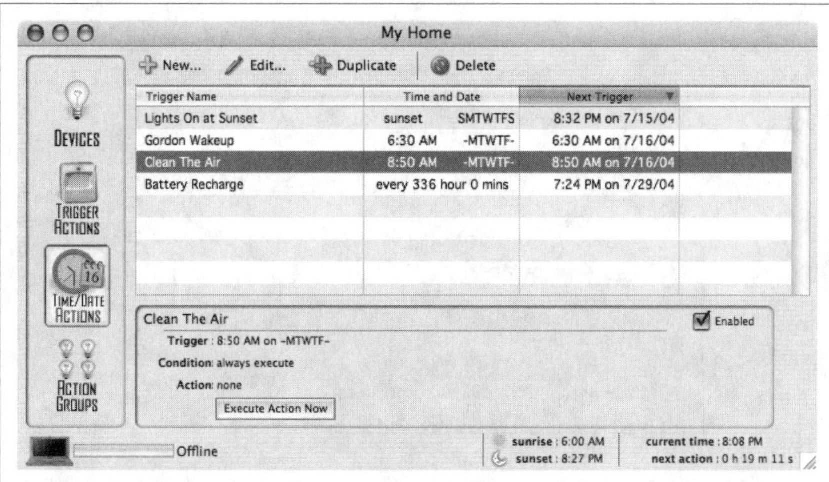


*Figure 1-41. Scheduled events*

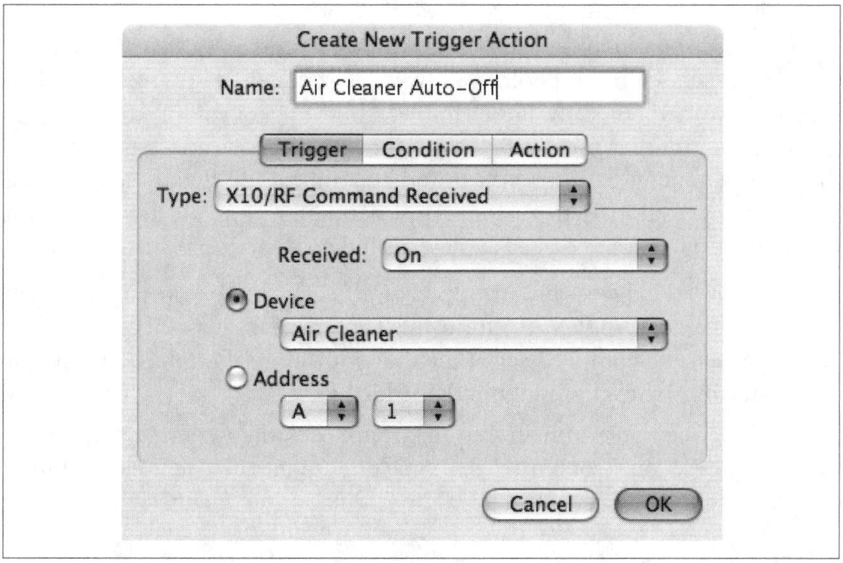

*Figure 1-42. New Trigger Action dialog box*

To specify the actions Indigo will perform for this trigger, click the Action tab. In Figure 1-43, we're specifying that 90 minutes after this trigger is activated, Indigo will send an `Off` command to the `Air Cleaner` unit. This ensures that whenever the air cleaner is turned on, it automatically will turn off later. If you wanted to specify that this action should occur only during daylight hours, or based on some other variable you're keeping track of, you can set this up using the Condition tab.

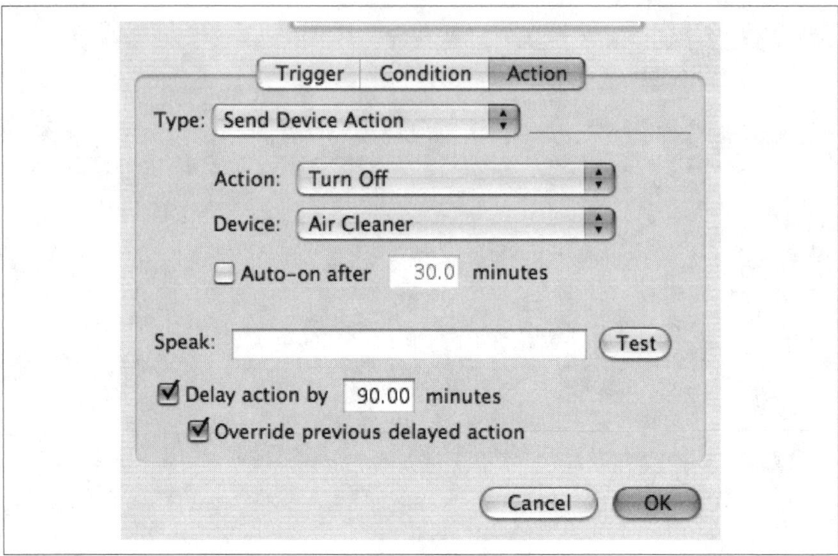


*Figure 1-43. Specifying the Indigo trigger actions*

Many of the hacks in this book work by using scripts that are triggered when a device is turned on or off. For example, a button on a Palm Pad might trigger an On script that turns off all the lights in the house and sets a variable that you've gone to bed [Hack #48]. In Indigo, you can do this by setting up a trigger whose action is to run an AppleScript, or you can use an Action Group to define a series of steps you want Indigo to perform:

- To have the trigger run an AppleScript, click the Action tab and then select Execute AppleScript from the Type pop-up menu. Then you can either select a compiled script file that you've saved on disk, or type (or paste) an AppleScript into the Embedded field.
- To have the trigger run an Action Group, click the Action tab and select Execute Action Group from the Type pop-up menu. Then select a previously defined action group from the Group pop-up menu.

## Defining Action Groups

*Action groups* are, quite simply, a list of actions that Indigo performs in the sequence you specify. To define an action group, choose New Action Group from the File menu. In the dialog box that appears, enter a name and a description (if you want), and then click the New button to define the first step in the sequence.

Just as with action triggers, you can have Indigo send an X10 command, set the value of a variable, execute an AppleScript, and perform several other actions for each step in the sequence. You even can execute another action group, which enables you to modularize your action groups into discrete, reusable chunks.

After you've defined your first action, simply click New to add another. The action group shown in Figure 1-44 speaks using text-to-speech, turns off the lamp in the guest bedroom, executes another action group that turns off the outdoor lighting, and then sets the `GoneToBed` variable to `true`.

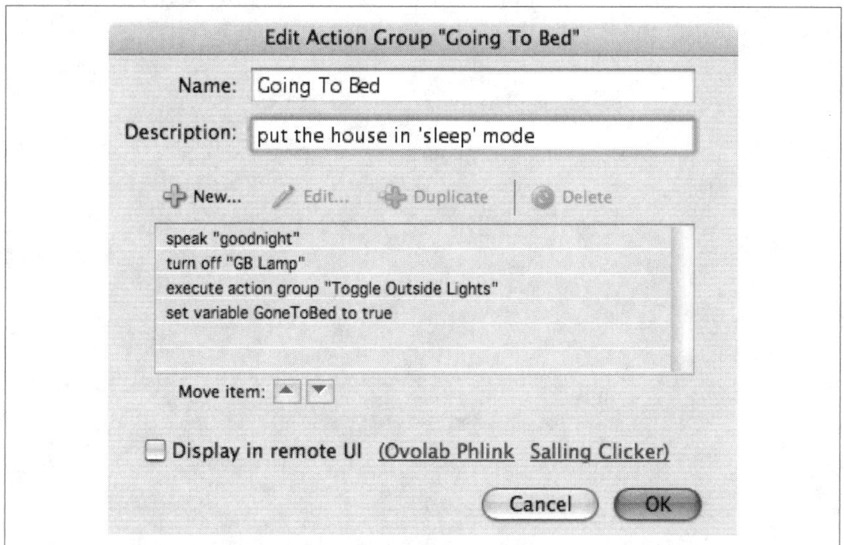


*Figure 1-44. An action group*

## Using Scripts

A *script* is a series of instructions, written using AppleScript, that can be executed by a scheduled event or even called by another script. Often, a script is the glue that ties your whole automation system together and uses the logic necessary to make your home behave more intelligently.

Earlier in this discussion, we created a scheduled event that turned on an air cleaner every weekday, and then we used a trigger action to turn it off after three hours. That's neat, but it's not particularly smart; it doesn't take into account anything else that might be happening in your home. To make it smarter, let's use a script to make sure the air cleaner comes on only when the office window is closed because it doesn't make much sense to try to filter the whole outdoors. Our script will also make the air cleaner come on only when someone is at home, so the air filter doesn't run when there's nobody around to enjoy its results.

You don't necessarily need a script to do this—you could just have the trigger action check a variable that indicates if a window is open. But using a script is often more straightforward and flexible for making logical tests, and learning how to use it through this simple example will serve you well as you expand your system.

Let's get started. Create a new action group by choosing New Action Group from the File menu. Enter a name for the group, and then choose Execute AppleScript from the Type pop-up menu. Click the Embedded button and enter your script in the field provided, as shown in Figure 1-45.

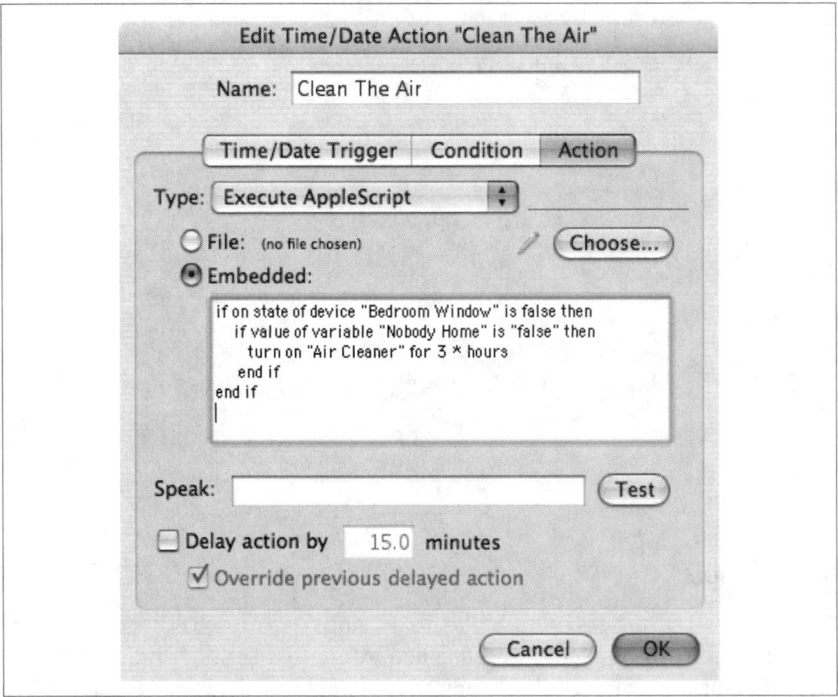


*Figure 1-45. An embedded script*

Click OK to save the script, and then click OK again to save the action group. Now you have an action group with a single step that executes your script. To call this script from anywhere else in Indigo, such as from a time/date action or from another action group, choose it as part of an `Execute Action Group` command.

For example, set up a repeating event. Choose New Date/Time Action from the File menu, and define an event named `Air Filtering` that executes the action group you just created every day at 8:00 a.m. When this event is triggered, the script embedded in the action group will turn on the air cleaner if the bedroom window is closed, and the NobodyHome variable indicates that the house is occupied. If you want to clean the air at other times, simply set up additional date/time actions that refer to the same action group.

## Using the Indigo Log

Indigo keeps an extensive record that enables you to see exactly what's going on with your home automation system. Not only does the log keep track of everything the program does, but also, because it's listening to the power line, you'll see X10 commands that are sent from other sources as well, such as minicontrollers [Hack #4]. Figure 1-46 shows just a few minutes of activity from a typical day. To show the log, choose Log Window from the Window menu.

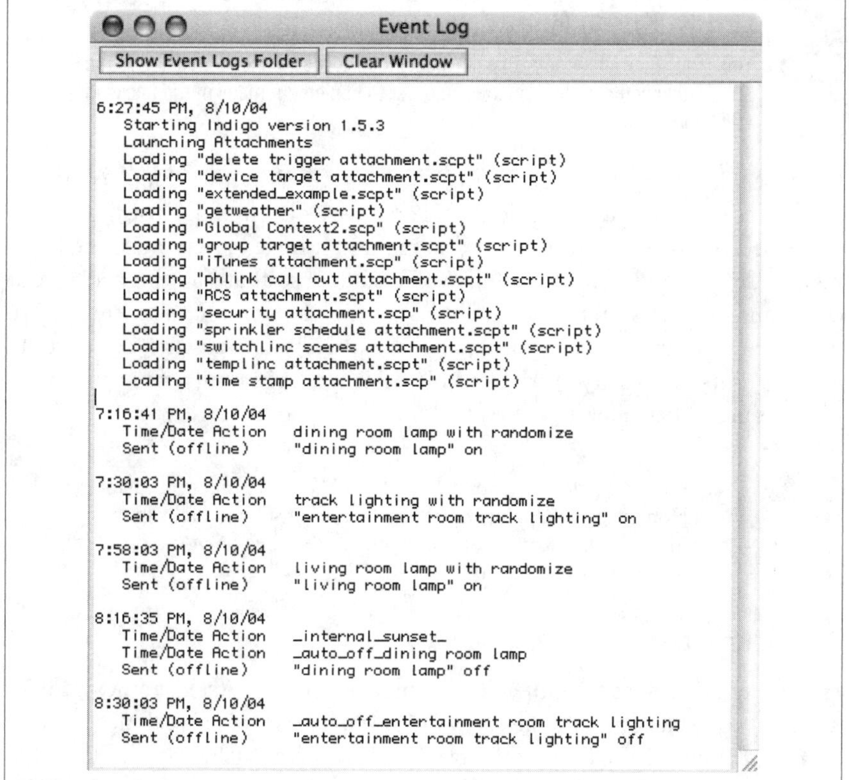


*Figure 1-46. A typical Indigo log file*

The Indigo Event Log window shows only the last several hundred log entries, or only the entries that occurred since the last time you clicked the Clear Window button. To view the full log file, or the log from a previous day, click Show Events Log Folder to open a Finder window with each daily log Indigo has created. These are plain-text files, so you can open them with TextEdit or something similar to view all their detail.

## Learning More

Now you know enough about Indigo to begin building your system, but be sure to visit Perceptive Automation's extensive web site (*http://www.perceptiveautomation.com*) and peruse the technical notes and tutorials. You'll also want to check out the product's discussion board (*http://www.perceptiveautomation.com/indigo/forum.html*) for the latest information and tips from other Indigo users, as well as the library of user-contributed scripts (*http://www.perceptiveautomation.com/indigo/scripts.php*).

# Get to Know HomeSeer

**You'll need to know some basic techniques to using the HomeSeer home automation software for Windows in order to implement many of the hacks in this book.**

HomeSeer is a popular program for Windows XP, Windows 2000, Windows 98 Second Edition, and Windows NT. By way of a familiar user interface, HomeSeer enables you to view and command a huge variety of home automation devices and modules, offers voice control so that you can speak your commands to the system, and includes a built-in web interface so that you can access your system from anywhere there's a network connection. Thanks to its popularity and extensible design, a large number of third-party plug-ins are also available that add additional capabilities.

For more details and the latest skinny, be sure to consult HomeSeer's help system. It's available from the Help menu in the application or online at *http://HomeSeer.com/products/homeseer/WebHelp/homeseer.htm*.

## Getting Connected

HomeSeer adeptly can handle all the home automation tasks you can throw at it, but to send commands to X10 modules, it needs a power-line interface [Hack #16]. HomeSeer works with several controllers, including the CM11A, LynX-10, and LynX-10 PLC, which are standard serial devices. If your computer has a USB port, you can use a SmartLinc USB or another controller that works with USB.

After you've connected the power-line interface to your computer, you'll need to configure HomeSeer. Choose Options from the View menu and then click the Interfaces tab. Click the Device list and select the interface you're using, such as a CM11A, and then select the Port where you've connected the power-line interface, as shown in Figure 1-47.

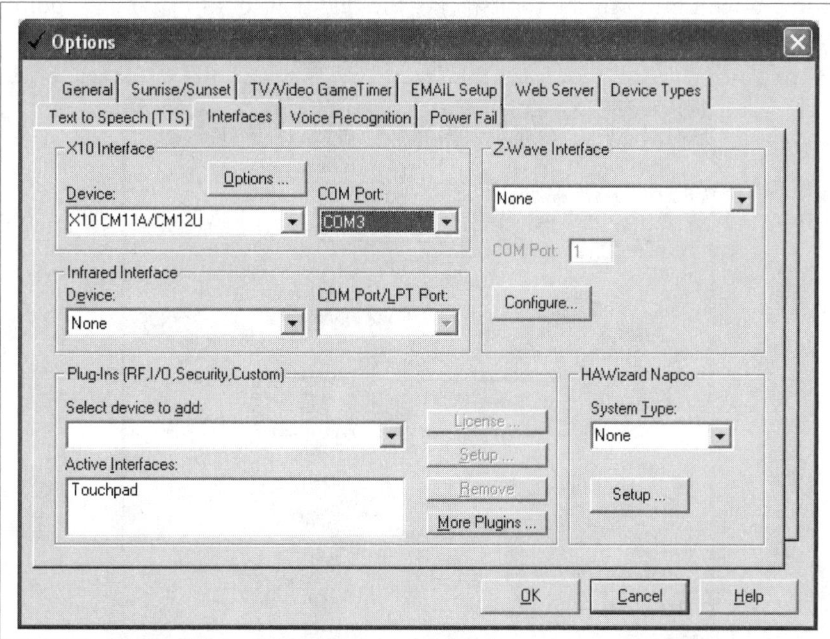


*Figure 1-47. Selecting a home automation interface*

HomeSeer also works with wireless X10 receivers **[Hack #5]**. These are excellent ways to improve your wireless motion detectors and the like, but wireless X10 is not a replacement for the power-line interfaces discussed earlier. If you're just getting started, don't worry about wireless X10 yet. Just know that it will work with HomeSeer when you're ready to go there.

HomeSeer's support for different types of devices is impressive. In addition to several X10-compatible interfaces, it also works with infrared and other types of controllers. Plus, it supports Z-Wave (*http://www.zen-sys.com/*), which is a wireless technology that someday might supplant X10.

Now that you've got HomeSeer set up, let's start adding the details it will need to run your home.

## Adding a Unit

HomeSeer needs to know about the X10 devices you have installed in your home so that it can send commands to them in response to events or to your actions, or by executing scripts on a scheduled basis. To add a device to HomeSeer, click the Devices button in the Views pane and then click the Add Device button in the toolbar (it's the one that looks like a light bulb). The Device Properties dialog appears, as shown in Figure 1-48.

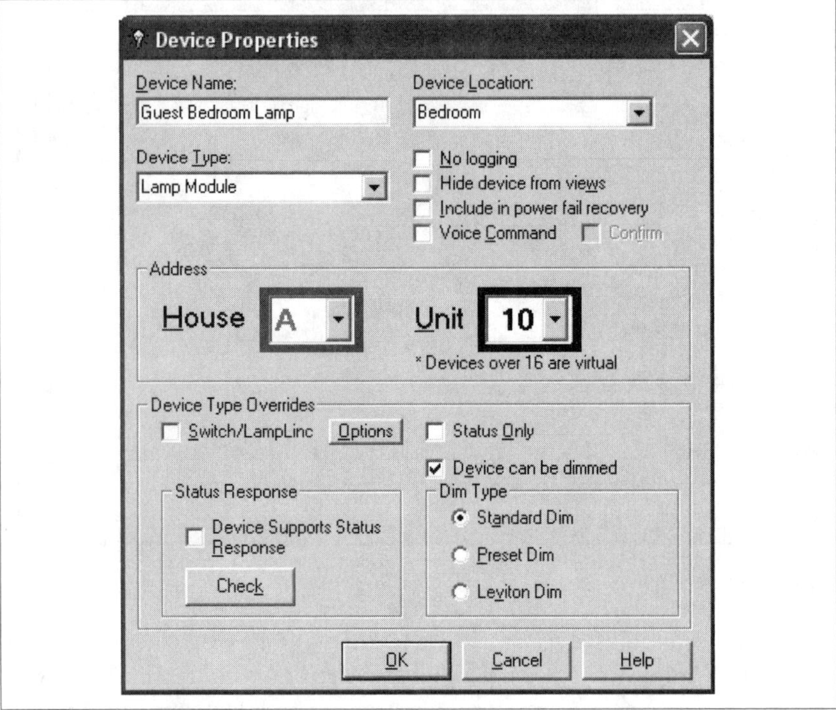


*Figure 1-48. Adding a new X10 device to HomeSeer*

Every device needs a name, so fill in the Device Name field with something meaningful. Because you might be referring to this name often in your scripts, choose something that is easy to type and reminds you of the device it represents. Because HomeSeer enables you to control your home using voice recognition, choose device names that are easy to say, too, such as *Bedroom Lamp*, *Overhead Light*, or *Side Table Reading Lamp*. Also keep in mind that HomeSeer will add the device's location, which you'll get to specify in a moment, to the name—for example, *Master Bedroom Overhead Light*, which can help you to distinguish between what otherwise would be similarly named devices.

The next important field is the Address [Hack #1]. Here, enter the address you assigned to this X10 module. In this example, the lamp in the guest bedroom is plugged into a lamp module whose address is set to `House Code A`, `Unit Code 10`—a.k.a. `A10`.

To ensure that HomeSeer knows about the capabilities of the new device, select its type from the Device Type list. When you do this, it sets several useful default options—such as the Status Only checkbox for motion detectors, which indicates that the device sends commands but does not receive them—but you can change the default options that HomeSeer elects if you need to.

If you're adding a lamp module and you want to be able to set the brightness level, make sure "Device can be dimmed" is selected. If you don't select this, `Dim` commands will not work as you expect, which can be quite perplexing to troubleshoot. When "Device can be dimmed" is selected, the device list will show the current brightness level in the Value column. In Figure 1-49 the `Guest Bedroom Lamp` device is dimmed to 70%.

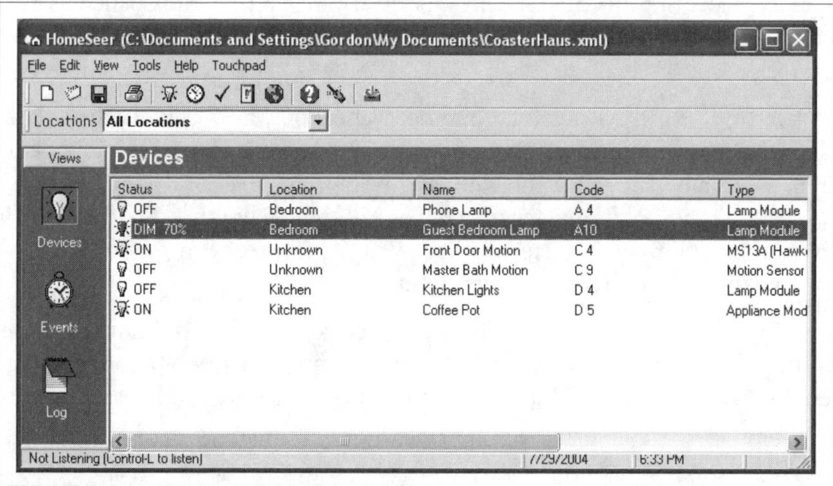


*Figure 1-49. The master list of devices*

The Device Location field is optional, but you might find it handy because it helps make the names unique, as described earlier. You'll be able to see it in the master list of devices (see Figure 1-49), which you can sort by location, to more easily locate devices in the list. It's also a good idea to change the location from its default value, `Unknown`, to avoid confusion when activity for the device is shown in the HomeSeer log file as `Unknown Guest Bedroom Lamp`.

## Adding Virtual Devices

Just as you're getting used to the idea of devices, let's throw in a measure of possible confusion for the fun of it. It turns out that a particular device doesn't necessarily have to correspond to a particular X10 module. You can, in fact, add devices with addresses that are invalid—that is, devices that have house codes beyond `P` or unit codes greater than `16`. These are called *virtual devices*, and you'll use them for tracking the states of things in your house, such as the temperature, who is at home, or the last time it rained.

Some hacks in this book, which are described as they'd be implemented in the XTension application **[Hack #17]**, refer to pseudo units. These are the equivalent of HomeSeer's virtual devices. Keep this in mind if you're adapting a hack that was described using XTension for use in HomeSeer.

To create a virtual device, click the Devices button in the Views pane and then click the Add Device button in the toolbar, as you do when adding an actual device. But this time specify an out-of-range X10 address, such as `A19`. Figure 1-50 shows a virtual device used for storing the temperature and sky conditions.

Notice that the "Device can be dimmed" checkbox has been checked. This is so that we can use the value of the device to store the outside temperature. Normally, the value indicates the brightness level of a lamp module, but in reality, it holds any negative or positive whole number. So, even if you live in chilly Chicago, or steamy Key West, HomeSeer can more than adequately track the current conditions **[Hack #64]**.

You can also use the Status field of a device to store bits of information, if you need to keep track of additional metainformation **[Hack #61]**, for example. To do this, you can use HomeSeer's scripting language to add custom status values:

```
hs.DeviceValuesAdd "A19","~Partly Cloudy" & chr(2) & "76" & chr(1)
```

In this script, `chr(2)` and `chr(1)` separate the status text and the value of the unit. This command associates the `76` with the string `Partly Cloudy`. The tilde (~) in front of the status text indicates that this value will not be added to the device's pop-up status selector in the HomeSeer interface. But from this point forward, whenever `A19` is dimmed to 76%, its status will display `Partly Cloudy` instead of `Dim 76%`, as HomeSeer usually would display. See the discussion of `hs.DeviceValue` in the Scripting section of HomeSeer's onscreen help system for more information.

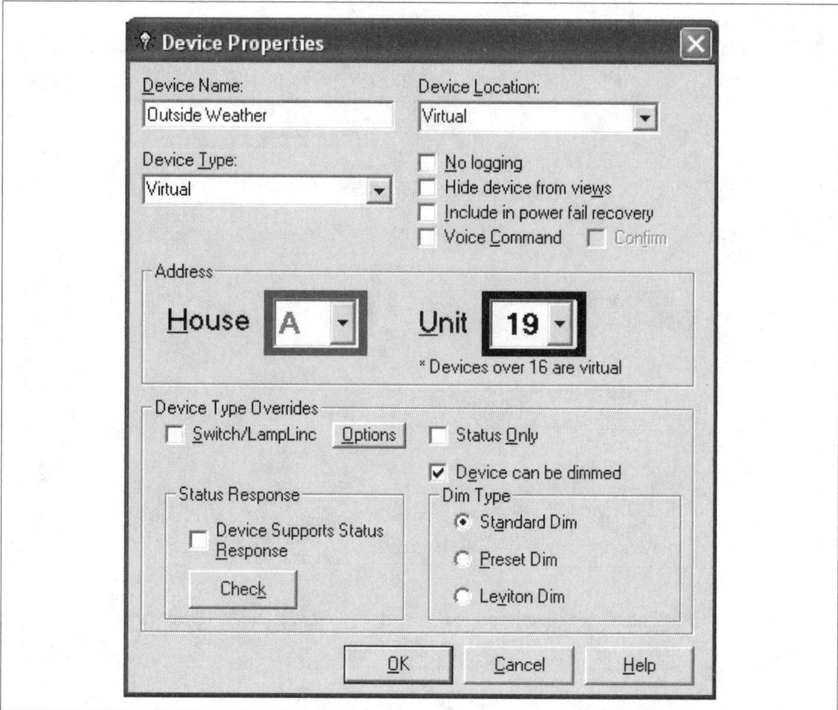


*Figure 1-50. Creating a virtual unit*

## Scheduling Events

If you want something to happen at a specific time in the future, either once or repeatedly, you set up a *scheduled event*. For example, you might want the air cleaner in the office, which is connected to an appliance module, to come on for three hours every other day. Or, perhaps you need to get up early tomorrow and you want the coffeepot to be started at 4:30 a.m.

To create a new scheduled event, click Events in the Views pane and then click the New Event toolbar button. The Event Properties dialog, shown in Figure 1-51, is where you specify what you want to happen.

Enter a name for the event so that you can identify it easily in the Events list. Then, select "Absolute time / Date" from the Type list and enter a date and time for the event to occur. If you want it to occur more than once, select Reoccurring from the Type list, type in how many times you want it to repeat, and then choose the interval between repetitions, such as 60 minutes, from the pop-up menu.

Next, to specify what you want this event to do, select the Device Actions tab. To send a command to a device, drag it from the Device Selection list to

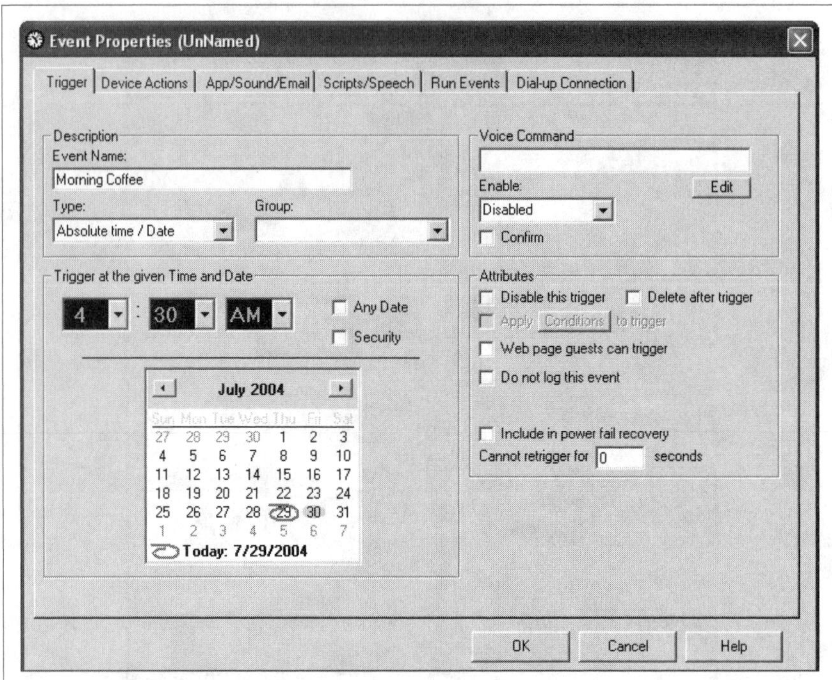


*Figure 1-51. Creating a scheduled event*

the Selected Devices list. Then, double-click the device in the Selected Devices list, specify an action in the dialog box that appears, and click OK. Figure 1-52 shows the result. The `Coffee Pot` device will receive an `On` command when this event triggers.

Why does the coffeepot show up in the Selected Devices list twice? It's a good idea to make sure the coffeepot isn't left on inadvertently, so the second action of this event sends an `Off` command at 5:35 a.m. To set this up, drag another copy of the device from the Device Selection list to the Selected Devices list and then double-click to edit the event. Change the Time Delay settings to have this command sent one hour and five minutes after the event is triggered initially, as illustrated in Figure 1-53.

## Responding to Events with HomeSeer

Although scheduled events are handy, having your home automation system respond to stimuli puts the "smart" in your *smart home*. Like the man behind the curtain in *The Wizard of Oz*, it's HomeSeer responding to events— motion detectors triggering, garage doors opening—which makes it all happen.

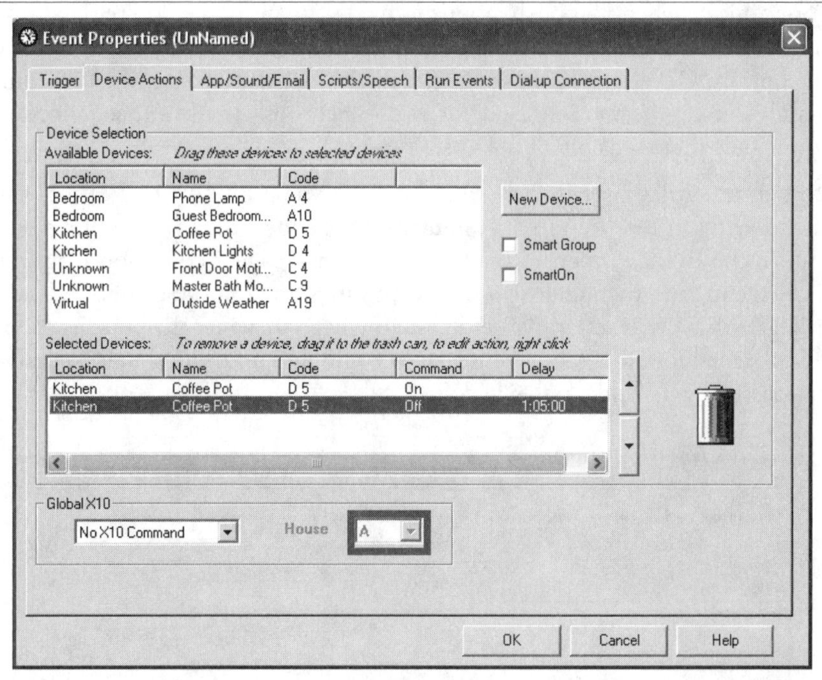


*Figure 1-52. Specifying Device Actions for an event*

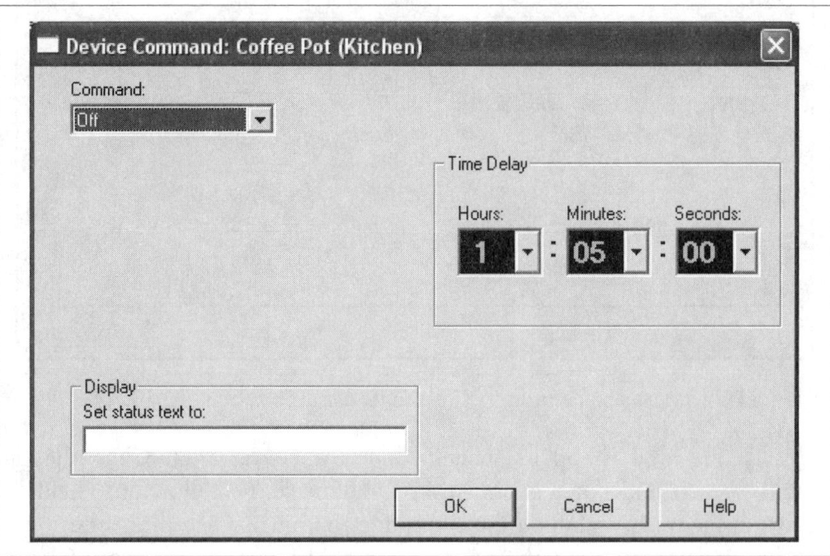


*Figure 1-53. Using the Time Delay settings to control sequences of commands*

For example, when you push a button that tells your home you're departing [Hack #70], HomeSeer runs a script when it receives the On command sent by the button. Or, after you've gone downstairs in the morning, the Off command sent by the motion detector in the bedroom triggers a sequence of commands that turns off all the upstairs lights.

To define a *trigger event* (i.e., what happens when a unit turns on or off), click Events in the Views pane, and then click the Add Event toolbar button. In the Event Properties dialog box, select a trigger condition from the Type menu, and then choose a device and its status. In the example shown in Figure 1-54, the action we're defining will occur when the Outside Front Door Motion motion detector sends an On command, indicating that someone has passed by its field of view.

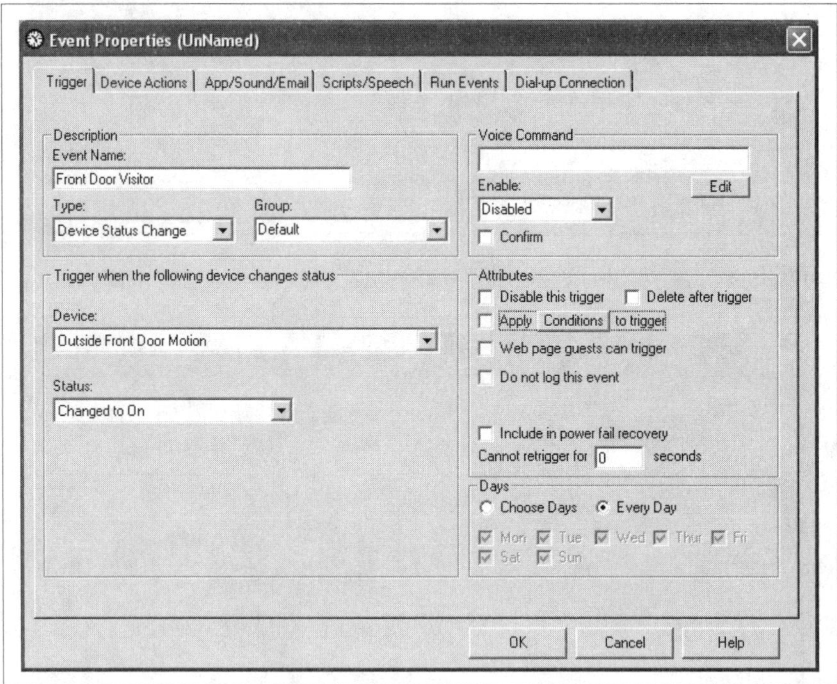


*Figure 1-54. Setting up an event trigger for a motion detector*

Next, define what happens when this event is triggered. Click the tab Scripts/Speech and enter a phrase in the "Speak the following:" field, as shown in Figure 1-55. When the computer announces the visitor, the sound will be heard throughout the house [Hack #28]. Then, the *frontdoorvisitor.txt* script will decide what other actions to take based on who is home and what time it is [Hack #74].

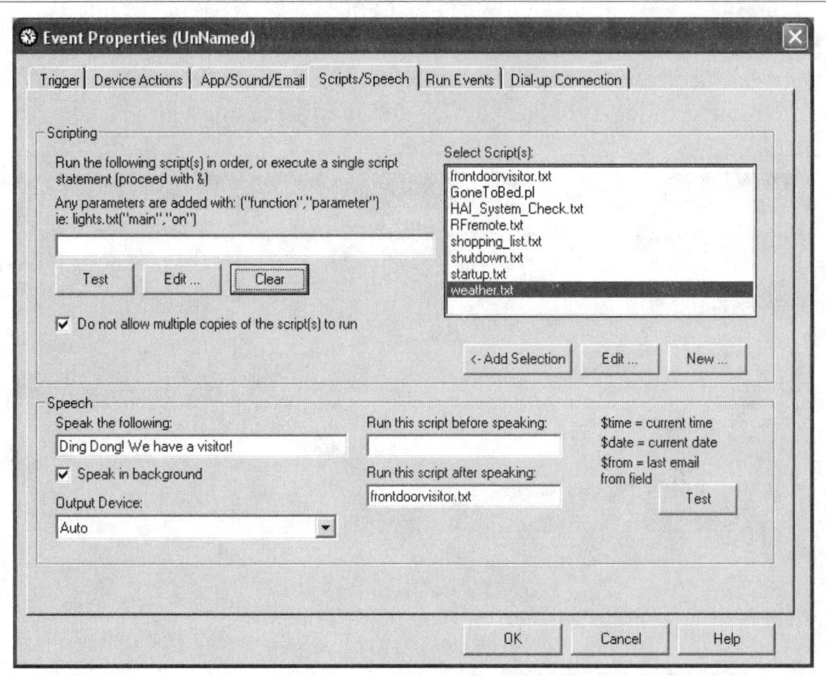


*Figure 1-55. Defining the actions for an event trigger*

That's the basic process of setting up an event. As you can see from the other settings in the Event Properties dialog box, you have plenty of choices for fine-tuning exactly how you want HomeSeer to react to any given command or situation. Let's look at a couple of additional techniques, based on what we've already set up, that will come in handy as you adapt the hacks in this book to your own liking.

## Triggering Sequences of Events

You can create your HomeSeer-based home automation system so that it's constructed of building blocks of command sequences. For example, you might create an event that turns off all the lights when you go to bed for the evening **[Hack #48]**. You also might find it useful to be able to turn off these lights at other times, such as sunrise, when you're cocooning in the den to watch a good movie, or when you're leaving the house. Let's create a single event that controls these lights so that you can call it whenever you need it, much like programmers use subroutines to keep their code efficient.

In HomeSeer, create a new event and set its Type to Manual Only. Then, click the Device Actions tab and drag all the devices you want to command to the Selected Devices list, as illustrated in Figure 1-56. Remember to set an appropriate command for each device (`Off` in this example).

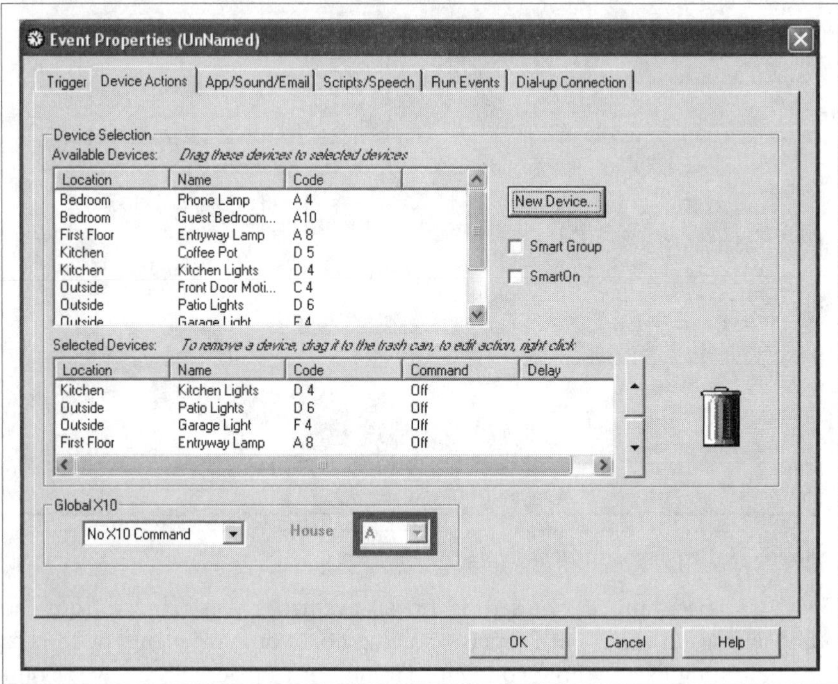


*Figure 1-56. Commanding several devices in sequence*

Because this event is defined as `Manual Only`, it won't be triggered until you trigger it using the user interface or via another event or script. To trigger this event from another, such as when you push a button on a Palm Pad **[Hack #5]**, set up a new event whose `Run Events` properties are set to call this one. For example, the event shown in Figure 1-57 will execute the `Downstairs Lighting` event.

As you can see, you can call any number of other events, in whatever sequence desired, to build new functions from other events you've created. This is a powerful and handy feature, but start with simple actions and keep an eye on the HomeSeer log file to help you debug any nested actions that seem to be misbehaving.

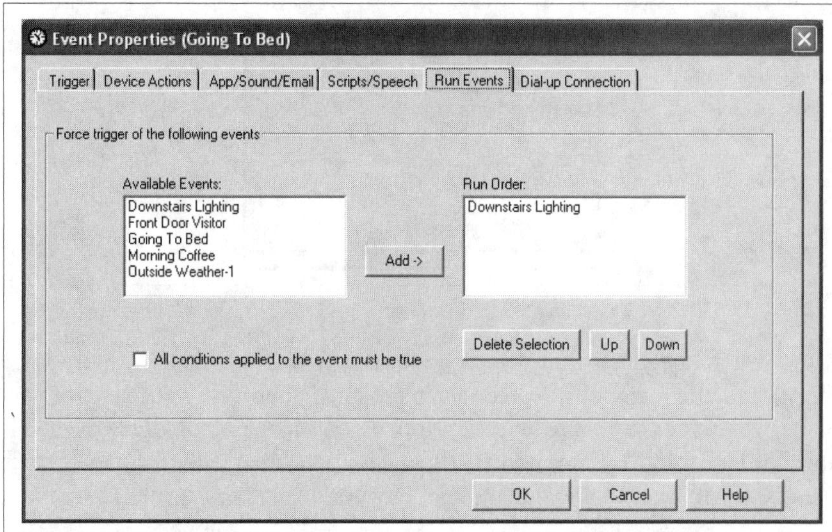


*Figure 1-57. Adding flexibility to your system with nested events*

## Triggering Scripts

You can do a lot with HomeSeer without using scripts, but you'll be missing out on some of the best features of the product. In addition, as you'll notice from the hacks in this book, scripting is often the only way to accomplish some of the smarter things you can do with your home. When it comes to scripting, HomeSeer is unique among home automation applications in that it supports three different scripting languages: VBScript, Perl, and JavaScript.

The best way to learn about scripting HomeSeer is to review its onscreen help, and to peruse the example scripts included with the product as well as the large library of user-contributed scripts available at the HomeSeer discussion board (*http://ubb.homeseer.com*). Assuming you have a script you want to use, whether borrowed or written from scratch, let's look at how you set up events to trigger it.

Several of the hacks in this book use scripts that are activated when you turn a device on or off. To set this up, you begin with an event that is triggered by the device changing state, such as turning `On`. Then, in the Event Properties dialog box, you click the Scripts/Speech tab. To create a script, click the Edit button. You're asked to enter a name for the script. The file extension you enter for the name determines which script interpreter will be used to execute the script, as shown in Figure 1-58.

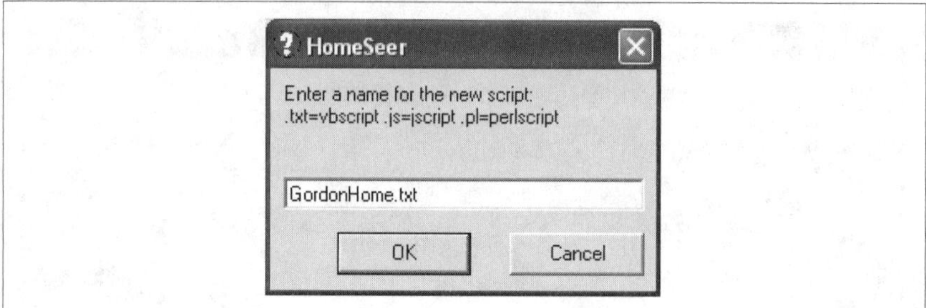


*Figure 1-58. Defining a new HomeSeer VBScript*

After you name the script, Notepad opens. Type, paste, or import your script, and then save the Notepad document. When you're done, the new script will be added to the Select Scripts list in HomeSeer and will be available for execution by any event. It's selected automatically for the current event and will be executed when the trigger conditions you specified are met.

Finally, notice the "Do not allow multiple copies of the script(s) to run" checkbox in Figure 1-55. This can be important if an event might be triggered multiple times in rapid succession—such as with some motion detectors—because new instances of your script might be started before already-running instances have finished. Depending on what your script does, selecting this option can save you some debugging headaches.

Similar to how you can use events as building blocks for more complex procedures, you can also have HomeSeer execute multiple scripts in a sequence. To add another script, select it in the list and click the Add button. The scripts will execute in the order shown in the Scripts field.

## Using the HomeSeer Log

HomeSeer keeps an extensive record that enables you to see exactly what's going on with your home automation system. Not only does the log keep track of everything HomeSeer does, but also, because the application is listening to the power line, you'll see X10 commands that are sent from other sources, too, such as minicontrollers **[Hack #4]**. To view the log, click the Log button in the Views pane, as shown in Figure 1-59.

The detail shown in the log makes it a valuable debugging tool, but it also means it can be overwhelming to sort through if you have a lot of devices and events. To reduce some of the clutter, you can set some devices so that their activity isn't logged. This can be handy for things such as motion detectors, which can generate many log entries in an active home. To set this up, check the "No logging" checkbox in the Device Properties dialog for each device you want to exclude.

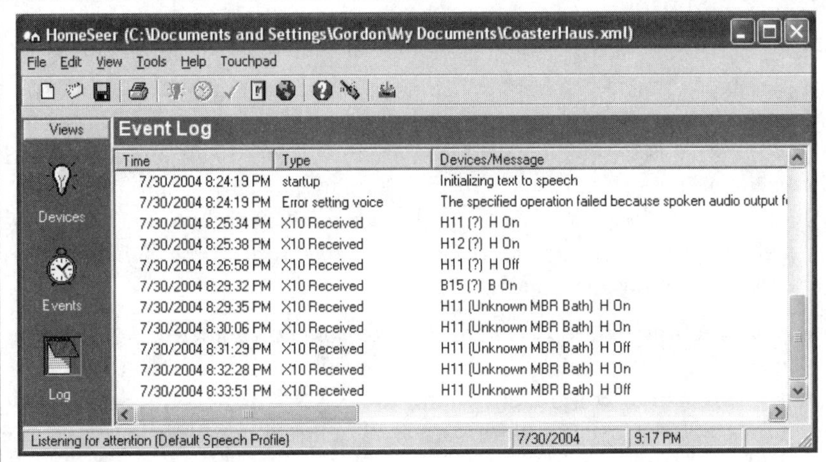


*Figure 1-59. A HomeSeer log file*

## See Also

Now you know enough about HomeSeer to begin building your system, but be sure to visit the product's extensive web site (*http://HomeSeer.com*) and peruse the technical notes and tutorials. You'll also want to check out the active, and friendly, community discussion board (*http://ubb.HomeSeer.com*). Even if you're not a HomeSeer user, this discussion board offers a daily dose of practical home automation techniques, news, and advice. It's well worth signing up for a free account so that you can read all the messages and post new ones.

## Sync with the Sun

**One of the best things about a smart home is that you never have to bother turning on lights when it starts to get dark or turning them off if you're an early riser and the sun is making them unnecessary.**

All the home automation software packages make it easy to control your lights automatically, or to perform virtually any other task, when the sun rises and sets. In fact, it's the perfect task for a computer-based home automation system; the computer can accurately calculate the sun times for your exact location and time of year, and then do your bidding with clockwork precision.

You need to make sure your computer's clock and location are set correctly so that the time calculations are reliable, but other than that it's simply a matter of configuring your home automation software and deciding what you want done. The latter bit is the hardest part: what do you want to do?

## Actions at Sunset

The most obvious sunset action is to turn on some lights in the home so that people on the inside can see what they're doing. You probably don't want the entire household to light up, so focus on ambient lighting. An efficient way to do this is to spend a couple of minutes every night and note which lights you and your family have turned on manually and which ones tend to stay in use until bedtime. For example, the small lamp in the family room, the light over the sink in the kitchen, and the front porch light might be good candidates for automation.

Next, identify the lights you might want to come on sometime *after* sunset. In addition to having lights turned on at sunset, it's simple to schedule events that occur a set number of minutes later. As the house gets darker, do additional lights provide comfort or safety? The lights over the stairway or the landscape lighting in the backyard are likely candidates.

Finally, consider more than just lighting. If you have a webcam on the birdbath outdoors, you probably want to turn it off after nightfall. Perhaps you want to mute the sound on your home automation system so that you're not bothered by the audible notification of incoming email and faxes. If you want to use the scheduler in your home automation software to run a script that backs up your hard drive, sunset might be a good time to do that, too.

## Actions at Sunrise

Typically, there are fewer things to do, automation-wise, at sunrise than at sunset. With a few exceptions, such as lights in darkened areas, you probably just want to turn off all the lights in the home at sunrise. In fact, many home automators use the sunrise trigger to check and reset the status of every device in their home, just to ensure that everything is in a known state as the day begins.

Other candidates for automation at sunrise include setting the sound volume on your home automation system computer so that audible alerts can be heard throughout the day, retrieving the day's weather report **[Hack #64]**, fetching the morning's email, or checking to make sure your broadband connection is still up by loading a URL or two. As with sunset scripts, you can use the sunrise event to schedule events that occur later, such as reminders to bring out the garbage **[Hack #25]**.

Now that you have a list of what you want to accomplish at each time, let's look at how you configure your system to handle your tasks.

**Sun events and XTension.** With XTension [Hack #17], you define what happens at sunrise and sunset by creating scripts that perform the tasks you want to occur. Simply create global scripts named Sunset and Sunrise, and XTension automatically will execute the appropriate script at the correct time for your location.

XTension knows when sunset and sunrise are in your location, by calculating the times based on your computer's settings. If you find they're not quite correct make sure the time and time zone are set correctly in the Date & Time pane of the Mac OS X System Preferences.

If you want to further adjust the sunset and sunrise times—perhaps because you live in a valley where it gets dark a little before sunset—you can offset the times XTension uses. Choose Preferences from the XTension menu, and then click the Suntimes tab. Enter how many minutes you want added to or subtracted from the calculated sun times, as shown in Figure 1-60. You also can set the exact latitude and longitude for your location, instead of using System Preferences' rough idea of your location based on your time zone, which also might increase the accuracy of the calculations of your sun times.

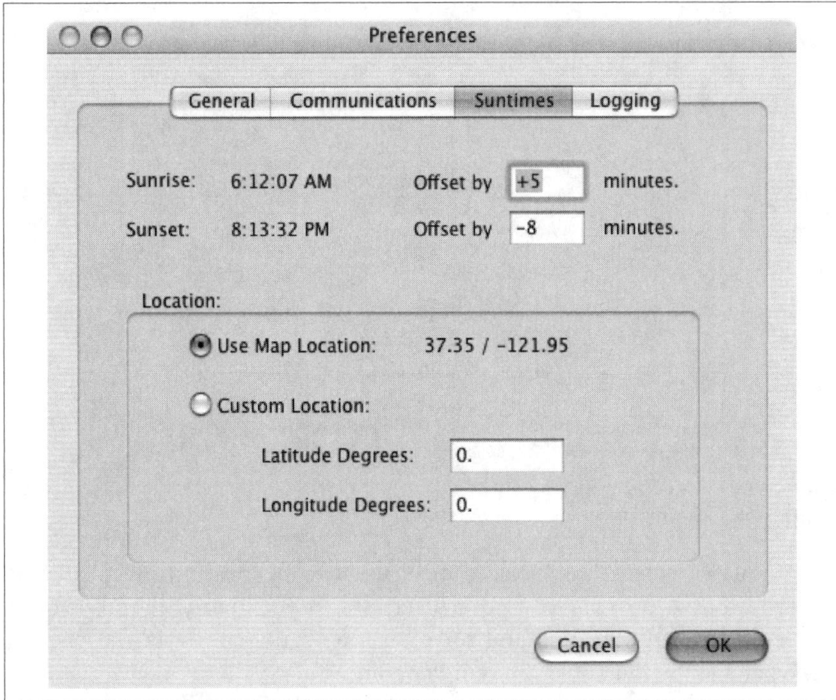


*Figure 1-60. Adjusting sunrise and sunset times in XTension*

After you've saved the adjustments, quit and restart XTension to trigger a recalculation of the sun times for the current day.

**Sun events and Indigo.** In Indigo **[Hack #18]**, you define what happens at sunrise and sunset by creating an action group that performs the tasks you want to occur. Then, create a Time/Date Trigger whose condition is `Sunrise` or `Sunset`, as shown in Figure 1-61.

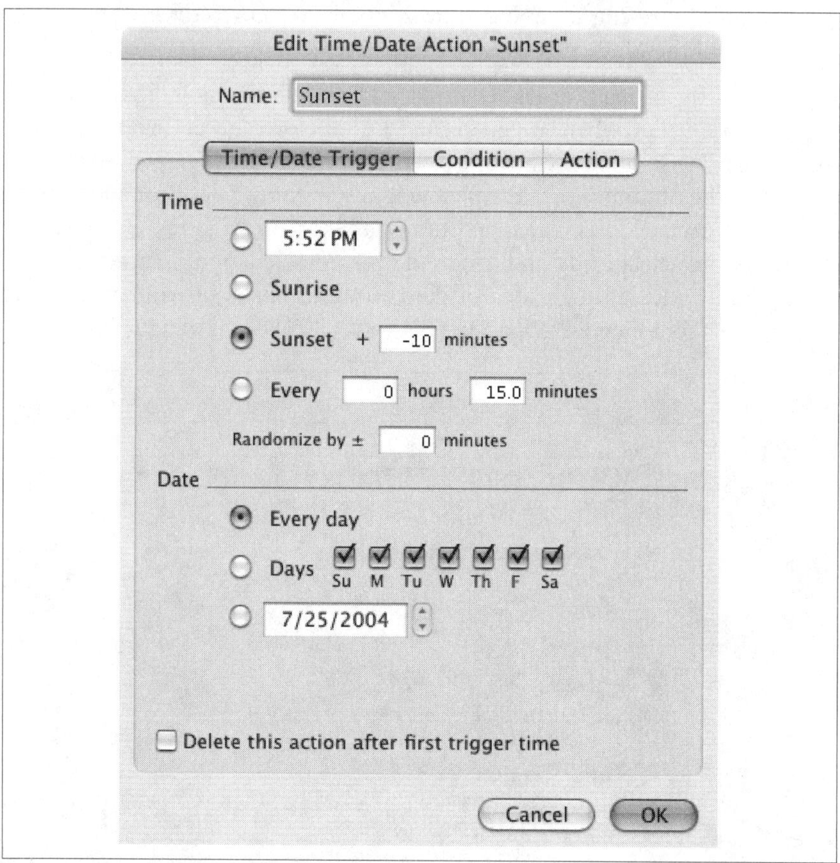


*Figure 1-61. Defining a sunset action in Indigo*

Indigo knows when sunset and sunrise are in your location by calculating the times based on your computer's settings. If you find they're not quite correct make sure the time and time zone are set correctly in the Date & Time pane of the Mac OS X System Preferences.

If you want to further adjust the sunset and sunrise times—perhaps because you live in a valley where it gets dark a little before sunset—you can offset the times Indigo uses. When you define the event, enter the number of minutes before or after sunset or sunrise when you want the event to occur. In the previous example, shown in Figure 1-61, the action will occur 10 minutes before sunset because –10 was entered in the Offset field.

You also can set the exact latitude and longitude for your location instead of using System Preferences' rough idea of your location based on your time zone. Doing this might increase the accuracy of the calculations of your sun times. Choose Preferences from the Indigo menu, and then click the Sunset & Sunrise tab. Click Override System Location and enter the longitude and latitude for your location. If you're not sure what to enter, there's a handy link in the Indigo window to a web site where you can look up your approximate values by city or Zip Code.

**Sun events and HomeSeer.** In HomeSeer [Hack #19], you define what happens at sunrise and sunset by setting up a trigger event that executes each day at sunrise or sunset. If you need to adjust the sunset or sunrise time—perhaps because you live in a valley where it gets dark before sunset—simply provide a positive or negative offset time when you define the trigger event.

For HomeSeer to calculate the sun times correctly, it needs to know your location. It will use the time zone that's set in Windows, or you can specify a more exact location. Choose Options from the View menu, and then click the Sunrise/Sunset tab. Select a nearby city from the Pick Location list, or enter the longitude and latitude in the fields provided. Click the Calculate button to tell HomeSeer to recalculate the sun times for the current day.

## Hacking the Hack

If you've implemented a state machine [Hack #24] so that your home automation system knows about the various conditions of your home, such as the weather and who is at home, you can take these factors into account when automating tasks for sunrise and sunset. For example, if the house is unoccupied at sunset [Hack #70], you might want to turn on a different set of lights or start a script that makes it looks like someone is at home to discourage burglars [Hack #72].

# Choose the Right Controller

**To some extent, the home automation controller you use defines your system. It also determines which home automation software you choose because not all programs work with every controller.**

Your choice of home automation controller determines the expandability, flexibility, and reliability of your system, and it might be the single most expensive item you have to buy to automate your home. This hack helps you make the right decision for your particular needs.

A *smart controller* is a software-programmable X10 transmitter and receiver. The word *smart* in the term distinguishes these controllers from devices such as minicontrollers **[Hack #4]**, which can transmit commands but cannot be controlled by a computer. Let's take a systematic look at the most common smart controllers that are available and rate each of them based on these criteria.

*Price*
: This will tell you quickly if you should consider a particular controller, depending on your budget.

*Features*
: Does the controller offer a comparative richness of features when compared to others in its category?

*Flexibility and expandability*
: How scalable and modular is the device?

*Reliability*
: Are there any known bugs or issues that you should be aware of or that are not commonly documented?

*Support and application base*
: This is a popularity measure, which indicates bang-for-buck. A better-supported controller will give you more options to choose from.

This hack rates each controller with a score from 1 to 5 against each criterion, with an overall score that's an approximate, opinionated, and weighted sum of the controller's score in each area. The result should be helpful in deciding which controller to use when compared against your needs and requirements.

Now that we've laid the ground rules, let's get started.

## Teasers and Appetizer Controllers

Controllers in this category will enable you to get a feel for what home automation is like, but their limited capabilities prevent many basic techniques.

**The FireCracker (CM17A/CM18A).** The CM17A, also called the FireCracker (*http://www.x10.com/automation/firecracker.htm*), was once X10 Incorporated's $6 postage-only giveaway to new customers, but now it sells for $40 as part of a kit that includes a transceiver, lamp module, and remote control.

*Praise*

At $6, the price was hard to beat, but it's not such a great deal at $40. The low price and small size of the FireCracker prompted people to write software for it. You can find an ActiveX control for commanding it via a web browser, and other utilities that can work with it are but a quick Google search away.

*Gripes*

The FireCracker can only *send* commands. It cannot receive commands and thus cannot trigger your home automation software to react to events in your home.

*Overall*

The FireCracker kit will give you no more than a taste of controlling lights from your PC. This is very different from anything that falls under home automation in my dictionary. I think you soon will grow tired of using it and you quickly will look for a better controller. In fact, even as a demonstration system, it gives the false impression that a home automation system is of limited use. If you want a much better taste of home automation, buy a HomeDirector kit for $20 on eBay instead; it includes a CM11A controller, described later in this hack.

Table 1-1 rates the FireCracker based on the criteria outlined earlier.

*Table 1-1. FireCracker rating*

| Category | Rating |
|---|---|
| Price | 5 |
| Features | 1 |
| Flexibility and expandability | 2 |
| Support and application base | 3 |
| Reliability | 1 |
| Overall score | 2 |

**The CP290.** The CP290 is the oldest controller in the game, but surprisingly enough, it has some distinct features that are not found in higher-priced devices. Though the CP290 is discontinued and no longer manufactured, you still might find one on eBay.

*Praise*

For starters, it's both a controller and a console. The console has On/Off buttons that you can use to manually control up to eight devices in a house code [Hack #1]. Another unique feature is that you can program a macro into the CP290 and execute it on a schedule or at randomized times to simulate a lived-in look. Finally, it has a battery that saves programmed macros in case the power goes out. Of course, if the power is out, no transmitted event will get through the power line, but at least when electrical service is restored, the events still will be stored.

*Gripes*

The worst shortcoming of the CP290 is that, like the FireCracker, it's a one-way (transmit-only) device. Also, because it's discontinued, most of the newer home automation software doesn't work with it. However, several DOS-based programs—and I'm sure, if you look hard enough, Windows-based programs—work with it. A pesky problem with the CP290 is that its clock drifts badly (more than a minute per month), although there is a remedy for this that involves replacing the clock chip with a better one.

*Overall*

For the price, there is no reason to buy this old timer (no pun intended) today.

Table 1-2 rates the CP290 based on the criteria outlined earlier.

*Table 1-2. CP290 rating*

| Category | Rating |
| --- | --- |
| Price | 1 |
| Features | 2 |
| Flexibility and expandability | 2 |
| Support and application base | 2 |
| Reliability | 2 |
| Overall score | 2 |

## Mainstream Controllers

These are the mainstream workhorses of home automation. They work with many home automation applications, and they are well supported and used in the home automation community.

**ActiveHome and HomeDirector (CM11A).** The CM11A (*http://www.smarthome.com/1140.HTML*) is the controller behind most residential home automation systems today, and for good reason. It offers a sensible balance between features, price, and software support.

*Praise*

The CM11A is the first controller we've looked at so far that sends and receives X10 commands, which makes it a lot more flexible for use with smart home automation. The CM11A strikes a good balance between price and features, and it enjoys the widest application support. I am not aware of any recent general-purpose home automation software that doesn't work with it. It also comes in versions that work worldwide: the CM11A is made for the 110V electrical systems; the CM11K is for 220V systems; and versions are available that work with French, German, and British electrical plugs.

Like the CP290, the CM11A can store macros for use without being connected to a host computer, although not all the software programs that otherwise work with the controller support this ability. It also has the ability to report back to the host computer about commands that collide with each other, as can happen when multiple devices try to send X10 commands at the same time. The CM11A also works with extended X10 commands, a feature missing from some of the other controllers.

*Gripes*

There are plenty of known problems with the CM11A when it comes to reliability. For example, there are at least four different ways the CM11A can become hung and stop responding to commands. There also have been reports of the device overheating to the point of being too hot to touch.

*Overall*

The CM11A is a sensible choice, unless you're looking for a build-it-yourself kit or want to integrate infrared control into your home automation system. For most people, your search for a controller to use will end with this one.

Table 1-3 rates the CM11A based on the criteria outlined earlier.

*Table 1-3. CM11A rating*

| Category | Rating |
| --- | --- |
| Price | 2 |
| Features | 3 |
| Flexibility and expandability | 3 |
| Support and application base | 5 |
| Reliability | 1 |
| Overall score | 4 |

**The LynX-10 kit.** The LynX-10 (*http://www.marrickltd.com/products.htm*) is, in my opinion, a gem. It is not nearly as popular as the CM11A, although the LynX-10's build-it-yourself price is in the same ballpark, at about $50. You can buy an already-assembled version for about $120, or an ISA card version for about $180. There is no real reason not to buy the kit to save some money, unless you absolutely don't know how to solder and don't care to learn. Also consider that if you're going to pay $120 for the preassembled version, you're approaching the price range of the Ocelot (discussed later in this hack), which might be a better controller to consider. However, in the $50 range, the LynX-10 kit rules.

*Praise*

The LynX-10 kit is an ideal controller for an enthusiast with an electronics background. You can place the assembled board inside the PC, on the bottom, resting on a sheet of plastic to protect from shorting. In addition, you can use the extra space on the printed circuit board for something useful, such as adding a watchdog timer that resets the PC if it stops working. The LynX-10 has extensive error reporting and statistics counting, which can result in a more robust system, and most home automation software packages work with it, too.

*Gripes*

To communicate with the power line, the LynX-10 requires an additional module called the TW523 two-way interface. This adds $20 to the overall cost of using the LynX-10. An issue with the TW523 is that it does not support extended codes and does not receive `Dim` commands, which prevents you from using many of the more advanced devices that are available.

Unlike the CM11A and CP290, the LynX-10 cannot function unless it is connected to a computer with home automation software. That is, it doesn't have the capability of storing and sending macro commands. However, for a truly smart home, you're going to want the capabilities of a computer [Hack #16], so you shouldn't consider this to be a drawback except for the simplest of systems.

Marrick Ltd. (*http://www.marrickltd.com/*), maker of the LynX-10, also makes the LynX-10 PLC, an improved version of the LynX-10 that does not require a TW523 and supports extended X10 commands. It costs about $100, preassembled, and is also well worth considering.

*Overall*

If you don't need infrared capabilities and you're willing to build it yourself, the LynX-10 kit is, in my opinion, a sure bet.

Table 1-4 rates the LynX-10 based on the criteria outlined earlier.

*Table 1-4. Lynx-10 rating*

| Category | Rating |
|---|---|
| Price | 4 |
| Features | 3 |
| Flexibility and expandability | 4 |
| Support and application base | 4 |
| Reliability | 3 |
| Overall score | 4 |

**Ocelot.** The Ocelot (*http://www.appdig.com/ocelot.html*) is on the high end of the mainstream controller category, and it is priced accordingly at about $200. In addition to sending and receiving X10 commands, it works with a wide range of modular add-ons that provide analog (e.g., temperature) and digital (e.g., alarm sensors) data acquisition capabilities. It can also control relay modules, such as sprinkler valves, and infrared equipment, such as stereos, directly.

*Praise*
: The Ocelot has a relatively tremendous program space for storing macros—16 K (compared to 1 K for the CM11A). It also can work with touchscreen devices, such as the Leopard (*http://www.smarthome.com/73103.HTML*), and it is considered to be a flexible and expandable controller.

*Gripes*
: The Ocelot, like the LynX-10, requires a TW523 and shares the same limitations regarding extended X10 commands. Additionally, the cost of the various (albeit attractive) add-ons adds up quickly.

*Overall*
: The Ocelot makes a superb and expandable X-10 controller, if you don't mind the price.

Table 1-5 rates the Ocelot based on the criteria outlined earlier.

*Table 1-5. Ocelot rating*

| Category | Rating |
|---|---|
| Price | 2 |
| Features | 5 |
| Flexibility and expandability | 5 |
| Support and application base | 3 |
| Reliability | 4 |
| Overall score | 4 |

## Expensive Controllers

These are professional home automation controllers, and they are priced accordingly. They also have extra capabilities that lesser controllers can do only via their host computer, or cannot do at all. These controllers cost in the $600 to $1,000 range, so here's a brief word to the wise before getting to their reviews. The X10 protocol has some inherent limitations, such as its speed and occasional lost commands due to collisions or noise. Many people find it bothersome, but people who are willing to pay nearly $1,000 for a controller might find it irritating. What this boils down to is that when you enter this price range, X10 might be the wrong technology for you. It's an appealing low-cost solution, but if you're willing to spend this much just for a controller, it might be worthwhile to consider other alternatives, such as hardwired dedicated control lines or wireless systems.

**TimeCommander-Plus.** The TimeCommander-Plus (*http://www.jdstechnologies.com/jdstcp.html*) costs about $600 and is similar to an Ocelot with the digital and analog I/O modules already built-in. It has 8 analog inputs, 16 digital inputs, and 8 relay outputs. If that's not enough, you can add more with a $400 expander unit. It has macro support and logging, built-in timers, and some special enhanced commands for high-end wall switches, such as the PCS SceneMaster (*http://www.smarthomeusa.com/Shop/PCS-Wall-Switches/*).

*Praise*
: The TimeCommander-Plus is an integrated unit with powerful hardware features. It easily can control your HVAC system, handle alarm sensors, and actuate sprinkler relays, all from its built-in I/O modules.

*Gripes*
: There is no direct support for infrared; it requires the Infrared Xpander ($300). Surprisingly, given its price, it also needs a TW523 module to support X10 commands, which brings along the TW523's limitations as discussed previously.

*Overall*
: This is a good choice for high-end systems, if your plans call for many sensors and controls actuated from a central location.

Table 1-6 rates the TimeCommander-Plus based on the criteria outlined earlier.

*Table 1-6. TimeCommander-Plus rating*

| Category | Rating |
|---|---|
| Price | 1 |
| Features | 5 |

*Table 1-6. TimeCommander-Plus rating (continued)*

| Category | Rating |
|---|---|
| Flexibility and expandability | 4 |
| Support and application Base | 2 |
| Reliability | 4 |
| Overall Score | 3 |

**Stargate.** The ultimate in hardware control, the Stargate (*http://www.jdstechnologies.com*) is a TimeCommander-Plus controller, with voice response and telephone control, including voice mailboxes. Voice response (not to be confused with voice *recognition*) enables you to record up to 100 phrases and tie them to events and commands.

*Praise*
: A high-end controller that integrates many components that others sell separately.

*Gripes*
: Like its predecessors, this controller needs the TW523 for X10 support. At $1,000, the added value is, in my opinion, questionable. The expense of providing features in a dedicated piece of hardware rather than in a host PC is not warranted.

*Overall*
: An excessively expensive controller for X10 applications.

Table 1-7 rates the Stargate based on the criteria outlined earlier.

*Table 1-7. Stargate rating*

| Category | Rating |
|---|---|
| Price | 1 |
| Features | 5 |
| Flexibility and expandability | 3 |
| Support and application Base | 2 |
| Reliability | 5 |
| Overall score | 3 |

## Final Thoughts

There are other controllers besides the ones reviewed here. (See Matt Bendiksen's sidebar, "The PowerLinc Family," for example.) However, their capabilities and features are similar, and the evaluations given here provide you with a basis with which you can compare.

## The PowerLinc Family

Smarthome, Inc.'s PowerLinc line of smart controllers, all of which can send and receive X10 commands, are worth considering when selecting a controller to use. They offer some fantastic features at affordable prices, but unfortunately, as of this writing, they are not available for use with 220V electrical systems.

*PowerLinc Serial (http://www.smarthome.com/1132.HTML; $40)*
: Acts as both a computer controller and a two-way TW523 interface replacement. It does not support standalone operation, but it works well with a dedicated computer and home automation software.

*PowerLinc USB (http://www.smarthome.com/1132U.HTML; $40) and PowerLinc Controller (http://www.smarthome.com/1132CU.HTML; $70)*
: Both of these controllers have a USB connection rather than an RS-232 serial connection, which is great to use with a Macintosh but requires a USB-to-serial adapter to utilize other controllers. The two controllers have the same features, but the PowerLinc Controller also can work as a standalone controller, with some limitations, for those times when your computer is turned off [Hack #15].

*PowerLinc IP (http://www.smarthome.com/1132IP.HTML; $30)*
: The latest addition to the PowerLinc line, the PowerLinc IP is quite unusual. It's a standalone network device that you control over the Internet via a subscription service called Smarthome Live. Apparently, it is possible for other software (not just Smarthome Live) to communicate with and control the PowerLinc IP, but no third-party software to do so is available yet.

X10 is, by nature, modular and incremental. However, think your system through before choosing a controller because changing to a different one later is more involved than adding new modules to the system. Make sure the one you select works with the software you intend to use, and that it can support what you want to do with your system over the next few years. Also consider the requirements of each controller. For example, do you want to have a computer running continuously? Can you wire all your inputs to a central location? Answering these questions should steer you toward choosing the right X10 controller.

*—Ido Bar-Tana*

# Maintain an X10 Library

You'll save yourself some hassle, and the need to memorize esoterics, by keeping an up-to-date collection of X10 and related modules.

There's one thing you'll learn quickly about using X10: it's a world that's filled with hard-to-remember details that, if you forget them, result in hard-to-diagnose problems or bothersome procedures. For example, if you use a computer-based controller that's based on the TW523 two-way interface **[Hack #21]**, you won't be able to use modules that work with extended X10 commands. That's fine if you don't have any when you first set up your system, but if you later add a two-way light switch, you might spend a lot of time troubleshooting the problem before you remember why it doesn't work correctly.

More commonly, when it comes time to replace the batteries in your X10 motion detector **[Hack #6]**, you'll have to remember the correct sequence in which to push its buttons to restore its settings. And the options are numerous: you have to reset its address, select whether it ignores motion during daylight, and determine how soon after the last time it senses motion it sends an `Off` signal.

Similarly, some modules from Smarthome, such as the LampLinc lamp module, require you to send a special sequence of X10 commands **[Hack #13]** to set their address, as well as other options. If you want to change its settings or address because you moved it to a new location, you'll need to repeat the programming procedure.

Clearly, keeping your equipment's manuals, instruction sheets, and package inserts organized is one approach to solving this problem, but even if your organization gene is strong enough to accomplish this, sorting though all that paper when you want to find something isn't very much fun. A better approach is to keep an all-electronic library that you can search and organize as you see fit. Here's a list of resources where you can find most of what you need:

*X10 Incorporated (http://www.x10.com/support/support_manuals.htm)*
: X10's web site thoughtfully provides a handy listing of manuals for all the company's products.

*Smarthome, Incorporated (http://www.smarthome.com)*
: Smarthome's web site almost always includes a product manual, in PDF format. To locate a particular manual, search for the product by name, and then look for a download link near the bottom of the Purchasing Information section of the product's page.

*Leviton (http://www.leviton.com)*
: Leviton provides manuals for individual products in the tech support area of its web site.

Finally, many X10 devices that are similar in function actually have quite different capabilities. For example, the TM751 transceiver does not enable you to control its built-in outlet via X10, but the RR501 does. The best resource for sorting out quirks such as this is Sand Hill Engineering's overview of X10 modules (*http://www.shed.com/x10stuff.html*).

## HACK #23 Shop for Secret X10 Devices

**It's possible that you already have an X10 device or two in your home, and you don't even know it. In addition, sometimes you can find good deals on modules that work with X10 but don't admit it.**

X10 technology has been around since the mid-1970s, so it's had plenty of time to become commonplace, and in fact has, with relatively little fanfare. The modularity of the system has allowed it to grow over time, with various modules and add-ons being introduced in response to new markets and demand for new capabilities. For example, the selection of X10 light switches ranges from X10 Corporation's simple pushbutton switches (*http://www.smarthome.com/2031.html*; $10) to Smarthome's two-way designer dimmers (*http://www.smarthome.com/2380.html*; $70), to Leviton's sophisticated switches with LEDs that indicate their status (*http://www.smarthome.com/4289.html*; $75).

This bounty of consumer choice isn't limited just to light switches. In today's market, there are hundreds of X10-capable devices to choose from, but not all of them are readily identifiable. Some companies prefer to de-emphasize (and sometimes outright conceal) their use of the X10 protocol. This might be an attempt to shield customers from X10's high geek factor, avoid the stigma of early and less-reliable X10 equipment, or perhaps make their technology seem unique.

If you come across a product that offers remote control or automation, but doesn't mention X10, it's worth a little digging to see what protocol it uses. First, determine if it receives signals wirelessly or over the power line. Devices that use the power line often use power line carrier (PLC) to describe their capabilities. If you see that, it's definitely worth investigating further. But remember that although X10 is a PLC protocol, it's not the only one, so don't buy a device that you're unsure about, unless you can return it easily. Also, other PLC protocols might interfere with X10 signals—there's limited signal space to share on your electrical lines—so keep that in mind as you investigate unfamiliar devices.

Here are three of my favorite ways to root out secret X10-compatible devices. First, take a close look at the device and see if it looks familiar. X10 Incorporated frequently manufactures devices for other companies to sell under other names. But even with a different label, the devices look essentially the same. For example, the X10 ActiveHome minicontroller (*http://www.x10.com/automation/x10_mc460.htm*; $13) is identical to the RadioShack Plug 'n Power minicontroller (*http://www.radioshack.com/product.asp?catalog%5Fname=CTLG&product%5Fid=61-2677*; $15), but the latter doesn't refer to X10 in its product description.

In fact, X10 devices that are sold under other names are the most common source of secret devices. They're often a good deal, too, because some companies, such as IBM, have exited the X10-based home automation business and shed their inventory of modules. Here are some of the names under which X10 equipment is sold:

X10 Powerhouse
The BSR System X-10
Magnavox
Leviton Manufacturing Co.
HomePro
Advanced Control Technologies
Stanley
GE Homeminder
PCS
Safety First
IBM Home Director
Heath Kit - Zenith
Wesclox
NuTone
RCA
HomeLink
SmartLinc
Universal Electronics
One-For-All
Sears
RadioShack Plug 'n Power

If you come across a device that you don't recognize as being an X10 module, see if you can spot a way to set an X10 address **[Hack #1]**. It might not be described as such, but if you see the familiar `A–P` house codes, and a way to set `1–16` unit codes, you've almost certainly found a compatible device. For example, you might find a wireless replacement doorbell kit that includes a secret X10 chime module **[Hack #9]** at your local hardware store.

Finally, if you're determined to discover how a mystery device works, turn to the manufacturer's web site to see if a manual is available for download. Check the technical specifications, and look for instructions on how to set the unit's address, if it has one. Or, turn to Google and look for posts on the *comp.home.automation* newsgroup, where newly discovered X10 equipment often is discussed and dissected. Happy hunting!

# Welcome to the State Machine

Set up a simple state machine to keep track of key conditions and events in your home. Then, use this information to make your smart home even smarter.

One of the keys to having a truly automated home—that is, a home that can react intelligently and proactively to current conditions—is setting up a *state machine*. A state machine is simply a logical approach to keeping track of key conditions and events so you can decide how to react when something changes.

For example, if you keep track of whether the sun is up when somebody rings your home's doorbell, you can turn on the porch light if it's dark, or keep it off during daylight. Without keeping track of whether it's dark outside, the best you could do is always turn on the light in case it might be dark—not a very good example of a smart home.

Another technique that uses a state machine is to have a text message sent to your cell phone when your home phone rings. The message contains the Caller ID information of the call you missed. If you're at home, the message isn't sent to your cell; it occurs only when you're away **[Hack #70]**. To do this, your house must know whether you're at home.

Table 1-8 shows some of the states I track in my system and a snapshot of their status.

*Table 1-8. Typical states for an automated home*

| State | Value |
|---|---|
| `Gordon Home` | `True` |
| `Gale Home` | `False` |
| `Houseguests` | `False` |
| `Daylight` | `True` |
| `Speech OK` | `True` |
| `Gone To Bed` | `False` |
| `Holiday Today` | `False` |

This handful of simple states enables me to program my system so that automation decisions are made in a seamless and seemingly intelligent fashion. Let's look at how the states can be used.

## Who's Home?

For my family, three key states need to be tracked: whether my wife and I are home, whether one of us is home, and whether we currently have houseguests staying with us. Who is currently at home determines how important automation-related messages get delivered. When I am away, the system will send reminders to my cell phone via email [Hack #73]. When I am home, the reminders are announced using text-to-speech. Similarly, reminders for my wife are routed to email, or announced, depending on her *at home* status.

Who is at home also determines which wake-up alarms are triggered in the morning. If my wife is traveling, her early-morning alarm does not sound, allowing me to sleep until my usual time. When I'm gone, my alarm is similarly silenced [Hack #47].

Whether we have houseguests is tracked for two important reasons. First, it tells the system to stop controlling the lights that are in the guest bedroom. In my experience, houseguests can be startled when a lamp seems to have a mind of its own.

Second, the `houseguests` state is used to indicate that the house still might be occupied, even though both my wife and I are away. This is an important logical test for reacting to outside motion detectors and other security considerations. As much as guests dislike having the bedroom lights turned off unexpectedly, that's nothing compared to the sirens and whole-house flashing lights they'd trip if the home was in "secure" mode after my wife and I left.

## Environmental States

Most home automation software automatically calculates sunrise and sunset times for your location. By executing scripts at these times, you can set a `daylight` variable for use in many decisions throughout the day. (Most home automation software can maintain a `daylight` flag for you automatically.) In my system, `daylight` is used to decide whether to turn on lights when outdoor motion detectors are triggered, such as on the front porch. When combined with knowing that the house is empty (by looking at the `Gordon Home`, `Gale Home`, and `Houseguests` variables), someone coming to the front door can also start a series of events that makes the home look occupied [Hack #74].

Another environmental state is called `Speech OK`. This one is checked before all text-to-speech announcements are made. It enables me to silence the system while we're asleep, watching a movie, or having a dinner party.

Similarly, if Gone To Bed is true, when an inside motion detector sees movement, nearby lights are dimmed to only half-strength—just enough to ensure safe passage—and then turned back off automatically a few minutes later.

Finally, the Holiday Today state is used to silence our weekday morning wake-up alarms. The scripts that wake us up in the morning check this state first, and if it's true, they just exit silently, so we can sleep in on those rare days off. It's possible to have this state set automatically by teaching your system which days are nonwork holidays. But I find it easier to just toggle it manually because I also want it to apply when I'm home sick, when I'm just taking a regular vacation day, or if want a later alarm for some reason.

## Tracking and Changing States

Most home automation software enables you to easily create variables that you can use to store the status of the states you're tracking. For example, in Indigo you can track states by using a variable **[Hack #18]**. In XTension, use a pseudo unit **[Hack #17]** for each state. In HomeSeer **[Hack #19]**, they're called virtual devices. A pseudo unit (or virtual device) does not have an X10 unit address, but it can be commanded as if it were an X10 module. That is, to set a unit called Gordon Home to true, simply turn it On. To set it to false, turn it Off.

Regardless of which software you're using, you will generally use one of three techniques for working with the states you're tracking. For some states, you'll want to set them using scheduled events. For example, you can set the Daylight state appropriately by scripts that run every day at sunrise and sunset.

Another method is to have the states set logically based on other conditions. For example, when I manually turn on the Gone To Bed state, the state Speech OK is also turned off **[Hack #48]**.

Finally, some states are best set manually. Because I don't have an exact bedtime, the Gone To Bed state is set when I push a button on a Palm Pad remote control **[Hack #5]**. Similarly, the states that keep track of who is at home are controlled by push buttons near the front door **[Hack #70]**.

## Final Thoughts

To get started with your own state machine, identify just a few of the household conditions that are useful for your home. Don't worry about discovering all of them up front. As with most things related to automation, start small and build your state machine over time.

CHAPTER TWO

# Office

## Hacks 25–36

Look around a typical home today and you're sure to see a lot of equipment that, just a few years ago, would have belonged in an office building. Now, at home, we're surrounded by computers, printers, telephones, and who knows what else. Have you ever wished for an assistant to help you manage it all? The hacks in this chapter will help.

You'll learn how to teach your house to talk [Hack #28] so that it can announce the name of incoming callers [Hack #27] and remind you of important events [Hack #25]. And, if you've ever wanted to make sure you closed the garage door when you left in the morning, you'll discover how you can reach out and control your home from any location [Hack #33].

### Remember Important Events

**A truly smart home should be able to gently remind you of upcoming appointments, things you have to do, birthdays, and other events. This hack provides two different approaches to this common problem.**

Reminders: we all need them to remember important dates and trivial tasks, but their success hinges on how easy they are to manage and how effectively they get your attention when they become due. In an ideal world, once you create a reminder you can put it out of your mind until your home alerts you about it, at just the right time for you to efficiently take care of whatever it is you're supposed to do. And because you're not at home all the time, your reminder system needs to be able to adapt to different situations so that you can be notified in a variety of ways.

## Deciding How to Be Reached

You need to decide how you want to be reminded about items when they come due. Do you want to be told about them when you're away from home, or is it sufficient that your reminders come up only when you're at home?

Consider the types of events for which you'll be creating reminders. For me, the calendar software at work handles all my job-related reminders, so my system at home is only for personal tasks. I try to keep the two worlds separate, so with only a few exceptions (such as a doctor's appointment during the day) will I have my house try to reach me with reminders during the day.

Here are a couple ways you can have reminders sent to you. Because each method has its strengths, you might want to use each of them, depending on the type of reminder you need.

*An email is sent to one or more addresses.* Email reminders are probably most effective for those times when you're away from home. You can send yourself an email at work or send a message to your cell phone's email address. However, depending on how attentive you are to your email, this might not be the best approach. For example, if your cell phone is turned off when you're at work, you'll miss tightly scheduled reminder messages.

*A message is spoken aloud by your home automation computer.* Spoken or other audio reminders are useful for when you need just a nudge to remember to do something. Tasks such as watering your houseplants every third night or closing the garage door before you go to bed are good candidates for spoken reminders.

Implementing each of these will require a little setup on your part, but before getting into those details, let's look at the different methods you can use to schedule the reminders.

The heart of your reminder system is the scheduler software. It needs to be flexible so that you can schedule one-time reminders, such as a doctor's appointment, and repeating reminders, such as the neighborhood parents' club that meets every third Wednesday of the month. The scheduler also needs to work with a scripting language so that it's flexible enough to handle the notification methods you've chosen to use.

## Scheduling with Home Automation Software

A simple way to implement a reminder system is to use the events scheduler built into your home automation software [Hack #16]. The software is, after all, robust enough to manage your home, so it should be good enough for reminders, too. You can create one-time or repeating events that execute scripts to send email, speak announcements, or display messages. Figure 2-1 shows a scheduled event in XTension [Hack #17] that executes every Thursday at 3:00 p.m. It reminds me to put the garbage can out on the curb for pickup the next morning. The Indigo [Hack #18] and HomeSeer [Hack #19] applications can also schedule events such as this one.

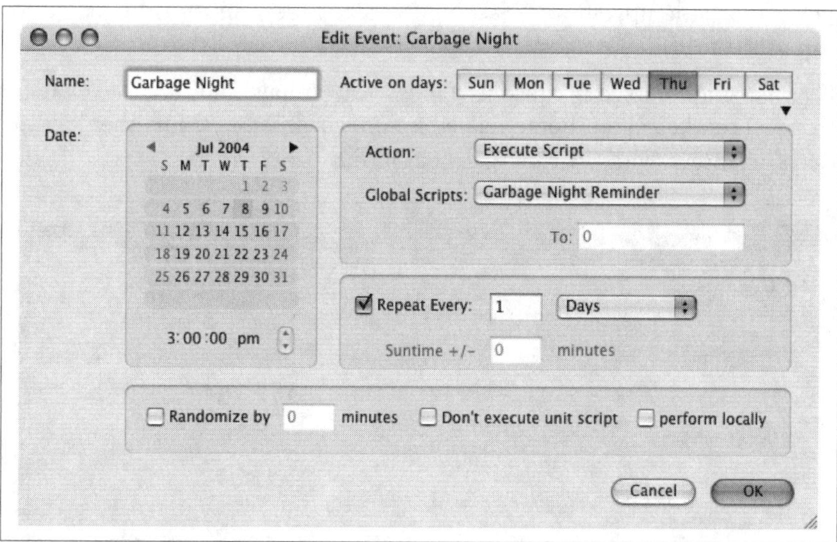


*Figure 2-1. Reminder to take out the trash*

Setting up a reminder event is easy enough, but scripting the reminder action can be trickier. One easy approach is to simply turn on a light or set off a chime module [Hack #9], but that isn't very useful if you have multiple reminders. You'll constantly be left wondering what the darn light or bell is signaling this time. For all practical intents, you'll need a script with a bit of sophistication.

If you've already implemented a general-purpose notification routine [Hack #73], it makes sense to use it here, too. The script that the event executes simply sets the text to be sent, then triggers the notification routine:

```
Set description of "Notify Gordon" to "It's Garbage Night!"
turnon "Notify Gordon" - causes description text to be delivered via email
```

As simple as this approach seems, it has drawbacks if you have more than a handful of reminders. First, you need to create separate events for each reminder, which can result in a lot of clutter in your home automation event scheduler. Second, because the reminder text is hardcoded into the script, as shown in the preceding example, you'll need to create a script for every reminder. That, too, grows unwieldy quickly.

## Scheduling with Calendar Software

Often, the best choice for a reminder system is to use an application designed for keeping track of dates and events. It'll make your reminders easier to manage, and provided that it can execute scripts when events become due, you can integrate it into your home automation system. Let's use Apple's iCal application (*http://www.apple.com/ical/*; free) as an example, but see the upcoming "Hacking the Hack" section for others that you can use instead.

iCal is a good choice because it enables you to have multiple calendars so that you can keep your home-related reminders separate from other kinds of dates, as shown with the *Haus* calendar in Figure 2-2.

*Figure 2-2. Home automation calendar in iCal*

In addition, iCal enables you to view your calendar on the Web, share it with others, and even synchronize it, using Apple's iSync application (*http://www.apple.com/isync*; free) with multiple computers. These features make it easy for you to check or edit your home-events schedule, even when you're not at home.

## Sending Email Reminders with iCal

iCal makes it easy to send reminders using email. When you define an event, simply turn on the alarm option and specify the email address to notify. Figure 2-3 shows an event with an email alarm scheduled to go off four hours prior.

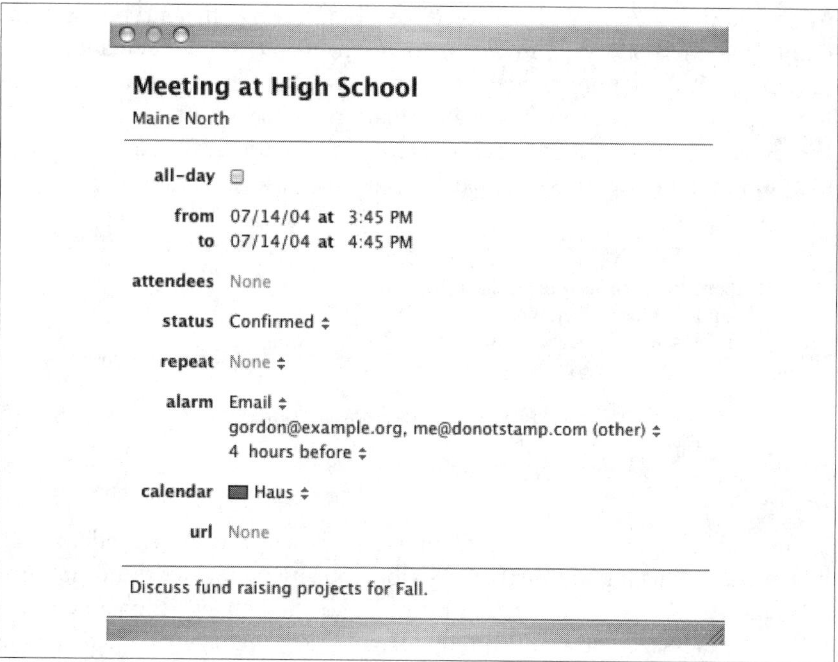


*Figure 2-3. An email alarm*

To send a reminder message to more than one address in iCal, add a new email entry to your "me" card in the Address Book. In the new entry, type multiple email addresses, each separated with a comma. When you schedule an alarm in iCal, select the multiple-address entry as the recipient for the reminder message. A copy will be sent to each address.

If you want to have the reminder communicate with your home automation system instead of sending an email, you'll need a script. To execute a script in iCal, use Apple's Script Editor to save the script as an application, and then set an *Open File* alarm and choose the application. iCal will run the script when the alarm is triggered. Here's a script that checks to see if anyone is home [Hack #70], then either announces the event or sends a reminder message as appropriate:

```
Tell application "XTension"
    if (status of "nobody home") is false then
        speak "Tonight is garbage night"
    else
        set description of unit "Notify Gordon" to "Tonight is garbage
night"
        turnon "Notify Gordon"
    end if
end tell
```

Although this script works fine, it's still not very efficient because the reminder text is specified in the script itself. You'd need a separate copy, modified with a reminder message, for each event. Unfortunately, at this writing, iCal does not pass any information to the script. Rael Dornfest, however, suggests a workaround. Here's a variation of the script that, instead of having the message text hardcoded, retrieves it from the name of the script file itself:

```
set fileInfo to info for (path to me)
set theMessage to displayed name of fileInfo
Tell application "XTension"
    if (status of "nobody home") is false then
        speak theMessage
    else
        set description of unit "Notify Gordon" to theMessage
        turnon "Notify Gordon"
    end if
end tell
```

To use this script, you'll still need a separate copy for each reminder alarm, but instead of having to modify the script, you simply make a duplicate copy of it, already saved in application format, and then change the name of the file to the message you want to send. Then, select the newly named copy as the action for an *Open File* alarm in iCal.

If you have an Audrey **[Hack #39]**, sending an email reminder to the account it uses is a great way to increase the visibility of the message.

## Hacking the Hack

A cross-platform alternative to using iCal is the Mozilla Calendar (*http://www.mozilla.org/projects/calendar/index.html*; free). It has the same ability to send emails for alarms (it does not run AppleScripts, however), and with a properly configured server, you can modify calendars from any computer, not just the one where you created the calendar, as is the case with iCal. Mozilla Calendar uses the same data format as Apple iCal, so if you have a Macintosh at home and a Windows computer at work, you can use either to update your calendar at home.

If you don't want a full-blown calendar system, plenty of good reminder applications are available for every type of computer. Find one that can execute scripts or send email, and give it a whirl using the principles described here. If you're a Mac user, consider using iMOnTime (*http://www.expersis.com*; $18), which offers several types of alarms and can be integrated easily with other applications using AppleScript.

# Keep the Lights On While You Work

Use only one motion detector to identify when a room is unoccupied and you might find yourself suddenly in the dark because you've been sitting a little too still.

Motion detectors are one of the most useful additions to your home automation system, but they also can be one of the most annoying. They're not particularly sensitive, so if you've entered a room and taken a seat you might not continue to trigger the detector. Depending on how you've set up your home automation system, this might signal the nearby lights to turn off. And nobody likes being left in the dark.

One solution to this problem is to wave your arms wildly every few minutes. This is almost certain to generate enough motion to trigger the sensor and keep the lights on. (Keep an eye out for the flashing red light on the front of the sensor to ensure you've been spotted.) But this is hardly something you want to be doing while quietly sitting and reading or sipping your tea—for one thing, the neighbors will think you're quite nutty.

Because motion detectors "see" by identifying patterns of heat and cold, when you're sitting motionless in a chair, your body heat becomes background noise to the sensor. And the further away you are from the detector, the greater your movements have to be to be seen as a significant change in the room's heat pattern.

One way to improve the situation is to use an additional detector in the room. Position one so that it sees you enter the room. Then put another directly under your desk where its field of view will be limited to just the area of your legs and the chair, as shown in Figure 2-4. As you move your feet, cross and uncross your ankles, and so on, these small movements will be more significant to the nearby and focused detector so that it will continue to signal your presence.

For a living room rather than an office, mount the detector on the ceiling pointing straight down at your easy chair. In the library, a motion detector midway up the wall where it will see the motion of your hands turning pages in a book might do the trick. You also can turn the detectors on their side to change their field of view, or construct a cardboard blinder to further focus it on the area where it's most likely to pick up your activities.

By using a second detector, you have a logical sanity check in your home automation software. That is, instead of relying on the state of a single detector to determine if the room is occupied, check both of them. If neither detector has fired in the last several minutes, more than likely the room is actually empty. But if one of them has fired, someone could still be present.

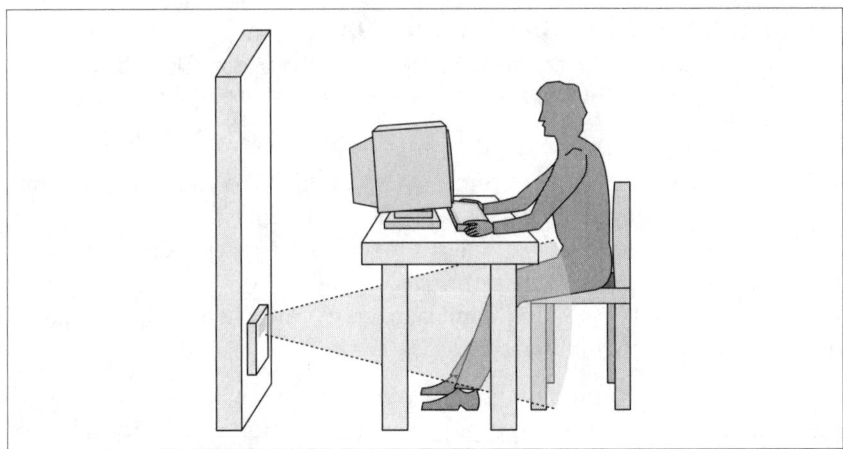

*Figure 2-4. An under-desk motion detector watching you work*

To do this logical check, you'll need to make sure the detectors are each set to unique addresses. But if you want to keep things simple, set both detectors to the same address. This will give you a single unit to check in your automation software, and in fact, you might not care which detector has fired lately, only that one or the other has. If you use this technique, however, you should ignore Off commands from the detector's address. Recall that motion detectors send an Off command after a period of inactivity. You determine the period when you assign the detector's address [Hack #6]. With multiple detectors set to the same address, you'll get an Off command when either of them decides the room is empty. If you don't ignore this signal, you might mitigate all the benefits of this hack.

## Hacking the Hack

If you're a HomeSeer user [Hack #19], one of the best techniques for keeping the lights on is to use the changed to on and set to on event triggers with a motion detector. The first trigger gets called only when the detector changes status from on to off. So, when you receive this event, execute a countdown timer for the period of time you want the lights to stay on. The set to on trigger is executed every time an On command is received, so use that trigger to reset the countdown timer; the subsequent On commands indicate the room still is occupied. When the timer finally does expire, turn off the room lights. You'll need to do a little scripting to set and reset the timer. See the Timer section of HomeSeer Help for details.

# Know Who's Calling

**Combine telephone Caller ID and voicemail with your home automation system to keep track of phone calls, ensure you don't miss messages, and even control your computer remotely.**

Just as home automation has gotten more sophisticated over the years, so has telephone automation. Two important inventions in this realm are the answering machine and Caller ID (CID). By combining these technologies with your home automation system, you can create an *automaton* that knows who is calling, takes messages for you, and passes this information to you in a variety of useful ways. Yes, life in the 21st century has its benefits.

Here are just some of the things you can do by combining telephony with home automation:

- Announce who is calling when the phone rings.
- Alert the first person home that calls were received while the house was unoccupied.
- Send an email when someone calls, perhaps with a recording of the message that was left.
- Display a pop-up CID message on every networked computer in your home, perhaps even your TiVo.
- Call home to take control of your house or check on its status.

To unite your telephone to your computer, here's what you need:

- CID service from your phone company
- Computer/telephone interface hardware
- Software that works with the interface hardware you've selected

Depending on how sophisticated you want the system to be, you might be able to use a modem as the computer/telephone interface. Let's start with the modem approach. For the most part, if you're going to use a modem, you'll be limited to CID functions. Although you can find modems that advertise their ability to record voicemail messages, the included software typically isn't well suited for integration with a home automation system. Also, if you're buying a new modem, consider getting one with USB so that you don't have to fuss around with USB-to-serial adapters to connect it to your computer.

This hack focuses on computer/telephone solutions for the Macintosh. But don't stop reading if you're a Windows user. The benefits apply to any platform, and on Windows, you'll find even more software available. In fact, HomeSeer Phone (*http://www.homeseer.com/products/homeseer_phone/*; $50), an add-in for HomeSeer **[Hack #19]**, offers powerful integrated telephony support and voice recognition of commands, which puts you ahead of Mac users in this regard.

If you have an old modem lying around that you want to use, check Lee Cohen's web site (*http://homepage.mac.com/maccallerid/modems.html*) for a helpful list of common Macintosh-compatible modems and their ability, or lack thereof, to understand CID.

Q: Hey, just about every Mac has a built-in modem! Can I use that?

A: Probably not. The vast majority of the modems that Apple builds in cannot decode CID information. The exceptions are some early iMac models that came with modems manufactured by Global Village and some clamshell iBook computers. There might be more exceptions, but Apple hasn't documented which models have CID-capable modems.

You'll need software that waits for the modem to send the CID information, which the phone company transmits between the first and second ring, and then passes on to your computer. The Mac OS X application CIDTrackerX (*http://www.afterten.com/products/cidtracker/index.html*; $10) can do this in spades. It works with a variety of modems, keeps an onscreen log of incoming calls, and works with AppleScript so that you can pass CID to Indigo or XTension **[Hack #16]** for further processing:

```
on run {CIDName, CIDNumber}
    Tell application "XTension"
        Set (description of unit "Last Call Rcvd") to CIDNumber
        Write log "Call Rcvd: " & CIDName & " " & CIDNumber
        Speak "Call From: " & CIDName
    End tell
    -- CIDName is passed as the Caller ID name that was sent
    -- CIDNumber is passed as the phone number that was sent
end run
```

You set the CIDTrackerX preferences to execute a script whenever a call is received. In this example, `CIDName` is the caller's name as provided by the telephone company (which is frequently `Unknown` or `Out of Area`), and `CIDNumber` is the caller's number. These values are passed to XTension to log and announce. You could also have the script increment the value of a pseudo unit [Hack #17] that keeps a running count of the calls you've received.

In addition to running a script, CIDTrackerX can operate as a CID server that broadcasts incoming CID information to other computers on your local network. Other computers that are running CIDTrackerX, or similar clients that work with client/server CID, can receive these broadcasts. See the "Client/Server Caller ID" sidebar for more information.

### Client/Server Caller ID

Like most programs that communicate with each other over a network, the client/server Caller ID (CID) applications use a special protocol. Unfortunately, there isn't a standard protocol, so some of the applications can't talk to each other. Here's the rundown of current popular network CID and the protocols they support:

*Network Caller ID (NCID; http://ncid.sourceforge.net/)*
: NCID is implemented using a command-line server and has clients that run on all major operating systems, including hacked Series 1 TiVo units. There isn't a scriptable Mac client for NCID, which limits its use for home automation.

*CallerID Sentry (http://home.houston.rr.com/jeffkohn/callerid_sentry.htm)*
: CallerID Sentry runs on Windows, but CIDTrackerX for Mac OS X can send and receive messages using the CallerID Sentry protocol. Additionally, the ACID application (http://www.timemocksme.com/acid) for the 3COM Audrey Internet appliance [Hack #39] also works with CallerID Sentry. Alas, there is no TiVo client for this protocol.

*Yet Another Caller ID (YAC; http://sunflowerhead.com/software/yac/)*
: YAC offers a server and client for the PC, and a client for Series 1 TiVo devices. CIDTrackerX is a YAC server and client, but no other Mac program supports this protocol.

If you want to do more than just capture and record CID, you'll need a specialized computer telephony interface and more sophisticated software. After a drought of good telephony software for Mac OS X, there are now three competitive choices:

- Phone Valet Message Center (*http://www.parliant.com/phonevalet/*) from Parliant Corporation
- Phlink (*http://www.ovolab.com/phlink/*) from Ovolab
- iAnswer (*http://www.matterform.com*) from Matterform

These three products have similar functionality. Each comes with hardware that connects to your computer via USB, and they all can decode and display CID information, answer the phone and play messages, record messages left by callers, and permit authorized callers to trigger AppleScripts on the host computer for automating tasks. This scripting capability is important; you can use it to control your home automation system **[Hack #34]** or to pause iTunes when someone picks up the phone **[Hack #36]**, to name just two examples.

The newcomer on the block, iAnswer, offers an interesting client/server mode. It supports the CallerID Sentry protocol for sending basic data to a client PC or 3COM Audrey, as CIDTrackerX does, but it has a more full-featured server implementation. You run the iAnswer Server application on the Mac that has the iAnswer hardware. The application runs silently and invisibly in the background, listening for CID information, running AppleScripts, and recording messages.

You run the iAnswer Client on any Mac OS X computer on your local network. (The server and clients automatically discover each other.) The client is the main method of interacting with the server. It will pop up windows with CID information as it comes in, display the entire CID history, and enable you to play voicemail messages the server has recorded. If you have set up your broadband connection so that you can access your home network from the Internet, you can do all of this remotely, too.

## Final Thoughts

Once you've gotten your Mac hooked up to your phone using the right hardware and software, particularly with a solution that can manage your voicemail and phonebooks, your home automation system quickly becomes an even more useful addition to your smart home.

*—Dean Davis*

# HACK #28 Broadcast Announcements to the Whole House

**If you use the text-to-speech capabilities or recorded messages to provide announcements, you need a whole-house audio system that will enable you to hear the announcements everywhere. If your house isn't wired with speakers, you can use your existing telephone lines to provide a voice-quality public address system.**

Most homes built within the last 30 years are wired to support multiple telephone lines. If you have only one or two phone lines, you likely have the capacity to support more lines. It's simple to use the spare telephone line capacity for an announcement system connected to your home automation system. The audio quality won't be good enough to play music, but it will be more than adequate for voice. It literally will be *telephone-quality* audio. Because you won't be able to achieve high quality, and it's not stereo sound, you can implement this system using inexpensive audio components.

## What You Need

To complete your whole-house public address system, here's what you'll need:

- An unused telephone line in your home's phone wiring
- One inexpensive speaker for each room in which you want announcements made
- A monophonic amplifier
- Several RJ-11 telephone connectors (one for each room)
- A telephone connector crimper
- Some telephone wire (a few feet for each speaker location)

## Planning

Decide which rooms you want to place speakers in to hear your computer's announcements. Make sure the room has a telephone jack, unless you intend to run wire specifically for this purpose. When deciding on the rooms, consider the type of announcements you plan on using.

If you're not sure you'll like whole-house announcements, or if you just want an easier method, see Jerry Prsha's "Voices from Everywhere" sidebar.

If you'll be announcing CID [Hack #27], you might want a speaker placed so that you can hear it from the dining room and television or family room, so you can decide if you want to interrupt your family time to answer the call. If you'll be having your system wake you in the morning [Hack #47], you'll definitely want a speaker in your bedroom. If you'll be using voice to provide status—such as when you leave the house [Hack #70] and to hear the weather report [Hack #64]—a speaker in the garage is a good idea.

In my system, I have four speakers: one each in the garage, kitchen, living room, and master bedroom. I purposely don't have one in the guest bedroom to spare my visitors the shock of a disembodied voice coming from under the bed.

## Voices from Everywhere

I've always wanted to have the home automation system alert me of things, but the cost of whole-house audio was too prohibitive. I was considering the possibility of running speaker wire to just a few rooms, via the heating ducts, when I suddenly had a technical epiphany. What if I were to install speakers, not in each room, but just in the air ducts?

I removed the humidifier and ran some long cables from my Mac to speakers that I placed on each side of the blower assembly. I reinstalled the humidifier and made the computer talk a bit using the Speech preference pane. It worked! I had sound throughout the home without running wire and with only two speakers!

After some experimentation, I learned several important things. The cheaper the speakers that you install in the vents, the better the sound will be. This is because expensive speakers are better at reproducing bass, which causes too much vibration. Some sound dampening, such as wrapping air ducts with insulation, also helps keep the vibration under control. The voice you choose can make a difference, too. The Trinoids voice works well and gives the sound a creepy omnipresence.

Use low-profile speakers to reduce the airflow restriction; large speakers will cause your blower to work overtime and make your system less efficient. Place speakers near the blower and facing in the same direction as the airflow, so the air can help carry the sound.

If you have someone out to repair the furnace, don't forget to alert him to what you've done. The last repairman who I forgot to tell about the system nearly soiled himself when the computer announced the hourly time and temperature.

Now that I have the speakers, motion sensors in and around the house alert us when raccoons are in our outdoor pond trying to eat our fish, and when our cats are in the litter box **[Hack #45]**. The Mac even wakes us up in the morning and reminds us again, like a snooze alarm **[Hack #47]**, if we're being lazy and sleeping that extra 10 minutes.

As I was writing this, the alarm system just spoke that it would be activating itself in five minutes because of inactivity in the house **[Hack #69]**. Sometimes, these handy verbal reminders, spoken through the vents, will also keep the homeowner from soiling himself by tripping his own alarm system.

The final step in planning is to determine which of the unused telephone lines you'll use for your speaker system.

## Examining Your Wiring

Telephones use only two wires (a pair) to operate. Standard telephone wiring consists of eight wires, enough for four lines, in a single bundle. Unless you have four separate phone numbers active in your house, you almost certainly have an unused pair you can use for this hack. Your task is to figure out which of the four pairs is not being used.

By convention, most homes use pairs 1 and 2. On a standard telephone connector (the end of the line that you plug into the wall), the pairs are numbered from the center out. That is, the two wires next to each other in the center of the jack are the first pair. The second pair is one wire over to the left and right, and the two outermost wires are the third pair.

In my home, I use pair 3 for the public address system. Most home telephone equipment doesn't even include connectors for the fourth pair; only pairs 1 and 2 are connected. You'll find in most cases that no wires are connected to the outermost pair.

## Wiring the Speakers

You'll need a single speaker for each room you identified earlier. Inexpensive, small, unamplified bookshelf or computer speakers work best. The type of speakers that often come bundled free with entry-level PCs work great. To get the best price, buy them in pairs and then split them up between rooms. Ideally, the speakers' wires should be removable so that you can more easily replace the wire later. But if they're not removable, that's OK. You'll just have some soldering to do.

Decide where you want the speaker positioned in each room. You might want it out of sight, such as under a bed or tucked under an end table. In the garage, I have the speaker mounted on the wall. Once you've decided, cut a length of telephone wire to run from the phone jack to the speaker. Don't make the wire any longer than necessary; doing so adds impedance and affects sound quality. Remember the old adage: measure twice and cut once.

Strip away a bit of the telephone wire's covering to expose the wires inside, and then connect one to each terminal on the speaker. Alternatively, if the speakers are so cheap that they don't have terminals, snip the built-in speaker wire close to the speaker, and solder the speaker wires to the telephone pair.

On the other end of the telephone wire, crimp an RJ-11 jack that connects the pair of wires you connected to the speaker to the unused telephone pair you identified earlier. In my system, I'm using pair 3, so I crimp the connector so that the outermost pins are connected to the speaker. Next, plug the wire into a nearby phone jack. Do this for each speaker and location you selected. Once you get the hang of crimping, it won't take but a few minutes for each connection.

But, what if a phone is already using the jack in the room where you want to install a speaker? Just buy a duplex phone adapter so that you can plug in both the phone and your speaker connection, as shown in Figure 2-5.

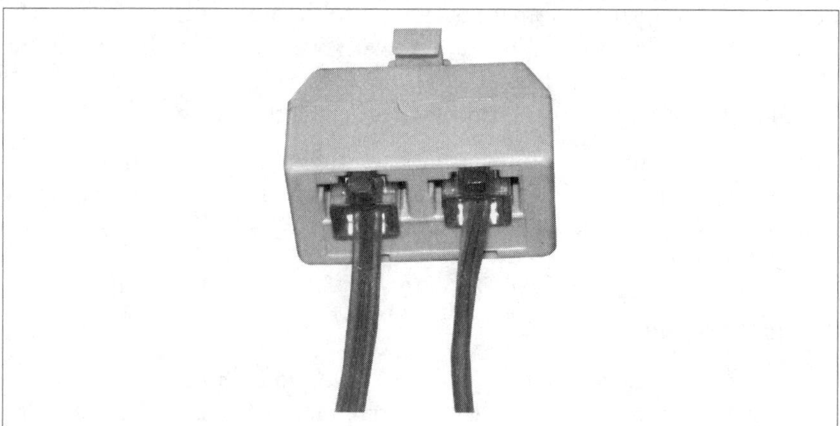

*Figure 2-5. Duplex telephone adapter*

## Amplifying the Announcements

You need a low-power monaural amplifier to send the sound across the phone lines to each speaker you've installed. RadioShack's (*http://www.radioshack.com*) Mini Audio Amplifier (Part #277-1008; $14) works great for this purpose. Get the AC Adapter (Part #273-1767; $14) for it; otherwise, you'll use up 9-volt batteries at an alarming rate because you'll be leaving the amplifier on around the clock. If you have another amp available, that's great, but a low-wattage amp works best. You'll find it easy to overdrive the telephone line and speakers, resulting in distorted sound, if you use something too powerful.

Use the telephone wire left over from wiring the speakers to create a wire that has an RJ-11 jack on one end, wired to connect to telephone pair 3, and a mini-audio plug on the other end. The plug connects to the earphone jack on the amplifier.

## Making the Computer Connection

Connect the amplifier to your computer by plugging it into your CPU's headphone jack. Now, turn on the amplifier, adjust the volume to a low setting, and have the computer speak a phrase or play an alert sound. If everything is connected correctly, you'll hear the sound throughout the house! It likely won't sound as you expect, so the next step is to fine-tune your audio settings.

When you connect the amplifier to your computer's headphone jack, your computer's built-in speaker might be muted automatically. If you want to also hear the announcements using the computer's speaker, look for a control panel setting that enables sound to play through both the headphone and speaker at the same time.

You'll need to find the best balance between the computer's volume setting and the amplifier's volume setting to get the best-quality sound at each location. (Here's where a higher-quality speaker will really pay off.) To begin, either enlist a family member to repeatedly cause the computer to speak, or temporarily set up a unit whose On and Off scripts speak a test phrase. The sample phrase you use for testing should be lengthy so that you can better judge how understandable it is, and it should include some numbers if you'll be using the system to announce the CID of incoming phone calls:

```
Say "Phone call from Mr. Esjay. 408 996-1010."
```

Set up a Palm Pad **[Hack #25]** to trigger the test unit, then wander the house and trigger the test phrase so that you can judge the quality and volume of the announcement. Adjust the volume controls until all distortion is eliminated at each speaker.

After you've gotten a clear sound, try each voice your computer's text-to-speech software offers. You'll probably find that most of them work well—the telephone wiring is designed to carry voice-quality sound—but one or two might be better understood than others. Try adjusting the rate of speech, if your OS supports doing so, to further tweak the voice you like best. Solicit the opinion of your housemates; they'll have to live with the voice you select, too.

You might start out selecting one voice and then later decide to switch to another after growing accustomed to spoken announcements. I once used Ralph (a voice available on Mac OS X) but now use Zarvox, which provides a wider tonal range that, for me, is more easily understood when playing over background noise, such as the television.

## Putting the Voice Announcements into Action

Now that you have the system set up, it's time to put it to good use. Here are some tips for getting the most benefit from your new public address capabilities:

- Consider prefacing all your announcements with "Pardon me" or another throwaway phrase. You'll find this will enable you to mentally tune into the announcement and not miss the first word or two.
- If you like two voices in particular, you can use them to distinguish between the type of announcement or the intended audience. For example, perhaps Victoria announces incoming calls, but Junior wakes you up in the morning. Doing this adds another layer of information that permits you to anticipate what's being said and evaluate its importance.
- Implement a `speech OK` state variable **[Hack #24]** that your scripts check before making an announcement. By turning this variable off, you can prevent announcements during movie time, in the middle of the night, or during other moments when you want some quiet.
- If you just can't find a synthesized voice you like, use prerecorded announcements **[Hack #29]** instead. They're not as flexible, but the quality can be much better.

# HACK #29 Announce Events with Recorded Announcements

**If your computer doesn't have text-to-speech capabilities, or if it does and you're not satisfied with the quality of its voices, you can use prerecorded announcements instead.**

Human speech is a great way to get someone's attention. Unlike a simple bell or "eep!" sound, human speech can convey detailed information, and although we often filter out the beeps and boops of modern life, the spoken word causes our ears to perk up and pay attention. Instead of having your computer play a sound or a song, have it speak a prerecorded announcement. The advantage of prerecording is that the vocal quality can be higher than synthesized speech, and the processing power needed to play a sound file is very small. Another advantage is that you can offload the playback to another application, such as QuickTime Player, to ensure that your home automation system can continue doing other things while the sound is played.

One drawback to prerecorded announcements it that you'll need to figure out in advance the announcements you want to make. But that's not too tough; for many homes, only a handful of events warrant the intrusiveness of an audio alert or notification. In fact, you might want your home automation system to be almost always silent so that when something does need your attention it's quite unique and conspicuous. Here are a few announcements you should consider:

- Someone's at the front door [Hack #74].
- The garage door has been left open [Hack #56].
- The mail carrier has opened your mailbox [Hack #62].

## Creating a Recorded Announcement

One way to obtain prerecorded announcements is to record your own. Many computers have either direct sound input capability or a built-in microphone. You'll need some software that enables you to record and then edit your speech. On a Macintosh, iMovie (*http://www.apple.com/imovie*; $50) will do the trick.

If you have an iPod and its microphone adapter, you can record a voice memo, then copy the track from your iTunes library for use in your automation system.

If you're satisfied with the quality of the text-to-speech function of your computer, but don't want to use it on-the-fly because of performance concerns, you might be able to create a recording by redirecting the output from the speech engine to a sound file. Doing this with Mac OS X Panther 10.3 is as simple as entering a single command. This script uses the voice Ralph to record the phrase "Welcome Back, Gordon!" to an AIFF file on the desktop:

```
say -v Ralph -o ~/Desktop/gordonhome.aiff "welcome back, gordon!"
```

Another approach is to use the Macintosh application Speechissimo (*http://www.speechissimo.com*; $49). Speechissimo has higher-quality voices than those built into Mac OS X and its voices are fluent in English, French, German, and Spanish. Recording a phrase to a sound file is very simple with its straightforward interface.

Finally, turn to the Internet and AT&T's speech synthesis demonstration web site (*http://www.research.att.com/projects/tts/demo.html*). You can type in the text you want spoken, choose from several high-quality voices for U.S. English, UK English, German, Latin American Spanish, and French, and then download a WAV file of the results to your computer.

### Playing the Announcement

Whichever method you choose to create the sound file, after you've created your recording, you'll need a little scripting in your home automation software to play it at the appropriate time. On Mac OS X and Windows systems, you can use QuickTime Player. This AppleScript, for use with XTension [Hack #17] or Indigo [Hack #18], tells QuickTime Player to open and play a sound file, then close the file when it has finished:

```
tell application "QuickTime Player"
        set theMovie to open file "Users:Home Automation:Sounds:Front Door
Motion.aiff "
        set close when done of theMovie to true
        play theMovie
end tell
```

You can use this script as part of an On script for a motion detector, or you can create a global script to call from multiple units.

### Hacking the Hack

If you're feeling really ambitious and want a lot of flexibility with your recorded announcements, consider creating several small recordings that you can play sequentially to create customized messages. For example, a script could play a recording that announced "Motion detected in" followed by a separate recording that said "Front yard" or "Back yard," as appropriate, to create what sounds like a single announcement.

## HACK #30 Send Pager Messages

**Although not as common as they once were (thanks to the cell phone), pagers remain useful for relaying messages from your smart home, particularly when you consider their long-lasting batteries, inexpensive nationwide coverage, and ability to receive signals where cell phone coverage is limited.**

Have you ever arrived home for the day and discovered that you've missed a phone call or a package delivery, or that the troublesome bathroom sink has leaked all over the floor, *again*? If you carry a pager, your home automation system can notice events such as these and transmit them to you using simple numeric codes. One of the advantages of sending notifications via pager rather than via email [Hack #73] is that it's simpler to set up a paging system. Another advantage is that sending the message is often much faster because your computer doesn't have to establish an Internet connection to send the message. Also, pager messages often are delivered nearly instantaneously, and they don't have to fight for attention in your inbox with the rest of your email.

To send a message to a pager, you will need a computer that has a modem connected to a phone line. To send a message to your pager, the computer must place a short phone call, so you'll want to brief others in the household [Hack #14] that occasionally they might hear the computer pick up the phone and dial whenever there's a message to send.

Using a pager to send messages is a good approach if you're setting up a home automation system to help out an elderly person [Hack #81] or to keep an eye on a building in another location [Hack #76]. You won't have to pay for an Internet connection or worry about keeping the system online and protected from network threats.

The Mac OS X application Simple Pager (*http://www.sentman.com/simplepager/*; $7) is easy to integrate with home automation systems such as XTension or Indigo. You simply enter the pager's phone number, the numeric message, and then click Send. Simple Pager dials using the computer's modem, sends the message you entered, and then hangs up the phone. As you can see in Figure 2-6, the interface couldn't be much simpler.

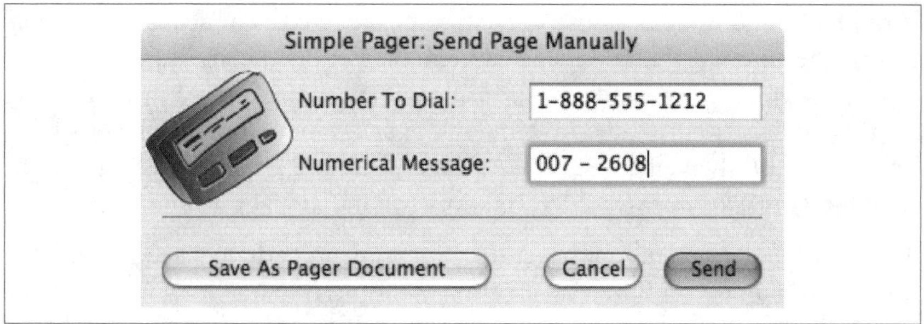


*Figure 2-6. Simple Pager for Mac OS X*

It's rare, however, that you'll be sending messages manually. After all, the point here is to have your system do so automatically when you're not at home. Luckily, Simple Pager makes automated operation nearly as easy as sending a page manually. You can create pager messages dynamically using AppleScript, or by creating *pager documents* that send messages you've defined already.

Sending messages using AppleScript is the most flexible approach because it enables you to determine the numeric message to send on-the-fly. For example, you might want to send a message about the current outside temperature, or how many visitors have come to your front door while you were away [Hack #74]. You can't know these values in advance, so you'll need an AppleScript that derives the information at the time the message is created.

For example, this AppleScript sends a pager message with the last outdoor temperature that was reported to XTension by WeatherManX **[Hack #64]**:

```
Set theMessage to (value of "Outside Temperature") as integer
Tell Application "Simple Pager"
        Page "1-888-555-1212" message theMessage
End tell
```

To receive a weather report when you're away from home, save this as a global script and then create a repeating event **[Hack #17]** that executes the script every couple of hours. In fact, you might want to enhance the script so that it sends the message only when your home automation system knows you're away **[Hack #70]** to avoid barraging you with messages when you don't need them:

```
If (status of "Gordon Home" is false) then
        Set theMessage to (value of "Outside Temperature") as integer
        Tell Application "Simple Pager"
                Page "1-888-555-1212" message theMessage
        End tell
End if
```

But not all the messages you send need to be created dynamically. If you just want to be notified that an event happened, such as the arrival of the daily mail **[Hack #62]**, you can set up *pager documents* with predefined messages. To do this, create a message in Simple Pager, as illustrated previously in Figure 2-6, and then click the Save As Pager Document button. Save the document in your *Documents* folder, or another handy location.

When Simple Pager opens a pager document, it immediately sends the message the document defines. By creating multiple documents with the messages you need, it's easy to have a variety of canned messages you can quickly send in reaction to events. Here's a bit of AppleScript, added to the On script for the unit that is triggered when the mailbox is opened:

```
tell application "Finder"
        open document file "Macintosh HD:Users:gordon:Documents:mailbox
pager"
end tell
```

One advantage of this method is that by passing off to the Finder the task of opening the document, XTension doesn't wait for Simple Pager to respond that it has accepted the message before it continues on with its next step, as it does when you use the dynamic message script described earlier. A disadvantage, however, is that the name and location of the pager document are specified in the script, and the script won't work correctly if you move or rename the document.

Simple Pager has some other nice features, too, such as the ability to execute an AppleScript before and after sending the page. This enables you to extend its capabilities, such as writing the message sent to a log file or speaking it aloud. Be sure to check Simple Pager's documentation for details and information about setting it up to work with your modem.

Now that you know how to send pager messages, you might be wondering how to best convey information using numeric data. It's straightforward if you're sending a phone number derived from the CID of a call you've received **[Hack #27]**, or a temperature as described earlier, but how do you send a message that the mail carrier has delivered the mail? This is probably the single biggest drawback to sending pager messages; you must resort to using coded messages.

If you have an alphanumeric pager that accepts text messages, consider communicating with it via email, which nearly every alphanumeric pager supports. See "Send Notifications of Home Events" **[Hack #73]** for information about sending email using your home automation system.

The coded messages I use are quite simple. To distinguish between messages that originate from my home automation system and those from people who are paging me, I preface each home automation message with `007`. That's easily distinguishable from a phone number, so I immediately can discern that I've received an automated message. (Sure, it's the international dialing prefix for Columbia, but that doesn't confuse *me*.)

Next, I add a hyphen to the message, followed by a three-digit code that indicates the message type. For informational messages, such as the temperature or the CID of missed phone calls, I use `411`. For events that are important, such as a water leak **[Hack #44]** or an inside motion detector that has detected an intruder **[Hack #71]**, `911` is my predictable choice.

Finally, following the message type is any data the message requires. In the case of a motion detector, my front door is designated as `701`. The complete message sent when it is triggered, then, is `007-411-701`. A message indicating that the outside temperature is 78 degrees Fahrenheit is `007-411-78`. A missed incoming phone call results in `007-411-408-555-1212`, where `408-555-1212` is the CID data received when the call arrived **[Hack #27]**.

Clearly, it's a good idea to limit the types of messages you send so that you don't have to memorize a long list of numeric gibberish. But that's a good idea anyway, because the novelty of receiving messages from your home quickly wears off and grows tiresome if you overuse this capability. And, like the villagers to whom the boy cried wolf, you're likely to ignore messages from your home and miss those that you truly care about.

## HACK #31 Broadcast Messages on Your Home Network

**Macintosh users can use LanOSD to send lovely and unobtrusive onscreen messages to every computer on the network.**

The LanOSD application (*http://oomz.net/Lanosd/*; free) is a clever *onscreen display* notification manager for your home or office network. Install a copy of LanOSD on every Mac in your house, and you can have one Mac broadcast messages that instantly appear on the screens of all the others. LanOSD is nearly invisible when there is nothing to display—it adds an item to your menu bar and not to the Dock—but it's beautiful when it has a message to deliver.

When a message arrives, it is displayed in a semitranslucent window that floats above whatever you're working on. After a few seconds, the message fades from view. Even in Figure 2-7, it's impossible to capture how nice this looks, so visit the product's web site and view the animated examples for yourself.

*Figure 2-7. LanOSD delivering an onscreen message*

You can control every aspect of LanOSD using AppleScript, or by using the command-line application that comes with it. You can specify the appearance of the message, including selecting which icon is displayed, the size of the text, and where the LanOSD window appears onscreen. You even can set it to play a sound file, or to speak the text of the message, which can come in handy for ensuring users don't miss its arrival.

I have LanOSD set up on the Mac I use for home automation, and I use it to send messages to all my other Macintosh computers when certain devices are triggered. For example, I use it to notify everyone when an outdoor motion detector has gone off, or when my garage door is open, which signals a family member to return home. To do this, I use this attachment script **[Hack #88]** with Indigo **[Hack #18]**:

```
using terms from application "Indigo"
        on LanOSD(DeviceName)
                tell application "Indigo"
                        set MessageKind to action type of device DeviceName
                        set PostText to ""
                        if supports dimming of device DeviceName is true
then set PostText to brightness of device DeviceName & "%"
                        set theText to DeviceName & " is " & (on state of
device DeviceName) & " " & PostText
                end tell
                tell application "LanOSD"
                        message kind MessageKind text theText icon "tv"
quadrant 4 fade delay 4
                        delay 4
                end tell
        end LanOSD
end using terms from
```

This script adds the LanOSD handler to Indigo's script processing so that it's easy to send a message to all the computers running LanOSD. But before covering how to send messages, let's look at some of the important things this script does.

In this line, the variable MessageKind is set to the device type property of the named device that you pass to this script:

```
set MessageKind to type of device DeviceName
```

For example, the device type for a coffeepot might be `Appliance Module`. A few steps later, the `MessageKind` variable is passed to the LanOSD application when the message is defined:

```
message kind MessageKind text theText icon "tv" quadrant 4 fade delay 4
```

When LanOSD receives multiple messages of the same `kind`, it displays them sequentially in the same alert window instead of creating new windows for each of them. This allows for a neater display should multiple messages arrive at nearly the same time.

Here, the script again uses the `DeviceName` parameter, passed to the script when you call it, to determine if the device supports dimming:

```
if supports dimming of device DeviceName is true then set PostText to
brightness of device DeviceName & "%"
```

If it does, its current brightness level is added to `PostText`, which is the text of the message that will be sent by LanOSD.

To send a message using LanOSD, set up an Indigo trigger action for the devices you want to monitor. For example, to monitor the brightness of the lights in the kitchen, add an action that is triggered when the `kitchen lights` device is dimmed or brightened, as shown in Figure 2-8.

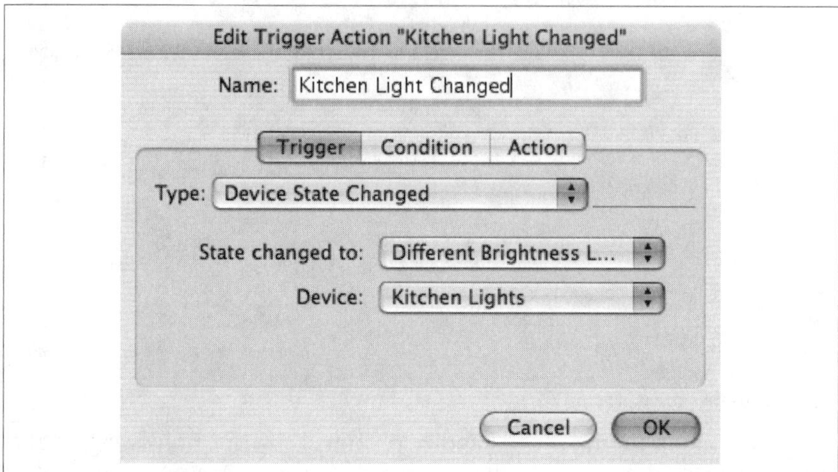


*Figure 2-8. An Indigo trigger action*

Define the trigger's action to run an AppleScript. Thanks to the attachment script you added earlier, all you have to do is pass the device's name to the LanOSD handler, as shown in Figure 2-9.

When the handler is called, it will query the state of the kitchen lights and construct a message for distribution by LanOSD. In this example, assuming that kitchen lights have been dimmed to 45%, the message will appear as shown in Figure 2-10.

## Hacking the Hack

Instead of passing the device name to the LanOSD handler, you could modify the script to retrieve the text of the message to send from an Indigo variable. Also, you might modify the handler so that it accepts more than one parameter, which would enable you to specify the icon and the length of time the text stays onscreen for each message you send.

*—Greg Smith*

*Figure 2-9. Calling the LanOSD handler*

*Figure 2-10. A message from LanOSD*

## HACK #32 Control Your Printer from Afar

**Have you ever noticed how your laser printer usually is switched off when you want to use it? This simple hack will save you from having to leave your comfy chair just to switch it on. Ahhh, Nerdvana.**

Laser printers are wonderful things, except they tend to be noisy and generate too much heat, even when they're not in use. As a result, unless you do a lot of printing, they're probably switched off most of the time. And they're almost always switched off exactly when you want to use them. If you have a laptop, turning the printer on might mean having to climb the stairs to the office just to flip the switch. Do this enough times, and you'll wish for a small child you can convince to do it for you, or an automated solution. Let's focus on the latter, and leave reproduction for another book.

As luck would have it, the X10 appliance module **[Hack #3]** is perfect for this kind of task. You'll want to get a model that accepts a three-pin power cord because that's what your laser printer will have. Also, pay attention to the module's power rating, compared to what your printer requires. If your printer is a real monster, you might need a heavy-duty module to handle the load.

You'll also want to make sure your printer has a power switch that stays in the on position, even when the printer has been unplugged. If it has a switch that you physically move to turn the printer on or off, as most laser printers do, you're all set. However, if it has a switch that doesn't stay in one position or the other, you won't be able to use this hack. With these kinds of switches, cutting off the power to the printer—which is how the appliance module does its magic—will cause it to forget that it's supposed to be on when the power is restored. If you're not sure, simply turn on your printer and then pull its plug. Wait a few seconds and plug it back in. If it stays off even after you reconnect the power, you won't be able to use an appliance module to control it. At least you can take solace in the extra exercise you're getting every time you need to get up and turn it on manually.

Once you're over the power hurdle, the rest is quite easy. Set the X10 address **[Hack #1]** of the appliance module to one that you're not otherwise using, such as `P1`, and then plug the printer into the module. In turn, plug the module into the wall outlet. When you want to turn on the printer so that you can use it, send the X10 command `P1 On`, wait a few seconds for the printer to complete its initialization sequence, and then print away. When you're finished, you can turn it off remotely by sending `P1 Off`. In fact, you'll need to teach yourself not to turn the printer off at the switch. If you do, the appliance module won't be able to turn it on again when you request, and you'll be back to where you started—turning it on manually. You'll make this mistake only a couple of times before you remember not to use the switch, but training your housemates likewise might take longer.

To send the `On` and `Off` commands to your printer, you can use a Palm Pad **[Hack #5]** or minicontroller **[Hack #4]**. Simply stash one or two near the locations where you most often use your computer—such as the TV area, kitchen, garage, family room, and bath—and you're all set. Or, if you use home automation software **[Hack #16]** to control your home, you can have the printer switched on automatically only during weekdays when you're at home [Hack #24] and turned off at night or when you leave to run to the store [Hack #70]. In these scenarios, the printer might be turned on when you're not really using it, but at the very least, it will be turned off when it's definitely not needed.

In Indigo [Hack #18], for example, you can define a trigger action that toggles the printer whenever the `isDaylight` variable changes, as shown in Figure 2-11. Indigo automatically maintains the `isDaylight` variable, which changes conditions with the rise and fall of the sun [Hack #20].

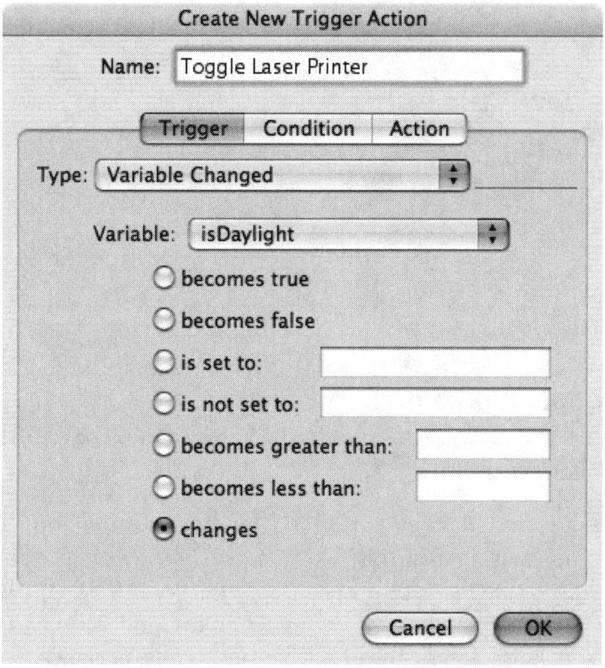

*Figure 2-11. A trigger action in Indigo*

After you've defined the trigger condition, click the Action tab, and select Toggle On/Off from the Action pop-up menu. Then select the printer's appliance module in the Device list, as shown in Figure 2-12. The toggle action simply reverses the current state of the unit; turning it off if it's currently on, and vice versa. This works quite well for our purposes, saving us from having to create separate events for turning it on and off.

Other techniques to consider are to simply set up a repeating event that turns it off every night, or to set up web-based remote control [Hack #99]. The best approach is the one that works with your system and, in the end, is easier than the manual method that it replaces.

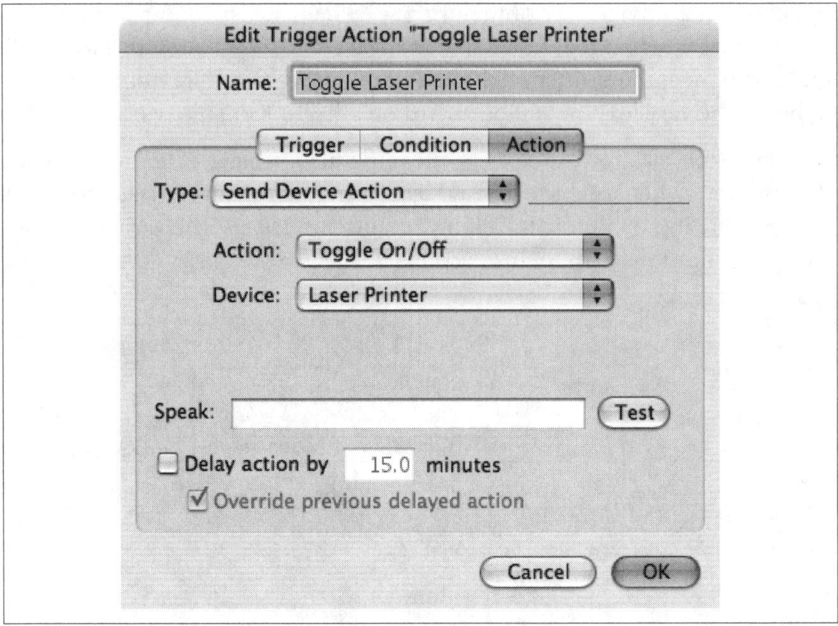


*Figure 2-12. The printer's appliance module in the Device list*

## HACK #33 Phone Your Home

**Control your home automation system by picking up the phone and dialing a few numbers.**

One of the handiest devices I've found for home automation is the TeleMaster from Home Controls (*http://www.homecontrols.com/cgi-bin/hci.pl?pgm=co_disp&func=displ&sesent=00&carfnbr=161*; $145). It's a small device, about the size of a couple of matchboxes, and you can tuck it away anywhere you have access to an AC outlet and a phone jack. The TeleMaster uses a two-way TW523 interface to send X10 commands to the power line, so you'll need an outlet with two free plugs. You need one plug for the AC adapter that powers the unit, and another for the TW523. Once you have the TeleMaster installed, you can send X10 commands by picking up an extension and dialing a simple sequence of numbers.

Don't confuse Home Control's TeleMaster with X10 Incorporated's Telephone Responder (*http://www.smarthome.com/5000.HTML*; $50). There's a bit of overlap in what they do, but they have different capabilities. The Telephone Responder, for example, works only when you dial into your home from the outside.

The TeleMaster works by listening to your phone line whenever an extension is taken off the hook. When it hears the right sequence of touch tones it beeps in response, and then sends corresponding X10 commands on the power line. You can choose between four different operating modes, depending on how many X10 addresses you need to be able to control (9, 90, 160, or 256 addresses).

Selecting the 60-address mode, which is what I use, requires that you dial four numbers to send a command. Dial * for *on* or # for *off*, followed by a single digit for the house code (`1=A`, `2=B`, and so on), and two digits for the unit number. For example, to turn on address `B7`, just pick up any phone and dial `*207`. To set the brightness level of a unit, hold down the last button in the command sequence.

Some sequences have reserved meanings, such as `*200`, which sends `All Lights On` for house code `B`, and `#200` for `All Units Off B`. One thing to be aware of is that some dialing sequences can conflict with custom calling features you might have on your phone, such as three-way calling. For example, my phone company uses `*70` to cancel the call-waiting function it provides. This means I can't send commands to house code `G` (`7` is the touch-tone digit for house code `G`), units `1` through `9` (each are dialed as `01`, `02`, and so on), without accidentally canceling call waiting. I can, however, command units `G10` through `G16` without any conflict.

You also can use the TeleMaster to notify your home automation system that the phone is ringing. It optionally will send `On` and `Off` commands to `P16` for each ring. If you have a lamp module set to this address, it will flash when the phone is ringing, which can be nice for the hearing impaired. Another interesting feature is the ability to send commands to `P15` when the phone goes on and off the hook. You could use this, for example, to stop voice announcements [Hack #28] while you're on the phone. Or, if you have a teenager in the house, you can trigger a script that checks the status of the phone after several minutes, and then begins flashing a light in your kid's room if he's been talking too long. If it's a weekend, however, you could give him extra time to speak.

In addition to using the TeleMaster to send X10 commands when you're at home, you also can use it to send commands while you're away. Simply call your home, wait until your answering machine answers, and then dial the TeleMaster commands. (It requires a pass code first.) If you don't have an answering machine, you can set the TeleMaster to answer the phone after it has rung unanswered for 50 seconds.

Around the home, I use the TeleMaster to trigger XTension [Hack #17] scripts that carry out multiple tasks. For example, when I retire for the evening, I just pick up my bedside phone and dial a command that executes a script that turns everything off for the night. I've had mine for almost 10 years—I use it every day—and the TeleMaster has never even hiccuped. My wife loves it, too [Hack #14], because she can control the lights in our henhouse without having to go out in the cold.

—*Rob Lewis*

HACK #34

## Control Your Home with Phlink

**The Phlink Telephone Adapter for Mac OS X provides a way to control and monitor your home over the phone, and it's easy to integrate with Indigo.**

How would you like to check on your house while at the office or on the road? There are a few ways to do this, such as using a web browser [Hack #99] or having your home send you email [Hack #73]. One of the most useful methods, especially in a situation in which you don't have access to a computer, is to control your home over the phone using touch-tone commands. But beyond just sending commands to your home [Hack #33], it's even better when you can have the home speak its status using voice synthesis.

This hack uses Indigo [Hack #18] and the innovative Ovolab Phlink (*http://www.ovolab.com/phlink/*; $150). The Phlink, which connects to your phone line and computer (via a USB telephone adapter), can answer telephone calls, identify callers based on CID information, record voicemail messages, and make outgoing telephone calls. Additionally, Phlink has terrific AppleScript support, which allows for integration with other Mac OS X applications. Perceptive Automation has written a set of AppleScripts that integrate Indigo and Phlink, giving you a speech-synthesized menu system that enables you to control Indigo in the following ways, all over the phone:

- Perform complex actions by executing Indigo action groups
- Hear the status of devices (e.g., "Hot tub heater is on")
- Turn devices on or off
- Hear the brightness level of lamps and light modules
- Brighten or dim lamps and light modules

### Installing the Phlink Scripts for Indigo

After you download the Phlink scripts for Indigo from Perceptive Automation's web site (*http://www.perceptiveautomation.com/hackbook/phlink.dmg*), follow the *read me* instructions to place the scripts and folders in the correct locations. Copy folders *1* and *2* and the file *greeting.scpt* into the */Library/Application Support/Phlink Items* folder. If you already have folders of the same name in that location, create a new folder with a different single-digit name, such as *3*, and copy the *1* and *2* folders from the Indigo download inside there.

> Phlink uses the names of the folders to build a menu system that you navigate by dialing the associated number. For example, you'll dial `1` to access the items that are in the *1* folder, after you get everything set up.

Next, copy *phlink call out attachment.scpt* into your *~/Documents/Indigo User Data/Scripts/Attachments* folder. This attachment script **[Hack #88]** defines two functions, `MakeEmergencyCallToNumber()` and `MakeEmergencyCall()`, which can be called from any Indigo trigger or time/date action to have Phlink call your mobile or office phone in an emergency.

If Indigo is running already, choose Reload Attachments from the Scripts menu to make it recognize the new script you added.

## Phoning Home

Now that we have everything installed, let's call into the system to see what happens. Make sure the Phlink application is open and set to answer calls. Then, dial your phone number using a cell phone or another line.

When the phone rings, Phlink answers the call and immediately executes the *greeting.scpt* AppleScript that you installed in the *Phlink Items* folder. That script uses speech synthesis to speak to the caller. It says:

```
Indigo home control server.
Press 1# to execute an Action Group.
Press 2# to control a Device.
```

When the user presses `1#`, Phlink will traverse into the *1* folder and execute its *greeting.scpt* AppleScript. Likewise, when `2#` is pressed, Phlink traverses into the 2 folder and executes the script found in that location.

The greeting script in the *1* folder queries Indigo to get a list of the action groups you've set as being eligible for execution over the phone. Similarly, the script in the 2 folder determines which devices can be controlled over the phone.

To set which action groups and devices can be accessed you change their settings in Indigo. To edit an action group or device, check the "Display in remote UI" checkbox, as shown in Figure 2-13.

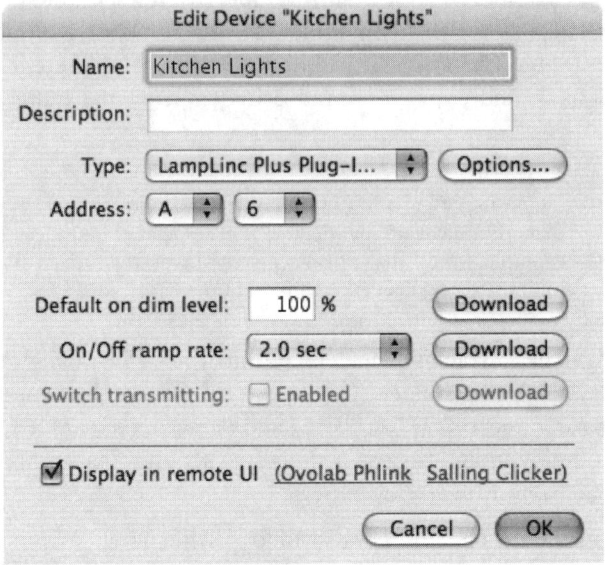

*Figure 2-13. Turning on phone control*

All action groups and devices that have this option selected will be available to the Phlink scripts. For example, if you select this option for the action groups `away from house`, `be home soon`, and `guest party arrival`, Phlink will speak the following menu:

```
To execute Action Group away from house press 1#.
To execute Action Group be home soon press 2#.
To execute Action Group guest party arrival press 3#.
```

Similarly, the device menu will include those you've selected for remote access:

```
To control device Porch Light press 1#,
To control device Entryway Light press 2#,
To control device Hot Tub Heater press 3#.
```

When you dial a number, Phlink passes the number to the *default.scpt* script in the current folder (inside folder *1* for action groups or inside folder *2* for devices). These scripts tell Indigo to execute the appropriate action group, or they get the status of the selected device. In our example, if the user presses `3#`, she hears:

```
Hot Tub Heater is off.
To turn on press 1#.
To turn off press 2#.
```

The script also handles the user's response, sending an `On` or `Off` command to the selected device.

## Custom Status Reporting

In addition to controlling devices and executing action groups, you can use AppleScript and Phlink to retrieve information about your home's status. For example, let's say you are interested in the number of times a motion detector on your front porch has been triggered today, to see if you've had any visitors while you've been out.

The first step is to set up Indigo so that it increments a variable each time `front porch motion detector` turns on. Create a new trigger action named `front motion trigger` and set the trigger so that it's executed when the device `front porch motion detector` is turned on, as shown in Figure 2-14.

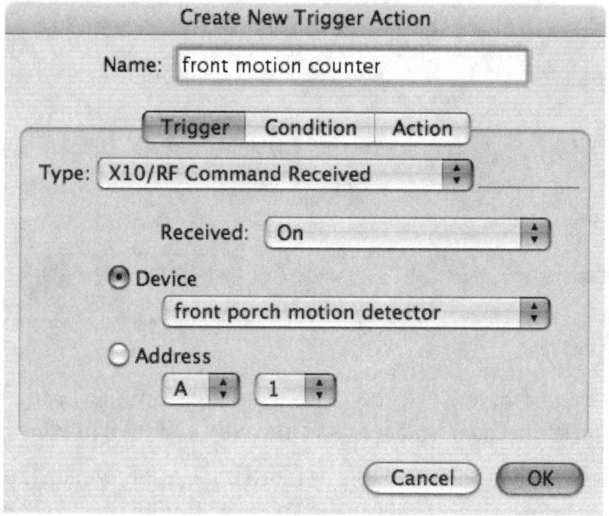

*Figure 2-14. Setting the Indigo Trigger Action*

Click the Action tab and choose Modify Variable from the Type pop-up menu. Set the options so that the variable `frontMotionCount` is incremented by `1` when this trigger is executed. If you haven't already created `frontMotionCount`, choose New from the Variable pop-up menu to create a variable, as shown in Figure 2-15.

This counter will continue to increment each time the motion detector is activated, but we're interested only in the count for the current day, so let's set up a time/date action that resets the counter every night. Add a new time/date action that executes every day, just after midnight, as shown in Figure 2-16.

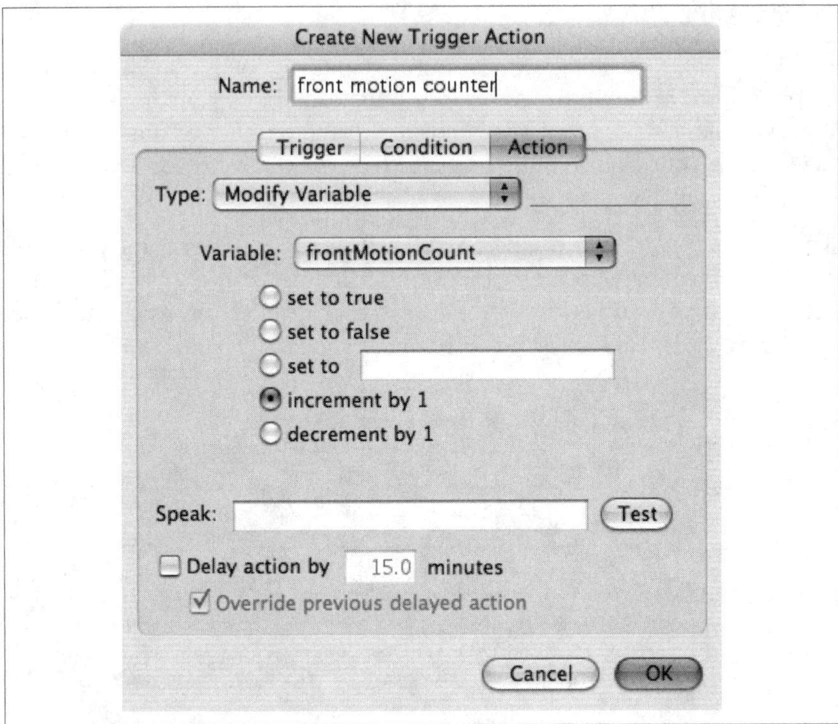


*Figure 2-15. Incrementing the counter*

Click the Action tab, and select Modify Variable from the Type menu. Then, choose `frontMotionCount` and set its value to `0`, as shown in Figure 2-17.

Now we have an Indigo variable, `frontMotionCount`, which automatically counts the motion triggers that occur over a 24-hour period. Next, we need a way to report the value of this variable, via Phlink, to the caller on the phone.

Open the *greeting.scpt* script, located in the *Phlink Items* folder, and insert four lines into the script:

```
on do_action given call:c
        tell application "Phlink 1.4.1"
                tell c
                        speak "Indigo home control server. Press 1 pound"

                        tell application "Indigo"
                                set motionCount to get value of variable
                              "frontMotionCount"
                        end tell
                        speak "front motion count is " & motionCount
```

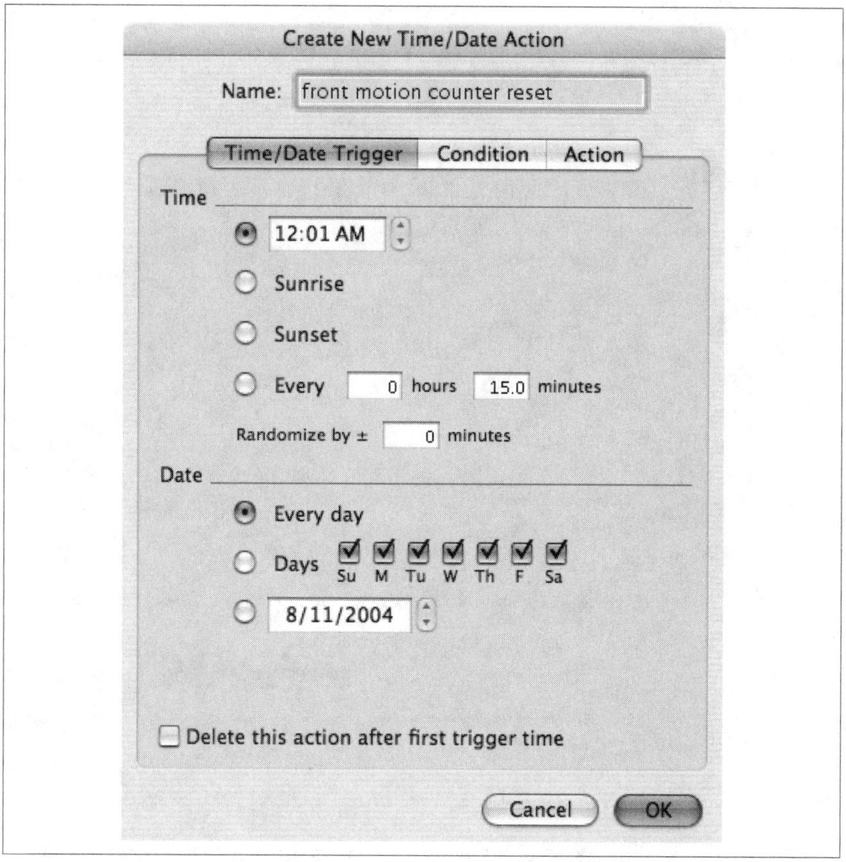

*Figure 2-16. Adding a midnight action*

```
                    set the require pound sign to true
                    set the email recordings to false
                end tell
            end tell
    end do_action
```

Now, we have a customized main greeting script that speaks the current motion detector trigger count every time it's executed. The next time you call your house you will hear:

```
Indigo home control server.
Press 1# to execute an Action Group.
Press 2# to control a Device.
Front motion count is 9
```

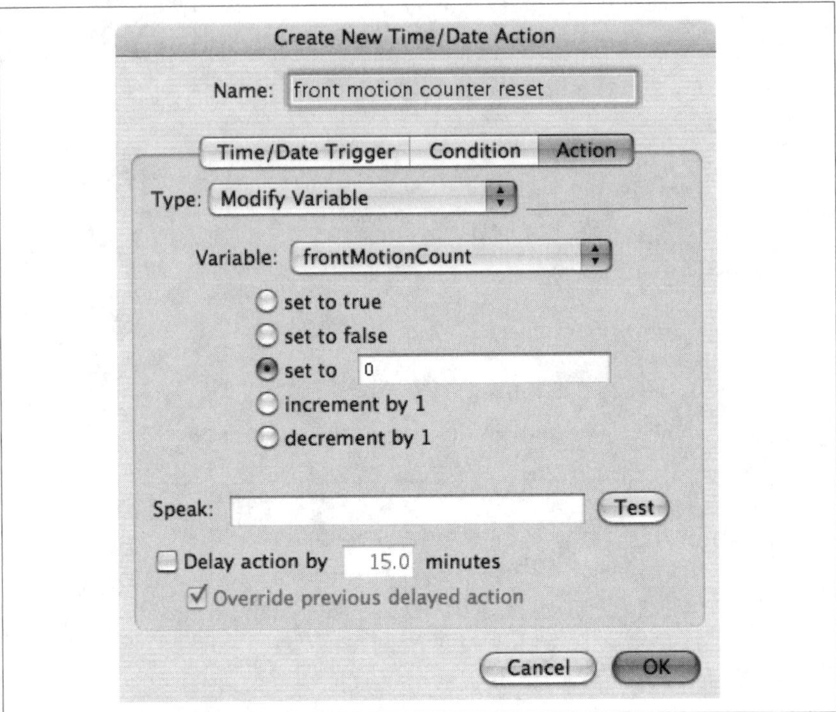


*Figure 2-17. Resetting the variable*

## Remote Notification

All the work thus far has created a system that you can call remotely to control and query the state of your house. But in addition to answering incoming calls, Phlink can place outgoing calls. From a home automation perspective, this is exciting because now we can have Indigo call our mobile or office phone whenever a critical trigger occurs, such as after a power failure, an intruder is detected, or a freeze/flood sensor has gone off.

The *phlink call out attachment.scpt* file that you installed earlier makes it easy. It's an attachment script **[Hack #88]** that defines two new functions you can use from any Indigo trigger or time/date action. The first function, `MakeEmergencyCall( )`, calls the emergency number you specify in the script. The second, `MakeEmergencyCallToNumber( )`, enables you to specify a number to call when you call the function. In both cases, you also provide the message you want Phlink to speak to whomever answers the calls.

Before you can use the AppleScript, you need to add your preferred emergency phone number to the script. Open the *phlink call out attachment.scpt* file in Script Editor and change the properties as shown here:

```
-- Default telephone number to call. Only used when MakeEmergencyCall()
function is called.
-- Use MakeEmergencyCallToNumber() to override this number for a particular
action.
property defaultEmergencyNumberToCall : "555 555 0101"

-- The number of times the message should be repeated. We repeat because
there currently
-- is not a way for Phlink to detect exactly when the call has been
answered.
property speakRepeatCount : 10
```

The `speakRepeatCount` property can stay at `10`, unless you find that having the phrase repeated 10 times does not allow enough time for the message to be delivered.

After you've saved your changes, choose Reload Attachments from the Scripts menu in Indigo to register your changes with the program.

Now, let's put it to use. Let's say you want to be notified when the outside temperature in your hen house reaches 32 degrees Fahrenheit. You have a freeze sensor (*http://www.smarthome.com/7193.HTML*; $25) connected to a Powerflash **[Hack #10]**, so Indigo will receive an X10 `On` command when the sensor detects that it's too cold.

Create a new Indigo trigger action that is executed when the command from the freeze sensor is received. The action for this trigger is to run an embedded AppleScript that consists of a single function call that calls the emergency function you updated with your phone number, as shown in Figure 2-18.

To test this trigger action, either put the freeze sensor in your icebox for a few minutes, or select the trigger action in the Indigo main window and click Execute Action Now. Phlink will call the phone number you defined and repeat the phrase "Freeze sensor triggered" 10 times.

If you want to have Phlink dial a number other than the one you defined in the attachment script, use the `MakeEmergencyCallToNumber` function instead. It accepts an additional argument that specifies the phone number to dial:

```
MakeEmergencyCallToNumber("freeze sensor triggered", "555 555 0102")
```

## Final Thoughts

This hack should give you a good foundation for integrating your home automation system with your telephone. You can modify the scripts you installed in the first few steps to do more. Examining how these scripts work is a great way to find hints on how to add additional functionality. If you

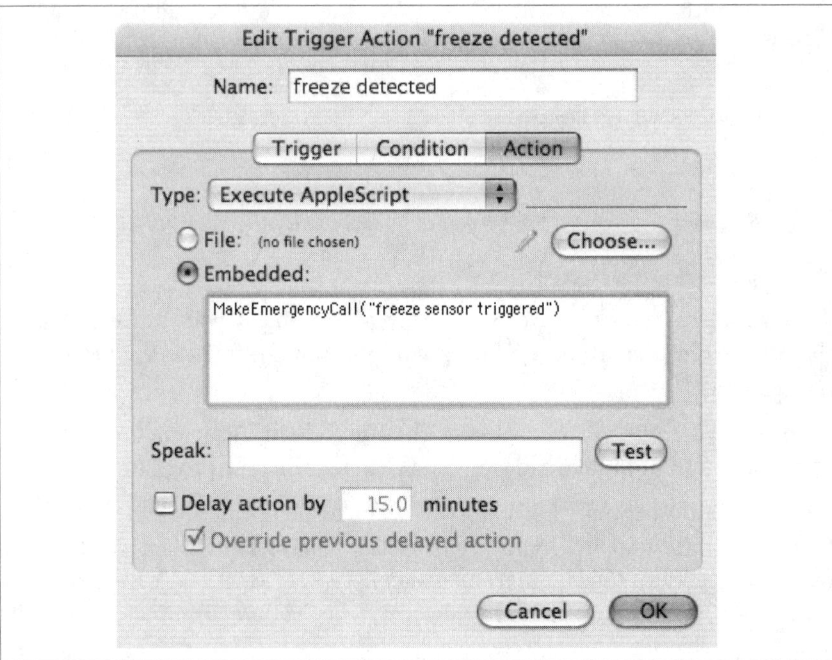


*Figure 2-18. Placing an outgoing call*

come up with an interesting script, visit the Indigo online support forum (*http://www.perceptiveautomation.com/indigo/forum.html*) and share it with other Indigo and Phlink users.

—*Matt Bendiksen*

## HACK #35 Forward Phone Calls

**The phone company's call forwarding options can be inconvenient to use, but with home automation techniques, you can get more value for your money.**

Did you order call forwarding from the phone company thinking you would use it a lot? Chances are, you don't use it very often because it's a pain to remember to activate when you leave the house. It's also a pain to remember to cancel it when you come home. Did you ever wish you could change your call forwarding number when you're away from home? Well, I get my money's worth from call forwarding by using my home automation system controller and telephone touch-tone interface.

I use a HomeVision-Pro (HV-Pro) automation controller (*http://www.csi3.com/homevis2.htm*; $600). I selected the HV-Pro because it supports multiple methods of input, output, and logic control. The best feature has been the ability to add new functionality that I did not envision when I purchased the HomeVision. I have successfully added two-way control to my Napco Gemini home security system (*http://www.napcosecurity.com/testframe.html?products-napco.html*), a Jandy Pool & Spa control system (*http://www.jandy.com/html/products/controls/onetouch.php*), and Powerline Control Systems' Universal Powerline Bus (UPB) products (*http://www.pcslighting.com/UPBProducts.htm*). These weren't on my original list of things to integrate with my automation environment, but HomeVision has enabled me to add them all.

When leaving the house, I press a button on the Palm Pad next to the door to initiate the call forwarding. I currently use one button to set the call forwarding to my cell phone (`M1 On`), another to forward calls to my wife's cell phone (`M2 On`), and a third to cancel call forwarding (`M1 Off` or `M2 Off`) that I've previously set. The general logic flow is as follows:

1. Send an X10 command (i.e., `M1 ON`).
2. The home automation controller receives the command and executes a predefined macro (i.e., `Forward to Jon's Cell Phone`).
3. The macro picks up the phone line and dials the sequence necessary to cancel any existing call forwarding.
4. The macro pauses and waits for the dial tone to return to the phone line.
5. The macro dials the code necessary to establish new call forwarding.
6. Pause to allow the command to take effect; then, hang up the phone.

To implement steps 3 through 6, the macro could simply dial the number `73,,,#72,555-1234,,,,,,,,,,,,`, where each comma introduces a two-second pause.

When I return home and want to turn off call forwarding, it would be simple to set up another Palm Pad button that triggers a similar macro to dial the sequence necessary to turn off the service. However, I don't usually do this; I prefer a smarter approach.

To automatically turn off call forwarding when I return home, I have my alarm system configured to send an X10 command (`M1 Off`) when I disarm the alarm upon returning home. When HomeVision receives this command, it starts a five-minute countdown timer. When the timer has expired, HomeVision checks to make sure the alarm is still disarmed. If it is, it runs a macro that cancels call forwarding, as described previously. This gives me a five-minute window to quickly run in and out of the house, disarming then arming the alarm, without disabling the current forwarding setting. This method of automatically turning off call forwarding works great, and it is handy when a second person returns home, which automatically cancels the call forwarding so that this person can receive phone calls.

Another handy, but more complex, technique enables me to change the telephone number that calls are forwarded to when I'm away from home. I call this *remote redirection of call forwarding*. If you have a second phone line for your fax or modem, you can use this technique to dial in and modify the call forwarding setting on line 1. You'll also need a *double pole, double throw* (DPDT) relay switch that's connected to your phone lines and configured so that changing the relay switches the HomeVision Phone/CID module between the two phone lines. The relay switch is connected to an output port on the HomeVision so that a macro can activate it and cause it to switch between the two lines.

In my setup, I rely on an output from the alarm system that signals HomeVision when the alarm is armed or disarmed. When armed (away), the Phone/CID unit is active on line 2, but when disarmed (home), it is active on line 1. Now, by dialing into line 2 and entering a pass code, HomeVision executes a macro that does the following:

1. After receiving the pass code and activating the macro, hang up line 2.
2. Activate the relay to switch the Phone/CID module to line 1.
3. Pick up line 1.
4. Dial the call forwarding code to set a new phone number.
5. Pause to allow the command to take effect; then, hang up the phone.
6. Activate the relay to switch the Phone/CID module back to line 2, to await another call.

## Hacking the Hack

If you don't have a phone control module for your home automation controller, you can implement a similar method by using a modem. You won't have the same control that HomeVision offers, and you'll need to find a scriptable terminal program to control the modem, but in general, the technique is quite adaptable to different systems and equipment.

*—Jon Welfringer*

# Silence the House when You're on the Phone

**Sometimes your smart home can seem dumb, such as when it starts making announcements or playing sounds while you're on the phone. Use this hack to teach it to know better.**

If your home automation computer announces important events when they occur, such as visitors coming to the front door [Hack #74] or the latest weather alerts [Hack #64], you'll occasionally find that it speaks at inappropriate times, such as when you're on the phone. When this happens, you'll be either annoyed or bemused at having to explain to your caller where that robotic voice is coming from [Hack #28]. In either event, you'll wish your home automation system would simply *shush* when it knows you're on a call.

Well, you can silence your house automatically, but first, you need the hardware necessary to detect when the phone line is off the hook. This hack uses Phone Valet for Mac OS X from Parliant Corporation (*http://www.parliant.com/phonevalet/*; $200). Phone Valet has many features, including voice mail, CID [Hack #27], and, most importantly for the purposes of this hack, the ability to execute AppleScripts in reaction to telephone events.

Windows users should check out HomeSeer Phone, an add-in module for HomeSeer that has similar capabilities (*http://www.homeseer.com/products/homeseer_phone/*; $50).

We're interested in the phone's *off hook status*. Phone Valet 2.0 makes it easy to (pardon the expression) hook into this event and add some automation. Choose AppleScripts from the Phone Valet menu, and then click the Hook Status tab, as shown in Figure 2-19. Use this dialog to select AppleScripts that will be executed whenever the phone line to which Phone Valet is attached is picked up, to either dial or answer or call, and then again when the line is hung up.

Parliant includes sample scripts that set your iChat status and mute iTunes, but you can add additional scripts and have more than one active at the same time. Install a new script in the */Library/Application Support/Parliant/AppleScripts/HookStatus* directory, and it becomes available in this dialog.

Let's look at a simple script that sets the computer's system volume when the phone line changes status. The sample scripts provided by Parliant are much more sophisticated—they work when you have multiple phone lines, for example—but if your needs are simple, as they are in this case, it's best to keep the approach simple, too.

*Figure 2-19. Phone Valet's hook status*

```
on handle_hook_status(isOffhook, lineName, numLinesOffhook)
    if numLinesOffHook is equal to 1 or numLinesOffHook is greater than 1
then
        set volume 0
    else
        set volume 5
    end if
end handle_hook_status
```

Every time the phone line goes off or on hook, Phone Valet runs the `handle_hook_status( )` handler in each script you've selected to receive this event. In this example, the script examines only the value of `numLinesOffhook`. If it's greater than `0`, at least one Phone Valet–equipped line is off the hook, so the computer's system volume is set to `0` (muted). Any sounds the computer generates (alarms, synthesized speech, etc.) will be silenced.

When all the monitored lines are hung back up, `numLinesOffHook` will be `0` and the script sets the computer's volume control back to the desired level. Ahhh, silence is golden.

## Hacking the Hack

If you want to allow some sound to still occur, perhaps higher-priority messages or reminders **[Hack #25]**, you might not want to adjust the computer's volume. Instead, set a state variable **[Hack #24]** in your home automation software that controls which types of audio events are allowed.

For example, to silence only voice-synthesized announcements, create a pseudo unit **[Hack #17]** to indicate whether speech is currently allowed. The Phone Valet script sets this unit instead of adjusting the volume control:

```
on handle_hook_status(isOffhook, lineName, numLinesOffhook)
    if numLinesOffHook is equal to 1 or numLinesOffHook is greater than 1
then
        tell application "XTension"
            turnoff "Speech OK"
        end tell
    else
        tell application "XTension"
            turnon "Speech OK"
        end tell
    end if
end handle_hook_status
```

Then, in your XTension scripts, check the value of `Speech OK` before making an announcement:

```
If (status of unit "Speech OK" is true) then
    Speak "Time to wake up, Gordon!"
Else
    Write log "Gordon wake up alarm silenced"
End if
```

If you decide to take this approach, you should adapt "Send Notifications of Home Events" **[Hack #73]** to create a single routine for announcing events so that you don't have to check `Speech OK` in multiple scripts.

Finally, if you don't need something as sophisticated as Phone Valet, look into the ELK-930 Doorbell and Telephone Ring Detector (*http://www.elkproducts.com/products/elk-930.htm*; $35). Connect it to a Powerflash module **[Hack #10]**, and you'll get an X10 signal whenever the phone rings. It won't signal when you pick up a phone to place a call, but it's a step in the right direction.

CHAPTER THREE

# Kitchen and Bath

## Hacks 37–46

Your kitchen and bathroom are the utilitarian workhorses of your home, and the hacks in this chapter reflect that character.

You'll find methods for keeping an eye out for water leaks [Hack #44], discouraging your teens from standing in front of an open fridge [Hack #42], and automating your furnace and air conditioner [Hack #41]. There's some fun here, too, such as warming a toilet seat [Hack #43] for those cold winter mornings and an easy-to-set-up alarm that will let you know if a repairman sneaks a beer [Hack #38] when your back is turned.

Come to think of it, this chapter also reflects that great tradition of having a kitchen drawer for keeping doodads and tools that you're not sure how you'd live without. If you look carefully, you're sure to find something that brings a smile to your face or a technique that you'll keep close at hand.

### Brew Your Morning Coffee

**Controlling a coffeepot is the home automation equivalent to a programmer's "Hello World" application; everyone needs to try it at least once, even if you don't drink coffee.**

The promise of a freshly brewed pot of coffee, timed perfectly to coincide with your morning routine, is the stuff from which marketing campaigns are made. I dare say, this task is second only to home security when it comes to home automation's *raison d'être*. Not only is it useful, particularly if you're a coffee drinker, but it's also a good starter project because it requires you to learn many of the key practices and concepts necessary for successfully automating other parts of your home. So, let's have a cup, shall we?

First off, you need to make sure your coffee maker has the right characteristics for automation. It has to work correctly when used with the X10 appliance module [Hack #3] that will be turning it on and off. The appliance module connects between the coffee maker and the electrical outlet, and it works by controlling the flow of electricity to the coffee maker. In other words, when you tell the appliance module to turn off the coffee maker, the module stops the electricity from reaching the machine. It's as if you unplugged the coffee maker from the wall outlet, without having to actually do so. When you tell the appliance module to turn on, it's the same as plugging the coffee maker back into the wall outlet.

Therefore, the coffee maker must be able to retain its settings after it's unplugged from the wall. To test this, turn on the coffee maker and set all the various options you like to use, such as the number of cups and so on. Turn the switch to brew (or whatever it is you do to begin making coffee), wait until it starts brewing, and then pull its electrical cord from the wall outlet. Wait a few seconds and then plug it in again. If it resumes brewing your coffee, it's a good candidate for automation. If it does not resume brewing, you can't use it for this hack.

The success of this experiment depends on the type of switches your coffeepot uses. What you need is a coffeepot that uses mechanical switches for its settings and, specifically, for the power switch. Otherwise, when the coffeepot is unplugged (or turned off by the appliance module) it will revert to its default setting, which is off. Unfortunately, most of the nicer coffee makers, particularly those that have built-in timers, do not use mechanical switches. If you want to put your home automation system in charge of brewing your coffee, you're probably going to have to shop for a more basic coffee maker.

The good news is that once you're past the power switch hurdle, you're almost home free. Automating your coffee maker can be as simple or as complex as you decide to make it. But there's one more thing to clarify before we move on. What you're automating is the starting and stopping of the coffee-making process. You're still going to be responsible for manually filling the machine with ground coffee and water, making sure the pot is in place, and cleaning and emptying the pot when you're finished. To avoid those tasks you're going to need a robotic housekeeper, or a kind and thoughtful partner, not X10.

Next, you need to decide exactly when you want your coffee to be ready, along with any exceptional conditions that would require a different start time or, heaven forbid, no coffee-making at all. This is a key differentiator between a smart coffee-making process that is controlled by a home automation system, and the not-smart, simple timer that's built into the coffee maker. Your computer has much more information available to it and can make much more sophisticated decisions. Sure, it's not a fair fight because your computer costs a thousand times more money, but who's going to complain?

Let's start out simple and say you want your coffee to be ready at 7:00 a.m. every weekday and at 8:00 a.m. on weekends. To accomplish this, set up two repeating events. One event is active Monday through Friday; the other event is active only on Saturday and Sunday. Figure 3-1 shows a weekend event defined in HomeSeer **[Hack #19]**.

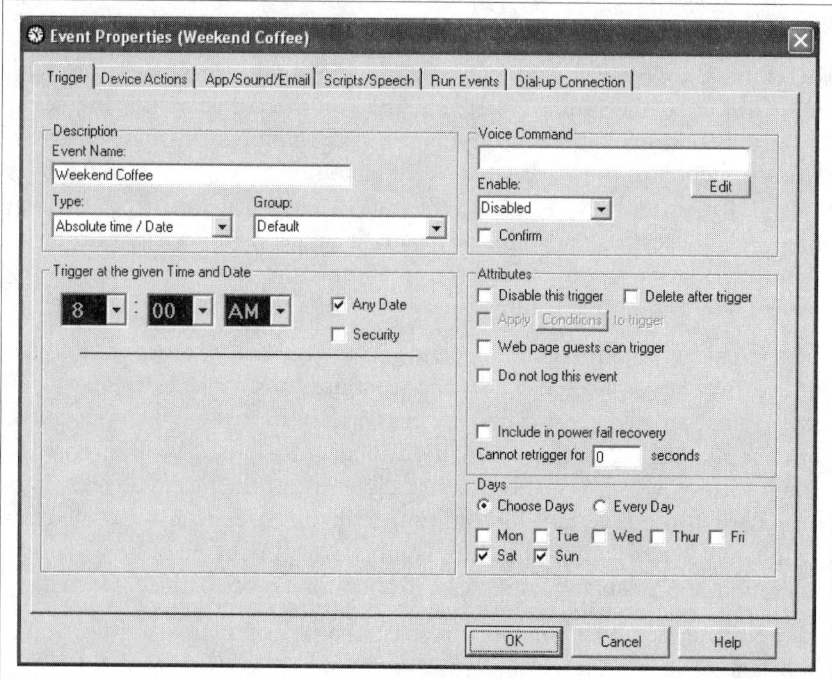


*Figure 3-1. Weekend coffee*

This event is set up to execute every Saturday and Sunday morning at 8:00. In the Device Actions tab, it is set up to send an `On` command to the appliance module that's connected to the coffee maker. Exactly one hour and five minutes after turning on the coffee maker, it sends an `Off` command to ensure it's turned off. You accomplish this by adding a second delayed command to the Selected Devices list, as shown in Figure 3-2.

So far, we haven't accomplished anything that a coffeepot with a built-in timer and calendar couldn't accomplish. So, let's up the ante and make the `weekend coffee` event smarter by having it turn on the coffeepot only if you're actually at home **[Hack #70]**. That way, if you go away for the weekend, the event will be skipped completely. To do this, open the `weekend coffee` event and choose By Condition from the Type pop-up list. This enables you to define the exact conditions under which HomeSeer will execute this

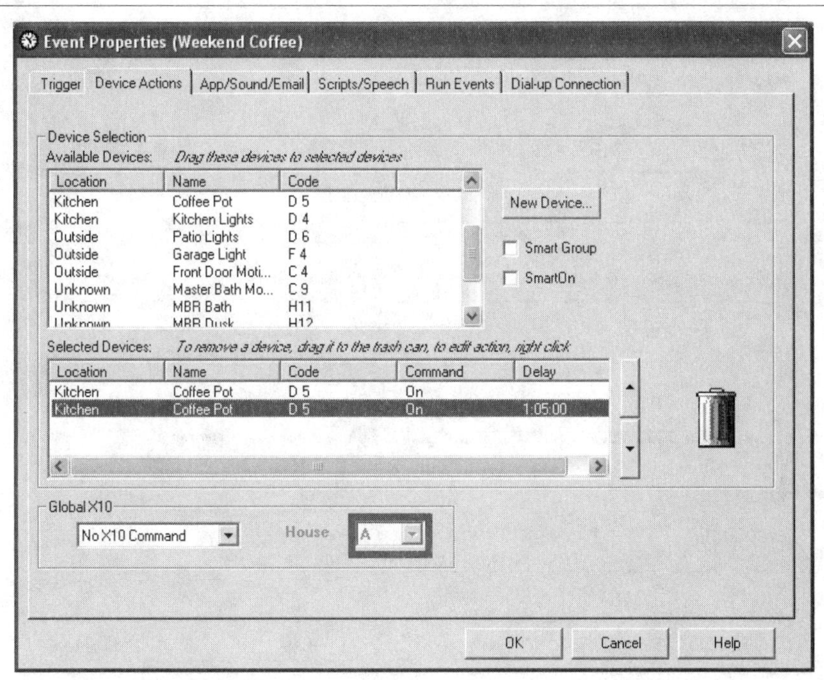


*Figure 3-2. A delayed Off command for safety*

event. First, choose Time Is from the If pop-up list; then, choose At from the pop-up list and enter 8:00 a.m. in the time fields, as shown in Figure 3-3.

Click the Add button to add the time condition to the list at the bottom of the window. Next, add another condition (`Gordon Home`, in this example) that checks the status of the virtual device you use to track whether you're at home. To do this, choose the device from the If pop-up list, choose Is On (as shown in Figure 3-4), and then click Add.

Both of the conditions you've defined must be true for this event to execute and turn on your coffeepot. If you're not at home at 7:00 a.m. on Saturday or Sunday, the coffee maker will not be switched on. Take that, Mr. Coffee!

A coffee pot that's inadvertently switched on, or left on, can cause a fire. For safety's sake, always make sure you physically turn off the coffeepot if it's not ready to make coffee and before you leave the house. Do not rely solely on X10 commands to manage this potentially dangerous situation for you.

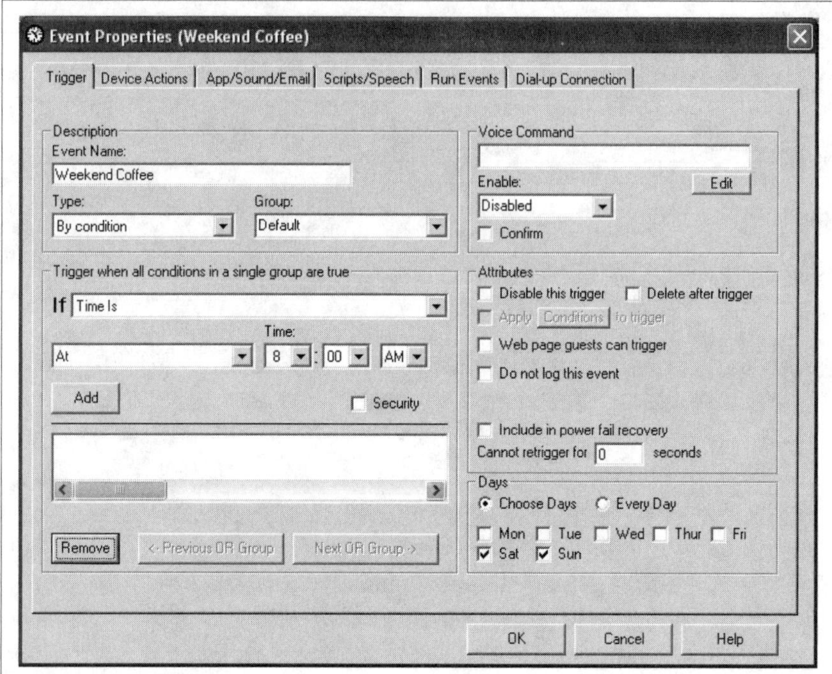


*Figure 3-3. Defining a time conditional*

## Hacking the Hack

You have the building blocks to create a more sophisticated solution by adding new events with more complex conditionals. For example, if you have guests visiting, you might want the coffee maker to stay on longer so that you can linger over conversation and still have a hot pot. Or, you might want to set up a virtual device that tracks whether the current day is a holiday or vacation day so that you can adjust the start time accordingly. The only limit is how closely your home automation system tracks the state of your home **[Hack #24]** and how much variance you need in your morning caffeine routine.

HACK #38

## Detect the Beer Thief

**Motion detectors are useful not just for finding out when someone is present, they can also tell you when a person is meddling where he doesn't belong.**

Michael Ferguson, of Sand Hill Engineering, had a persistent problem at his home. A minifridge used to keep a few bottles of beer cold for hot Florida days was raided frequently by visiting friends and workers, but Michael had a hard time catching them in the act.

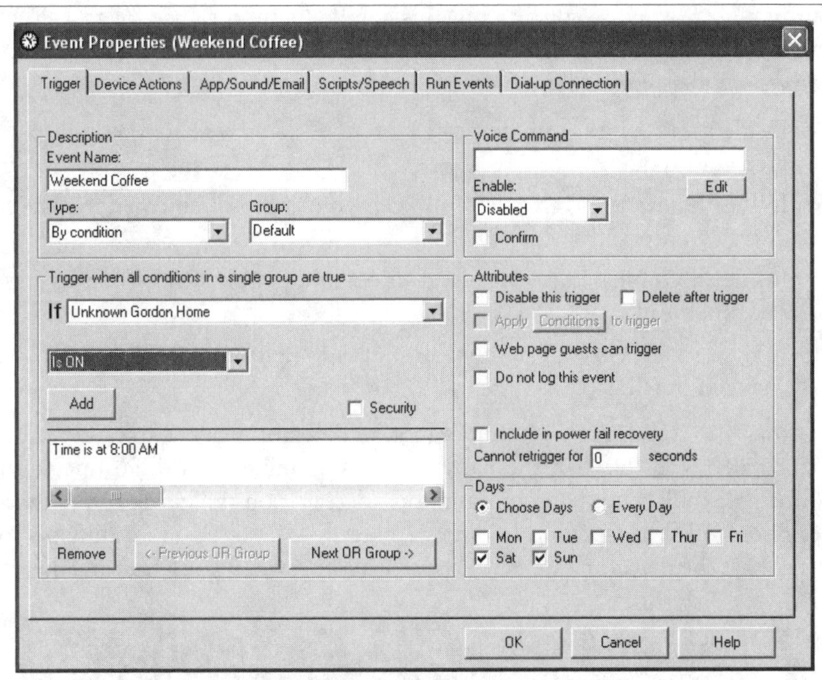


*Figure 3-4. Defining a device status conditional*

So, he good-naturedly created a solution that takes advantage of the small size and long battery life of X10 motion detectors **[Hack #6]**. It's a clever yet simple idea that has many possible applications.

Simply place an X10 motion detector in the fridge, and whoever opens the door will be caught red-handed. You might have to make sure the transceiver is nearby to receive the signal from the motion detector, due to the steel shell of the fridge. As a bonus, the detector's dusk sensor can help you discover that the door has been left ajar because it won't signal dusk until the light in the fridge goes off.

## Hacking the Hack

If you have home automation software **[Hack #16]**, your log will show that someone has been in the fridge, but for more fun, consider having the detector trigger a nearby chime or siren module **[Hack #9]**. Either way, you'll have the information you need to finger your thief. Here's a sample On script for XTension **[Hack #17]** that writes a special message to the log and activates a chime module:

```
write log "Someone is in the fridge again!"
turnon "chime module"
```

But in all seriousness, you can put this simple method to use for tasks that are more important. Consider other places where you can use a motion detector to alert you to activity:

- Inside a locked gun cabinet, to ensure the kids haven't secretly discovered your hidden key
- Inside the Ziploc bag where you keep your stash of emergency cash in case of a natural disaster
- Inside the tool shed, to alert you to neighbors borrowing your favorite rake or shovel
- In the furnace closet, to remind you how long it has been since you last changed the filter

Another useful technique is to put a motion detector inside your liquor cabinet, to alert you to when the babysitter is siphoning off a dram or two of your single malt. Simply mount a motion detector on the inside of the cabinet door, near the opening (as shown Figure 3-5), so it's more difficult to sneak past the detector's threshold of sensitivity, even if you know it's there.

*Figure 3-5. Keeping the babysitter honest*

After you've mounted the detector, add an On script, as described earlier, that logs the activity or sends you a notification **[Hack #73]**.

HACK #39

## Install a Kitchen Terminal

**As tempting as it might be to have a full-featured computer in every room of your home—even the kitchen and garage—if you have a home automation system, you can use small, dedicated terminals to accomplish many of the same tasks with less expense and clutter.**

During the Internet boom years, many companies bet their future on ubiquitous computing and tried to sell lightweight, network-savvy computers to provide web browsing, email, and other Internet applications in every room of the house. Alas, the idea never really took hold with average consumers. But enthusiasts loved the idea and quickly began snatching up these *thin-client* computers as soon as they were discontinued and sold at closeout prices.

One device in particular, the 3COM Audrey (shown in Figure 3-6), really struck a chord in the home automation community. It was introduced as an *Internet appliance* for about $500. Audrey includes email and a web browser, and can sync its calendar and address book with a whole family's worth of Palm devices. And it's small, so it fits nicely on a kitchen counter.

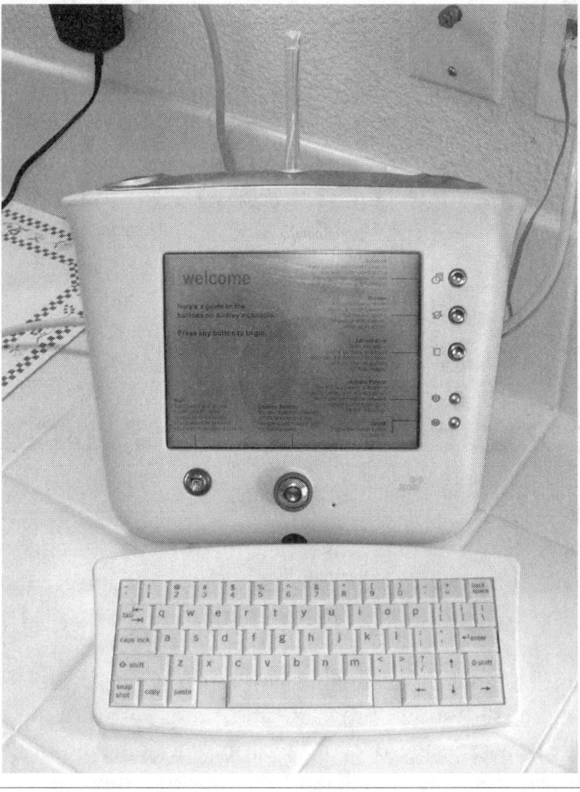

*Figure 3-6. The 3COM Audrey*

After 3COM discontinued the Audrey, less than a year after introduction, new units were dumped at less than $100. It wasn't long before like-minded people began getting under Audrey's hood and figuring out exactly how she worked.

Although the Audrey is no longer being made, it's readily available on eBay. But if they become scarce, remember that the techniques, and most of the software, described in this hack can be adapted to similar devices.

Integrating Audrey into your HomeSeer-based **[Hack #19]** automation system is straightforward. Thanks to HomeSeer's built-in web server **[Hack #99]**, which enables you to control your home remotely, you can use Audrey to access HomeSeer from any networked room in your house. In fact, HomeSeer's TouchPad feature (shown in Figure 3-7) is a perfect match for Audrey. It provides big, graphical controls that you can activate easily, simply by tapping Audrey's screen with your finger.

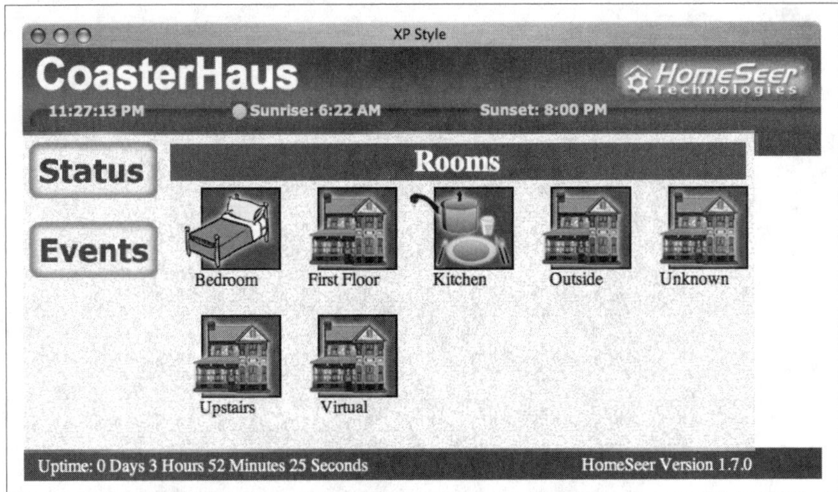


*Figure 3-7. HomeSeer's TouchPad*

Audrey owners also can take advantage of the MCS Audrey plug-in for HomeSeer. Although the TouchPad enables you to control HomeSeer from an Audrey, the MCS Audrey plug-in completes the circle by permitting HomeSeer to control Audrey's special features. It adds commands to HomeSeer's scripting language that enable you to turn on Audrey's LED status lights; activate its slideshow, browser, and other software; and even cause Audrey to speak a message over its built-in speakers. Overall, the $19.95 plug-in adds nearly 50 Audrey-related functions to HomeSeer.

If your home automation system is based on MisterHouse [Hack #16], you should consider using MrAudrey (*http://www.mraudrey.net/MrAudrey.php*; free). MrAudrey is a complete package of Audrey programs that provides tools for communicating with MisterHouse as well as other applications useful for home automation.

One of my favorite Audrey add-ons is Audrey Caller ID (ACID, *http://www.timemocksme.com/acid/*; free), a program that allows Audrey to function as a Caller ID (CID) display via its built-in modem. It comes in quite handy when deciding whether to answer a phone call during dinner or other times when you'd rather not be disturbed.

ACID also can send the CID data over the home network so that other applications can display the information, too [Hack #27]. This is great fun and very useful. You're working at home on your laptop, the phone rings, and a message appears on your screen telling you who's calling. Sweet.

## Final Thoughts

The Audrey is popular, especially due to its compact design and high SAF [Hack #14], but it's far from the only Internet appliance available. The home automation community is constantly looking for reasonably priced technology to fit this niche, as are other like-minded technoholics, and the I-Appliance BBS at the Linux Hackers web site (*http://www.linux-hacker.net/cgi-bin/UltraBoard/UltraBoard.pl?Session=*) is one of the best places to monitor for new developments. Who knows? You might just find the next must-have addition to your smart home.

# Install a Home TV Server

**Add a TV tuner and software to your PC to create a server that makes live video from your home viewable in a web browser.**

I have two web cameras [Hack #82] that I can access from the Internet, but I also have several closed-circuit television (CCTV) cameras in and around my home. I can view the output from the CCTV cameras from any TV in my home, thanks to a *channel modulator* from ChannelPlus (*http://www.channelplus.com*; from about $60) that broadcasts the cameras onto unused channels on my home's cable TV system. This enables me to tune any TV in the home to channel 110, for example, to view the image from the camera at the front door. Tuning to channel 111 shows me what's happening in the kid's room, and so on, with the other cameras. It's a useful system, but I wanted to be able to also view the cameras from the Web, not just on the televisions in my home.

To accomplish this, I needed something that could stream audio and video to a web browser and provide controls for selecting which channel to view. I tried several products, but the only one I found that fit my needs is Beyond TV from SnapStream Media (*http://www.snapstream.com*; $60).

Beyond TV works by controlling a television tuner card that's installed in the computer. My tuner card is the TV Wonder VE from ATI Technologies (*http://www.atitech.com/products/tvwonderve/index.html*; $50). It is inexpensive and readily available at consumer electronics stores. Another option that the Snap Stream web site highly recommends is the Hauppauge PVR-250 (*http://www.hauppauge.com/html/wintvpvr250_datasheet.htm*). It has a hardware-based MPEG-2 encoder, which relieves the computer of having to perform the computationally intensive task of compressing the video. But I am using Windows Media Player to view the streaming video on the Web. MPEG-2 isn't suitable for my use because Beyond TV cannot stream video in this format

Beyond TV's web interface works well for me. It requires Microsoft Internet Explorer, and it enables me to stop or pause the live video stream and to change channels. Because the ATI Wonder VE card is connected to my cable system, I simply load the URL for my home's web site, navigate the camera page, and click a link to view the streaming video from channel 110—my front door.

To view the cameras while you're away from home, you'll need to have configured your broadband connection and your home network to allow connections from the outside world. You'll also have to either know your home's current IP address or have arranged for a locator service, such as DynDNS (*http://www.dyndns.org*), to help you manage this process.

If your system already is set up for you to access it from the Internet, you simply need to open two ports on your router to allow access to the TV server. You'll need access to Beyond TV's administration port (`5129`) and the Live TV Streaming Server port (`8080`). If you need to change these ports, see "Changing Port Settings" in Beyond TV Help.

This hardware and software combination works well for viewing my CCTV cameras and, as a bonus, it also enables me to set up automated recordings of any TV channel in MPEG-2 format, similar to a TiVo digital video recorder. I can watch these on the PC's screen or I can burn them to DVD for later use. That's quite a nice bonus feature to have on hand.

—*Edward Cheung*

# Control Your Heating Remotely

Replacing your existing thermostat with one that you can control by X10 commands enables you to control your home's temperature remotely.

To control your home's heating and air conditioning using home automation equipment, you need to replace your existing thermostat with one that responds to X10 commands. The RCS TBX16 HVAC Control System (*http://www.smarthome.com/3045b.html*; $250) is the choice of many home automators. It looks like a regular thermostat (as shown in Figure 3-8) and you can control it manually, just like the one you're replacing. This gives the system a high SAF **[Hack #14]** because, for all appearances, it's just a normal thermostat. You'll like it, too, because it can seamlessly replace nearly any type of 24-volt thermostat you're currently using—for gas, electric, oil, or propane systems—and you don't have to pull new wires between the thermostat and the heater; the existing wiring will do the job.

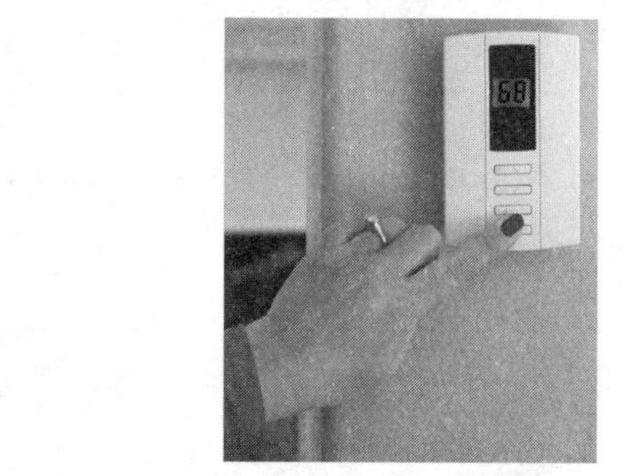

*Figure 3-8. The TBX16 thermostat*

What's different about this thermostat is that you can control it completely by sending it X10 commands on the power line. Any of the things you can do when standing at the thermostat and pushing buttons you can also do remotely with X10 commands. You can turn on the system, choose between heating and cooling, and select the desired temperature. After you've set the options, the TBX16 controls your HVAC system just like your old manual thermostat did.

Installing the TBX16 involves removing your old thermostat, mounting the new one in its place, and splicing into the thermostat's control wiring to insert the TBX16's control unit into the mix. The control unit is what sends and receives the X10 commands and tells the thermostat how to change its settings accordingly. You usually mount the control unit on the wall, next to your heating and air conditioning system, which is easy to do.

To hear commands from the power line, the control unit uses a TW523 two-way interface [Hack #21], which you need to plug into a wall outlet near the control box. You'll also need an outlet to plug in the power supply for the control unit, which uses a small transformer. To get details on the installation process, download the TBX16 manual from the Smarthome, Inc. web site (*http://www.smarthome.com/3045b.html*).

After you have the thermostat installed, you need to set its house code [Hack #1] by following the instructions in the manual. The TBX16 requires you to reserve an entire house code for its use. For example, if you set the thermostat to house code `N`, you send `N1 On` to turn on your HVAC system. To set the temperature, the thermostat uses a simple lookup table of temperatures and X10 commands. For example, to select a set point of 76 degrees Fahrenheit, send `N5`; for 77 degrees, send `N6`, and so on.

One of the most common uses of the TBX16 is to install it in a vacation home. You can use it to make sure the place is appropriately heated or cooled before you arrive for a visit. This saves everyone from, for instance, having to walk around in coats while the heater warms up the place.

Here's how to make sure your cabin is fully heated before you arrive. Install an X10 Telephone Responder device (*http://www.smarthome.com/5000.html*; $70) at your cabin. The Telephone Responder connects to the telephone line and plugs into a wall outlet to draw power and to allow it to send X10 commands. The Telephone Responder listens to phone calls for a certain sequence of touch tones that it translates into X10 commands [Hack #33]. Set the Telephone Responder to the same house code as your thermostat (`N` in our example), and you're all set.

Before you leave for the road trip, call the cabin; the Telephone Responder will answer the call. If you have an answering machine on the same phone line, it can work with most of them, too. You will need to dial in a security code (so that someone doesn't accidentally get into your system) and then dial in the temperature.

If you usually keep the cabin heated at a low temperature, such as 40 degrees (so that the pipes don't freeze), simply dial in the new temperature. The temperature you set is made up of X10 codes. For example, if you want the cabin heated to 75 degrees, dial `4` and `*` on the phone's keypad. The Telephone Responder will send the X10 signal `N4 On` over the home's electrical wires. When the thermostat receives that signal, it will change the temperature set point to 75 degrees and the heat will come on. Now, the family can enjoy a warm and toasty cabin upon arrival.

—*Smarthome, Inc.*

# HACK #42 Monitor the Refrigerator Door

Here's a solution to the problem of a refrigerator door left open too long.

If you have a refrigerator or freezer with a door that's hard to close, at the very least you're wasting energy when it's accidentally open. At the very worst, you risk spoiled food, which is what prompted me to come up with this hack. If you have kids who have a hard time remembering to close it properly, or stand there forever with the door wide open while they decide what they want, you have the same problem.

Here's a simple solution: use your home automation system to detect when the door has been left open for too long. If it has been left ajar inadvertently, the system can remind you to close it securely. But if the reminder is triggered because someone is standing there indecisively, the reminder should prompt him into action.

To set this up, mount an alarm system door and window sensor **[Hack #78]** to the refrigerator or freezer door. Place it at the bottom of the door to help keep it out of sight, as shown in Figure 3-9.

Connect the wires from the sensor to a digital input port on your home automation controller so that your system receives a signal when the fridge door opens. When this occurs, trigger an event that will sound an alarm 60 seconds later. When you receive the signal that the door has closed, cancel the triggered event so that the alarm does not go off.

If your home automation controller does not have digital inputs (if, for example, you're using the CM11A), you can use a Weeder Board (*http://www.weedtech.com*; $70) or a LabJack (*http://www.labjack.com*; $120) instead. Or, adapt "Know When the Mail Arrives" **[Hack #62]** to apply to your fridge door instead.

As for how to signal the alarm, you're limited only by your creativity. I use an inexpensive piezoelectric buzzer that's connected to a relay on my home automation controller. You could send an `On` command to an X10 chime module **[Hack #9]** or use your computer's text-to-speech capability to speak a reminder **[Hack #28]** (or reprimand, as the case may be).

I've found that the 60-second timer works well for my family. It has rarely gone off on my wife or me, but it gets the kids every time; they keep the door open just too darn long! It's a simple but effective solution to a common problem, and if you have a home automation system already, adding this to your home is worth the minimal effort.

—*Jon Welfringer*

*Figure 3-9. A door sensor on the fridge*

# Heat the Toilet Seat

Build your own heated toilet seat to take the edge off those cold winter mornings.

In a moment of whimsy, I decided I wanted a heated toilet seat. A web search revealed that heated seats seem popular in Japan, and that Kohler (*http://www.us.kohler.com/pr/n_seamless.html*) sells one in the U.S. for about $100, but it was the wrong size for our plumbing. As a result, I decided to build my own.

First, I acquired some foil heaters from the Minco Corporation (*http://www.minco.com*) that are about 20.1 ohms of resistance each (part #HK5427R20.1L12B) and draw about 1 watt per square inch. I also purchased a new toilet seat (shown in Figure 3-10) for about $15. I selected a molded wood seat, instead of a plastic one, because I suspect that wood has better thermal conduction. In addition, the wood is easier to machine and cut.

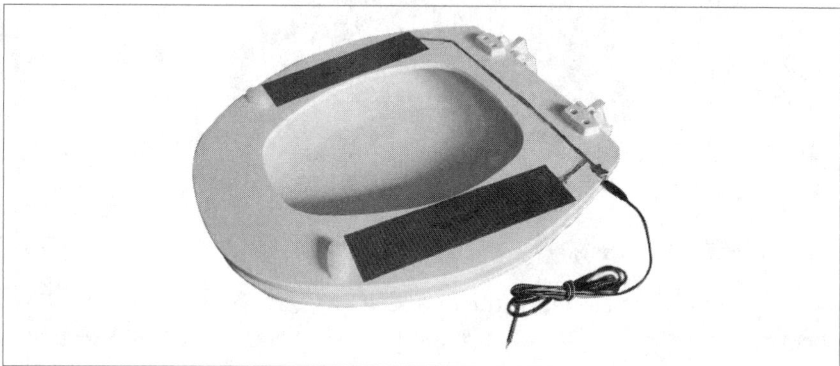

*Figure 3-10. The foil heaters installed*

I attached the two Minco foil heaters, wired in series, to the underside of the seat. I also added a small electrical jack at the back edge for connecting to a power transformer, as shown in Figure 3-11. I made a groove for the wires using a circular saw, and then filled the groove with epoxy to protect the wires and reinforce the seat's structural strength.

*Figure 3-11. Jack for the power transformer*

After I assembled this I spray-painted the whole face to restore the uniform appearance. This resulted in a clean look, with only a small silver jack visible from the side, as shown in Figure 3-12.

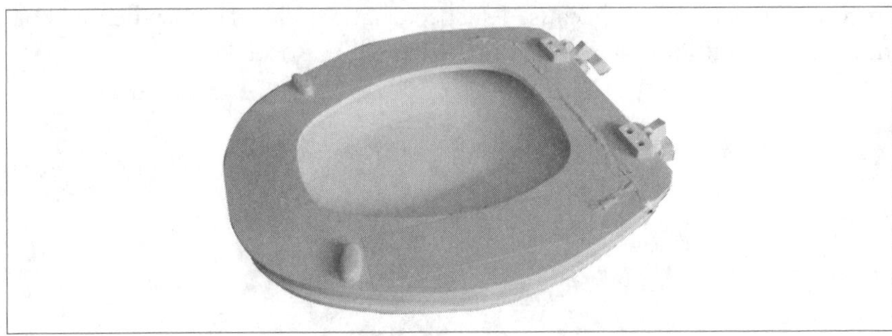

*Figure 3-12. The finished unit*

I power the seat with a transformer rated at 16V AC. With the 40-ohm load, this results in about 3.2 watts per foil heater. This small amount of power heats the foil side to a noticeably warm temperature (about 90 degrees Fahrenheit) but is cooler and more evenly distributed on the top of the seat. I measured the total power draw with my Kill A Watt power meter (*http://www.p3international.com/products/special/P4400/P4400-HG.html*; $30) and it was only 11 watts (about 4 watts lost in the transformer), or about one cent per day to run this device.

I also installed a GFCI power outlet near the toilet to power the device, as shown in Figure 3-13.

*Figure 3-13. The installed unit*

After a few months of trial use, I replaced the large transformer with a smaller one and hid the wires, for a much cleaner look.

## Hacking the Hack

Although the cost of operating the unit is low, you might want to use an appliance module [Hack #3] to turn off the heater when you go to bed [Hack #48] or are not at home [Hack #70]. You also could have your home automation system turn it on several minutes before your morning alarm [Hack #47] so that it's heated before you get up in the morning.

—*Edward Cheung*

# HACK #44 Detect Flooding

**This do-it-yourself sensor inexpensively, yet effectively, alerts you to water leaks in your home.**

A few years ago, the drain in the floor of the walk-up entrance to my basement door became blocked, and water started to enter the house. I didn't discover this problem until several hours later, after much of the basement carpet was soaked. After laboriously cleaning up the mess, I decided I needed to install a water intrusion alarm.

The sensor I use is simply a standard $5 battery-powered smoke detector. I soldered a pair of wires each to one of the connectors where the push-to-test button makes contact. In Figure 3-14, you can see the wires connected to the contacts, between the round buzzer and smoke detection unit, with the wires exiting via the mounting hole.

The other end of each wire is stripped bare and mounted in a corner, at floor level, where it will get wet when water begins entering the basement. You'll have to experiment with the positioning of the wires to ensure they provide an early warning when flooding begins. When the water gets high enough to touch both wires, the circuit will be completed, and the smoke alarm will sound.

I fastened the unit to a two-by-four stud on the ceiling of my mechanical room and ran the wire pair to three locations where water leaks are likely: at the outside entrance to my basement, near my sump pump, and next to my water meter. In each location, a small area of the wires is stripped bare to allow moisture to reach the core.

I added a small label with the date of installation and instructions on where to look for leaks if the alarm sounds. At the time of this writing, the unit has been installed for 10 years and numerous flooding incidents have been

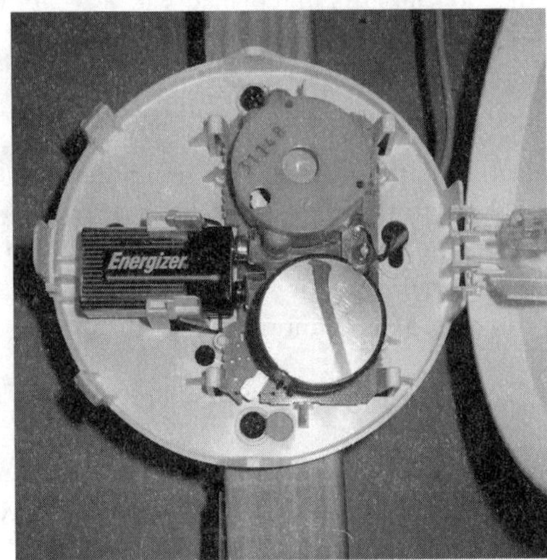


*Figure 3-14. A water intrusion alarm*

averted thanks to its watchful eye and very loud alarm. The only maintenance it has required is a new battery every couple of years. When the alarm goes off, which happens to sound very different from the smoke alarms we have, we know to clean the drain in the walk-up entrance and to check the other areas for leaks. In addition, the unit still functions as a smoke detector, which gives us a backup sensor in the mechanical room where our furnace and water heater are located.

## Hacking the Hack

Using a standalone, inexpensive, and loud smoke alarm is hard to beat, but you could consider using a Powerflash module **[Hack #10]** instead. Connect a pair of wires to the terminals on the Powerflash, as described here, and it will send an X10 command when moisture closes the circuit.

—*Edward Cheung*

HACK #45

# Monitor the Litter Box

**Keeping track of whether you need to empty the litter box provides a good way to learn several useful techniques for home automation.**

Jerry Prsha, an XTension user, uses his home automation system to monitor his cat's litter box so that he knows how many times it has been used and when it needs to be emptied again. This is a somewhat odd application of a technique that, when generally applied, is useful for other home automation projects.

Instead of paying attention to some mundane detail concerning your home life, such as keeping an eye on the litter box, use a sensor to monitor the area and have your system alert you to when a threshold point has been reached. For example, you might decide you want to empty the litter box every third time it is used, and you can ignore it until you're notified that it's time to pay attention to it.

To accomplish this, place an X10 Eagle Eye motion detector **[Hack #6]** near the litter box, positioned so that it is triggered only when the box is occupied. If you have a covered litter pan, such as the Booda Dome (*http://www.thecatconnection.com/Booda-Dome-Litter-Box.html*; $30), mount the detector inside the cover using double-stick tape, pointing down so that it will be triggered when a cat stands in the pan. If you don't have a covered litter box, position it on the wall, near floor level, and attach cardboard flaps to the side of the detector to limit its field of view (like blinders for a horse). Alternatively, use some duct tape to mask off part of its lens. The goal is to prevent the detector from seeing incidental motion and generating a false report. For more discussion about how motion detectors see the world, see "Keep the Lights On While You Work" **[Hack #26]**.

Next, you need to decide how frequently you want to be notified of activity. Jerry prefers to know every time one of his kitties uses the box, so he's defined an On script for the litter box motion detector that announces the activity over his home's speakers **[Hack #28]**:

```
Speak "there is a kitty in the litter box!"
```

This script is executed every time the motion detector receives an `On` command. However, because X10 motion detectors send an `On` command every few seconds, for as long as motion continues, it's better to have the script speak only if it hasn't seen any motion recently—say, within the last five minutes:

```
If (time delta of "litter box motion detector") > 5 * minutes then
    Speak "there is a kitty in the litter box!"
End if
```

This will work fine, but it's going to be up to you to remember how many times you've heard the announcement, assuming you heard them all, and decide that it's time to empty the box. That's not too much fun, so let's make the script keep track of this for you:

```
if (time delta of "litter box motion detector") > 5 * minutes then
    speak "there is a kitty in the litter box!"
    set theCount to (Get Unit Property "useCount" from unit "litter box
motion detector") as integer
    write log theCount
    set theCount to theCount + 1
```

```
        Set Unit Property "useCount" in unit "litter box motion detector" to
    theCount
        if theCount greater than or equal to 3 then
            speak "it is time to empty the box!"
        end if
    end if
```

This script stores the number of times the litter box has been used as a property of the motion detector unit. In XTension, you can use properties to store bits of data you want to keep track of. You access them programmatically as shown in the script, or you can view and edit them via the user interface by clicking the Edit Unit Properties button in the Edit Unit dialog box. It's not necessary to define properties in advance; your scripts can add and delete them as needed.

Once the useCount property has exceeded a value of 3, the reminder to empty the box will be announced every time until useCount is reset. You can do this manually using the XTension interface, but it would be better to use a Palm Pad [Hack #5] or minicontroller [Hack #4], kept near the litter box, to reset the counter after you've emptied the pan. This On script, assigned to the X10 address sent by the Palm Pad, does the trick:

```
Set Unit Property "useCount" in unit "litter box motion detector" to 0
Speak "Thank you for emptying the litter box!"
```

Alternatively, if you're in the habit of emptying the box every night, you can have the counter reset automatically in the morning. Add this snippet to XTension's sunrise script:

```
Set Unit Property "useCount" in unit "litter box motion detector" to 0
```

Given the consequences of forgetting to empty the box when it needs it, this might not be the best approach. Nevertheless, the sunset and sunrise scripts [Hack #20] are useful for resetting properties that you're using to accumulate daily statistics. For example, you might want to know how many times a day the litter box is used, so you can report back to your cat's veterinarian:

```
set theCount to (Get Unit Property "useCount" from unit "litter box motion
detector") as integer
Write log "Litter Box total use: " & theCount
Set Unit Property "useCount" in unit "litter box motion detector" to 0
```

Here, the sunrise script writes the usage count to the XTension log file, but you could have it saved to an external file instead so that you can import it into Excel or plot it on a graph [Hack #92].

### Hacking the Hack

The ways in which you can apply these techniques are virtually limitless. By storing information in unit properties, you have the ability to track the time of day or day of the week, the outside temperature, or who was at home at the time any event occurred.

Here are just two more examples to get you started: how many times a week is your liquor cabinet opened [Hack #38] and what time of day does the mail usually arrive [Hack #62]? The ultimate toy for compulsive personalities, don't you think?

Instead of announcing or logging the information you're monitoring, you might send it to your cell phone's email account [Hack #73].

## HACK #46 Avoid Battery Memory Problems

**Use home automation techniques to sensibly recharge your often-used battery-powered devices.**

If you have an electric razor, toothbrush, or other battery-powered device that you use for short intervals nearly every day, you've probably had it go dead at exactly the most inconvenient time. Who can remember to charge these things on a regular basis? If you're not paying attention, they'll run out of juice. If you charge them too often, you risk limiting their life by inducing the dreaded battery memory effect, exhibited by its refusal to hold a charge for more than a few uses. You can't win either way, but luckily your home automation system can rescue you from this mundane, and maddening, chore.

Here's what I do with my beard trimmer and electric toothbrush. Each device's charging stand is plugged into an appliance module [Hack #3]. I keep the trimmer and toothbrush in their charging stands when they're not in use, but most of the time the charger is turned off (via the appliance module) so that it simply is acting as a holder.

A scheduled event [Hack #16] turns on the appliance module for just long enough to charge the device's batteries. The event executes once every two weeks, during the afternoon, when it's assured that I won't be using the devices and they can charge undisturbed.

Because I use the toothbrush more often than my beard trimmer (you'll be glad to know, I'm sure), the charging times are adjusted for the usage difference. Instead of having separate events for each device, a script handles both:

```
turnon "Toothbrush Charger" for 3 * hours
turnon "Trimmer Charger" for 1 * hours
```

As I acquire more rechargeable devices, it's simple to add a line to the script for the new item, with its own recharging interval.

If you don't care to charge devices separately, plug each charger into a power strip, then plug the power strip into an appliance module. To turn on all the chargers at once, send an `On` command to the appliance module. You won't be able to charge each device separately, but you'll spend less on appliance modules.

CHAPTER FOUR

# Bedroom

## Hacks 47–53

You might think the bedroom is the last place in the house you want a computer, and almost everyone would agree, but the hacks in this chapter enable you to benefit from computer-based automation without having to turn your sanctuary into an office.

In fact, bedroom-related hacks are among the subtlest, and most interesting, in this book. You'll learn how your home can gently urge you from bed in the morning [Hack #47], help you navigate a dark house during the middle of the night [Hack #49], and play pleasing chimes [Hack #53]. These hacks are sure to delight your whole family.

### Educate Your Alarm Clock

**Create a set-it-and-forget-it alarm clock for each individual in your home. It goes off only on workdays, and even can nag a reluctant riser to make sure she gets up.**

How many times have you gone to bed, only to have to get up a few minutes later to double-check that you set the alarm clock? Or, how about that early-morning weekday alarm setting you always forget to turn off, disrupting your blissful Saturday morning slumber? And what about those days when you need to get up extra early? You find yourself setting and resetting the alarm clock several times a week.

Those days are in the past if you have a smart home, with just a few prerequisites:

- A house that knows who is at home [Hack #70]
- A whole-house sound system [Hack #28] or a computer in the bedroom
- A motion detector [Hack #6]

If you don't have all these pieces yet, you can adapt the hack to meet your needs and equipment. The key is a simple script that you schedule to execute at the time you want to wake up:

```
if (status of "holiday today") = false then
    if (status of "gordon home") is true then
        turnoff "Gone To Bed"
        say "Gordon.  Wake up! It's time to arise! get up! get up! get up!"

        if (time delta of "MBR Bath Motion") is greater than 5 * minutes
then
            execute script "Gordon alarm" in 4 * minutes -- reschedule
myself
        end if
    end if
end if
```

This script is called Gordon Alarm. You'll need a personalized copy for each person in your household for whom you want to set an alarm. Each script is set up as a repeating event. If you need to wake up earlier on certain days, simply add additional scheduled events that fit your schedule. For example, if you get up early for a morning swim every other day, schedule an event at 5:00 a.m. for Mondays, Wednesdays, and Fridays and another at 5:45 a.m. for Tuesdays and Thursdays.

To provide more sleep time on holidays, the script checks the `Holiday Today` variable before doing anything else. If this evaluates to `true`, the script exits silently. You maintain the state of the holiday variable manually. I have a Palm Pad **[Hack #5]** at my bedside to turn it on or off.

Instead of using the script to check a `Holiday Today` variable, another good technique is simply to suspend execution of the alarm scripts completely when holiday mode is activated. This makes the alarm scripts slightly simpler, but you have to remember to maintain the holiday mode script to suspend each of the appropriate alarms, so I prefer to have each script check the global variable.

You could use iCal or a similar calendar program that enables you to execute a script for events. When you add a holiday or vacation day to your calendar, set up an alarm action that runs a script. The script turns on the holiday variable in your home automation software. See "Remember Important Events" **[Hack #25]** for examples of running scripts using calendar software.

After checking to see if it's a holiday, the script determines who is at home **[Hack #70]**. If the individual is checked out of the house, the alarm exits silently. If your spouse travels, this is probably the best feature of the script—no more fiddling with the alarm clock to make sure you've turned it off when she's on the road.

Next, the script turns off the Gone To Bed script [Hack #48], which in turn enables voice announcements and turns on the appropriate lighting if it's still dark outside. If you don't have to do anything first thing in the morning, simply omit this step from the script.

Text-to-speech is used to announce the alarm. Have the computer say whatever it is that will get you motivated, and that it lasts long enough to ensure you wake up. If your speakers are controlled with an appliance module [Hack #3], be sure to turn them on before making the announcement:

```
say "Gordon.  Wake up! It's time to arise! get up! get up! get up!"
```

If you're a light sleeper, the alarm has done its job and you're ready to start your day. If you're slow to wake, like I am, you'll need a little more prodding:

```
if (time delta of "MBR Bath Motion") is greater than 5 * minutes then
    execute script "Gordon alarm" in 4 * minutes -- reschedule this script
end if
```

The script does this by watching the `MBR Bath Motion` motion detector that I have mounted on the ceiling of the master bathroom. It checks to see if this motion detector has been activated in the last five minutes, and if it has not, the script reschedules itself to run again four minutes later. In other words, the alarm goes off every four minutes until it sees that I have gotten up and into the shower. Note the logic in the script: because the motion detector doesn't discriminate about who is in the bathroom, I have it speak the alarm *before* checking for motion. This ensures that my alarm doesn't get skipped because my wife is up already and has triggered the motion detector recently. However, this logic also means the alarm nearly always goes off twice: the first time, when I get up, and again four minutes later. If that annoys you, you could set up another state variable to keep track of whether the alarm has gone off already today.

## Hacking the Hack

If you're interested in how much sleep you're getting at night, modify this script to calculate the elapsed time between turning Gone To Bed on and off. You can announce or log results during the alarm, or even create a graph [Hack #92] to show your boss how overworked you are. See "Calculate Elapsed Time" [Hack #97] for an easy way to calculate the elapsed time.

# Put the House to Sleep for the Night

When it's time to go to bed, have your house turn off all the lights, check to make sure things are secure, and ready your motion detectors to light the way should someone get up for a drink of water.

There's nothing like having your house shut itself down for the night. Instead of spending your final waking moments traipsing around turning off lights, have your computer do it for you. You can have it set your alarm system [Hack #80], check to see if you left any windows open [Hack #78], make sure the garage door is closed [Hack #56], and announce tomorrow's weather [Hack #28].

You do all this by writing a script that takes care of each task for you. The easiest way to trigger the script is to keep a Palm Pad [Hack #5] on your nightstand. For example, let's look at my XTension On script [Hack #17] for the unit the Palm Pad controls, named Gone To Bed.

## The Evening's Announcements

The first few steps in the script check some important conditions and alert me to things that I want to be aware of before drifting off to sleep:

```
if (status of "Holiday") is true then
    say "Tomorrow is a holiday, the wake-up alarms are turned off."
end if
```

First, if the holiday state [Hack #24] is turned on, an announcement is made to remind me that the morning's wake-up alarms won't be executed [Hack #47]. If the next day really is a holiday or vacation day, hearing this announcement always brings a smile to my face in anticipation of sleeping in. If it's not a day off, a button on the nearby Palm Pad enables me to turn off the holiday state so that the morning alarms will be triggered.

Another variable that can cause the morning alarms to be skipped is if someone is out of town. For example, if my wife is away on business, her alarm won't go off. By making another announcement about who is gone [Hack #70], I get to confirm that the alarm status is set correctly for the next morning:

```
if (status of "Gale Home") is false then
    set sayString to "Gale is gone."
    say sayString
end if

if (status of "Gordon Home") is false then
    set sayString to "Gordon is gone."
    say sayString
end if
```

Finally, just one more state to check; the oft-forgotten garage door [Hack #55]:

```
if (status of "Garage Door") is true then
    say "The garage door might be open!"
end if
```

If you've integrated weather information into your system, such as the current outdoor temperature [Hack #64], this would be a good time to make those announcements, too. It might seem like a lot of chatter is occurring with these steps, but by keeping the phrases short and focusing on what you really care to know when going to bed, you can limit the total announcements to just a few seconds of informative speech. Speaking of speech:

```
turnoff "Speech Allowed"
```

As the final step in this phase, a flag is set to indicate that no further announcements should be made. This is checked by the script triggered by the outside motion detectors, such as the one at the front door [Hack #74]. I've learned through experience that announcing "Front Door Motion" in the middle of the night because a cat has walked too close to the front door is a rather rude awakening, not to mention very unpopular with my wife [Hack #14]. If Speech Allowed is off, the detector simply puts an entry in the log instead of making a commotion.

## Lights Out

Now comes the part you'll quickly grow to love. The script begins turning off the lights and putting the home into nighttime mode:

```
turnoff "First Flr Lights"
```

All the lights on the first floor go off first, accomplished by sending the `Off` command to the group I've defined in my home automation software [Hack #95]. Several lights are in this group, so it takes a few seconds for the X10 commands to be sent to each light individually, but it's only one command in the script.

I want the lights in the garden to go off a little later because I enjoy the soft light they cast against the drawn window shades in the bedroom, so the following script schedules an event that will turn off the `Outside Lights` group after 25 minutes:

```
turnoff "Outside Lights" in 25 * minutes
```

That takes care of the yard and the first floor of my home. I handle the second floor a little differently:

```
if (status of "Guest Home") is false then
    turnoff "Second Flr Lights"
else
```

```
        turnoff "Loft Lamp 1"
        turnoff "Office Light"
        turnoff "Stairway Light"
        turnoff "MBR Globe"
    end if
```

On most nights, the second floor lights are also turned off by addressing a group. However, if we have houseguests [Hack #70], units are addressed individually so that the lamps near the guest bedroom are not turned off. The script leaves those lamps to be controlled the old-fashioned way: manually, by our guests.

Finally, the air cleaner in the master bedroom is switched on, providing some white noise and filtered air for us all night long:

```
turnon "Air Cleaner"
```

Additionally, because I know this is the last step in the script, hearing the air cleaner come on confirms for me that the house has finished transitioning to nighttime mode.

## During the Night

Now that the house is in nighttime mode, as tracked in the home automation software by the `Gone To Bed` unit, any inside motion detector that gets triggered will turn on a nearby light to quarter-brightness for five minutes:

```
If (Status of "Gone To Bed") is true then
    brighten "kitchen light" to 50 for 5 * minutes
end if
```

This provides a nice soft light for seeing your way to the bathroom or kitchen, or to check on the kids.

You might want to have other scripts check nighttime mode, too. For example, you can skip a script that gets the latest outdoor temperature [Hack #64] while you're asleep, as you won't need the information at that time.

## In the Morning

The `Gone To Bed` state needs to get reset in the morning so that your motion detectors, and other units or actions that use the variable, will behave correctly. On weekdays, the state is reset by the morning wake-up alarms [Hack #47]. For resetting the state on weekends or holidays, when the wake-up alarms are turned off, a script runs every day at 8:30 a.m. to reset the home to daytime mode:

```
If (status of "Gone To Bed") is true then
    Turnoff "Gone To Bed"
End if
```

The script first checks to see if an earlier alarm already reset `Gone To Bed`. This allows the script to run every day and properly reset the state on holidays and vacation days that occur during the middle of the week, as well as weekends.

When `Gone To Bed` is turned off, its Off script is executed. The first thing it checks is whether it's daylight, using the variable XTension maintains automatically based on local sun times:

```
if (daylight) is false then
    turnon "Stairway Light"
    dim "Garden Window" to 50
    dim "MBR Globe" to 60
end if
turnon "Speech Allowed"
say "Good Morning!"
execute script "Speak time & weather"
```

If the sun isn't up yet, a few lights are turned on or dimmed to half-brightness. Then, `Speech Allowed` is turned back on, and "Good Morning" is announced along with the current time and weather conditions.

Next, the script once again examines if we have houseguests before deciding which additional lights should be controlled:

```
if (status of "Guest Home") is true then
    turnon "Garden Window"
    turnon "LFT Lamp 1"
    turnon "MBR Globe"
    turnon "Outside Lights"
else
    turnon "First Flr Lights"
    -- turn on all upstairs lights except two
    block unit "Office Light"
    block unit "Phone Lamp"
    turnon "Second Flr Lights"
    unblock unit "GBR Lamp 1"
    unblock unit "Office Light"

    --turn on all outside lights, except the rope light
    block unit "Outside Rope"
    turnon "Outside Lights"
    unblock unit "Outside Rope"
end if
```

Take a close look at what happens if `Guest Home` is `false`. All the lights in the `First Floor` group are turned on except for the `Office Light` and the `Phone Lamp`, which are blocked from being controlled before the group is turned on **[Hack #95]**. Immediately after turning on the group, the units are unblocked so that they're ready for commands later in the day. This is a handy technique for when you want to control most of the members of a group, but not all of them.

With that, everything is reset and the house and home automation system are ready for another day.

### Hacking the Hack

If certain variables need to be reset daily, such as high temperatures or the number of phone calls you've received, the On or Off scripts for Gone To Bed are a convenient means for taking care of these kinds of housekeeping tasks. If you have an X10-controlled thermostat, add steps to the Gone To Bed scripts that set the temperature for nights and mornings **[Hack #41]**.

## HACK #49 Lighting for Insomniacs

**If you get up from bed in the middle of the night to get a drink, the only thing worse than feeling your way in the dark is turning on a bright light that might blind you and wake others. A better idea is to have your home automation system gently light your way, then turn everything back off a few minutes later.**

If you have indoor motion detectors **[Hack #10]**, you can use them to turn on lights that illuminate your way for late-night jaunts to the kitchen or bathroom. This is also a great way to help guests, children, or the elderly feel safer when they venture out of bed after the household has gone to sleep.

With a home automation system that knows when you've gone to bed **[Hack #48]**, it's simple to add a few lines to the detector's On script so that it can react appropriately when it's turned on by someone walking near it:

```
if (status of "Gone To Bed") is true and ((time delta of "Upstairs Motion")
is greater than 8 * minutes) and (daylight) is false then
        dim "Stairway Light" to 40 for 10 * minutes
end if
```

The script checks to see if the Gone To Bed flag has been turned on, if it has been more than eight minutes since this motion detector was last triggered, and if it's after sunset. If all these conditions are true, the unit Stairway Light is set to 40% brightness for 10 minutes.

The script examines the detector's *time delta*, a built-in property for any unit in XTension **[Hack #17]**, because the motion detector will send an On command nearly continuously while it sees motion. Checking this property allows the script to ignore all of these extra events instead of turning on the light every time one is received. This simple approach to time-filtering reduces the number of X10 commands sent, and in this case, prevents multiple events from being scheduled to turn off the lights 10 minutes later.

## Hacking the Hack

Due to the inherent delay in detecting motion [Hack #10] and sending the X10 commands, it's a good idea to illuminate areas before the person who is walking will reach them. You can use multiple motion detectors to determine where the person is heading [Hack #85], or you can guess and turn on lights for only a brief period so that they're not burning needlessly for a long time:

```
if (status of "Gone To Bed") is true and ((time delta of "Upstairs Motion")
is greater than 8 * minutes) and (daylight) is false then
        dim "Stairway Light" to 40 for 10 * minutes
        dim "Garden Window" to 45 for 5 * minutes
end if
```

In this example, the motion detector is located at the top of the stairway, so the light over the garden window near the bottom of the stairway is switched on, under the assumption that the person who triggered the detector is about to go downstairs. If you have a second motion detector downstairs, you could use that one to confirm that your guess was correct and to keep the light on longer, but certainly, having an unneeded light come on for just a few minutes won't be too bothersome.

HACK #50

# Adjust Lights as the Sun Rises

**Instead of turning on the lights at full brightness in the morning, have them come to a lower level that slowly decreases as sunrise approaches.**

It's 3:00 in the morning and my girlfriend has to get up for work, but I don't need to get up for several more hours. When she arises, the motion detector in the bathroom will turn on the light for her automatically, but at that hour, there's nothing worse than having a light snap on to full brightness. Neither of us wants that to happen!

This script provides for kinder, gentler morning light by starting out with less-than-full brightness, then gradually turning the lights off as sunrise approaches. So, if it's midnight, the light level is set to about 50%. Then it's slowly ramped down until, at dawn, it's no longer on.

The script that handles this process is an attachment script [Hack #88] for Indigo [Hack #18]. It adds a function named `DimADevice()`, which any Indigo AppleScript can call to calculate the appropriate brightness level for any light.

## The Code

The appropriate brightness level of the light is calculated every time the motion detector in the bathroom is triggered.

```
using terms from application "Indigo"
    on DimADevice(DeviceName)
        tell application "Indigo"
            set EarlyCurrentDate to (date string of (current date)) & " 12:
00:00 AM"
            set CurrentDateSunrise to (calculate sunrise for date
EarlyCurrentDate)
            set CurrentDateSunset to (calculate sunset for date
EarlyCurrentDate)
```

The script begins by creating a variable, `EarlyCurrentDate`, which can be used for time calculations. It does this by getting the system date and adding a timestamp of midnight. This creates a reference date that's used for Indigo's sunrise and sunset calculations. With the results of these calculations in hand, `CurrentDateSunrise` and `CurrentDateSunset`, we start deciding the appropriate brightness setting:

```
            if (current date) > CurrentDateSunrise and (current date) <
CurrentDateSunset then
                --It's daytime so lights are at 100%
                set DimPercent to 100
else
```

If the system's current time is not between the calculated sunrise and sunset times, the script continues:

```
if the day of (current date) is the (day of (calculate sunset for (current
date))) and (current date) > (calculate sunrise for (current date)) then
                    set theSunset to (calculate sunset for (current date))
                    set theSunrise to calculate sunrise for ((current date)
+ (1 * days))
                else
                    set theSunset to (calculate sunset for ((current date) -
(1 * days)))
                    set theSunrise to calculate sunrise for (current date)
                end if
```

This script might be running before or after midnight. If it's before, the sunrise time we care about is the next sunrise (i.e., tomorrow morning). If the script is running after midnight, the sunset time we need is yesterday's, not today's:

```
            set Totalsecs to theSunrise - theSunset
            set CurrentSecs to theSunrise - (current date)
            set DimPercent to (CurrentSecs / Totalsecs) * 100
            if DimPercent < 10 then DimPercent = 10
end if
log DimPercent
```

Some simple math calculates the percentage of darkness (nighttime) that's remaining and sets `DimPercent` to match. If less than 10% of the night remains, `DimPercent` is set to 10% because most lights can't be dimmed to such a low level and still remain visibly on. For your lights, you might need to set this to another value, such as 25%, if they're not particularly responsive to low brightness levels.

```
            if (on state of device DeviceName) is false and (value of
variable "isDaylight" as boolean) is false then
                    if (supports dimming of device DeviceName) is true then
                        brighten DeviceName to (DimPercent as integer)
                    else
                        turn on DeviceName
                    end if
            end if
        end tell
    end DimADevice
end using terms from
```

The preceding code checks to see if the device we're about to dim can, in fact, be dimmed. If it can't, it's simply turned on instead. It also checks the status of `isDaylight`, a variable maintained by Indigo that indicates if it's after sunrise. If the script finds that it's sunrise, `DimPercent` is discarded and no command is sent to the light.

To call this script, I set up a trigger action for the motion detector in the bathroom. The trigger's action is an embedded AppleScript that calls the attachment handler, passing the name of the bathroom light device that I want dimmed to the appropriate level, as shown in Figure 4-1.

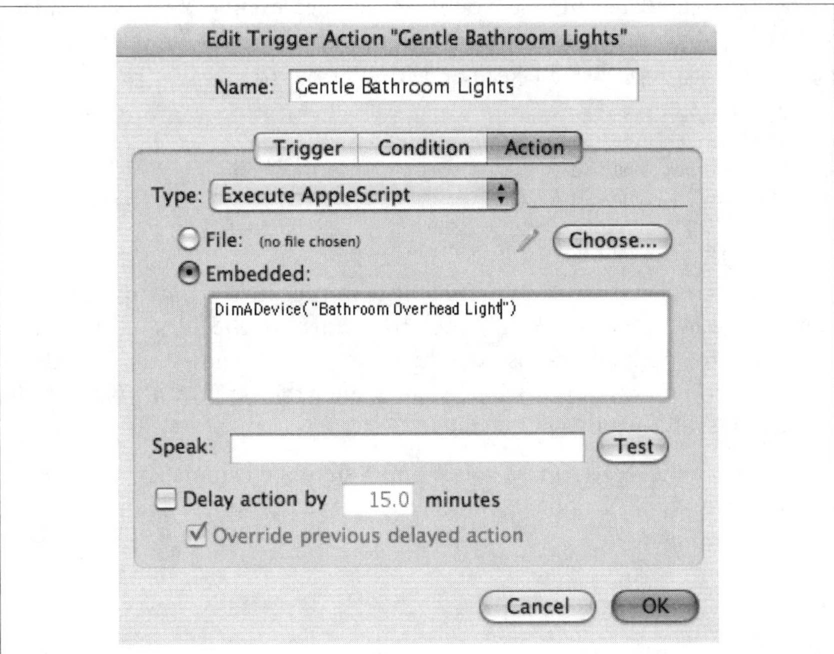


*Figure 4-1. Calling the handler*

If you have multiple lights that you want to control with this motion detector, simply call the handler again, passing the name of each device.

### Hacking the Hack

Though the gentle decrease of light is comforting, it might not be very noticeable, depending on the type of X10 light switches you use. In that case, adjust the percentage calculation so that it changes the lights in 10% or 20% increments.

*—Greg Smith*

## HACK #51 Simulate a Sunrise

**Instead of getting jolted from sleep with a rude alarm, wake up gently by gradually increasing the brightness of your bedside lamp.**

Some sleep research suggests that the best way to wake up in the morning is to do so slowly, with a light that gets progressively brighter, reaching full strength at the time when you want to arise. It's said that this can even eliminate the need for an alarm; you will just naturally wake up with the simulated sunrise. If this idea appeals to you, you can use your home automation system to try it out without having to buy a special alarm clock (*http://www.dreamessentials.com/a_clocks_sunrise.aspx*; $110).

To set up a simulated sunrise, you use scheduled events that send gradually increasing `Dim` commands to your bedside lamp at regular intervals. Here's a script that accomplishes this using XTension **[Hack #17]**:

```
dim "Bedroom Lamp" to 10
dim "Bedroom Lamp" to 20 in 3 * minutes
dim "Bedroom Lamp" to 30 in 5 * minutes
dim "Bedroom Lamp" to 50 in 7 * minutes
dim "Bedroom Lamp" to 75 in 9 * minutes
dim "Bedroom Lamp" to 100 in 11 * minutes
```

Save this as a global script and schedule it to execute 11 minutes before the time you want to awake. When the script runs, it schedules all the events necessary to gradually adjust the light. Figure 4-2 shows the Scheduled Events window as it appears immediately after the script runs, with all the associated events waiting in the queue.

If you use HomeSeer **[Hack #19]**, you set up a simulated sunrise by creating an event with multiple actions, each delayed by an appropriate amount of time, as shown in Figure 4-3.

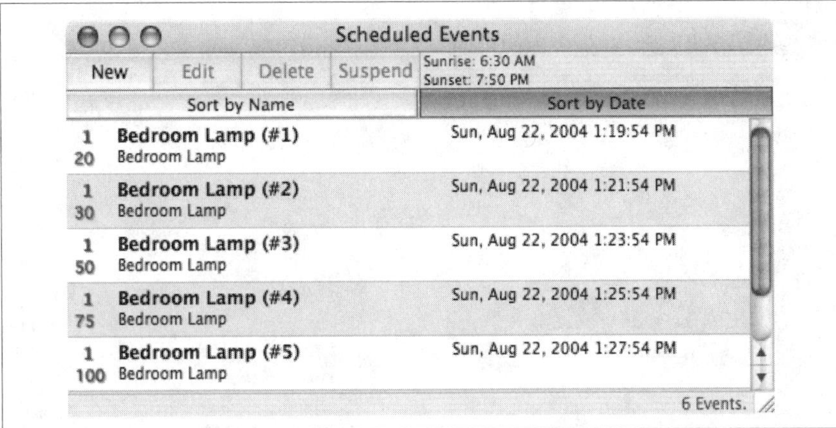


*Figure 4-2. Scheduled events in XTension*

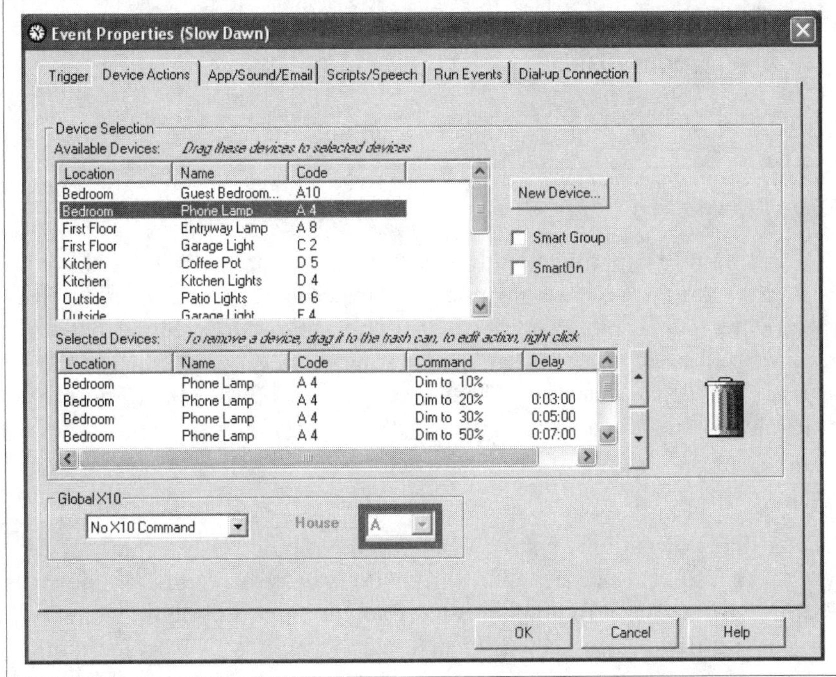


*Figure 4-3. Delayed events in HomeSeer*

Remember that when you define the event, you can select which days it is active, such as weekdays only (as shown in Figure 4-4), and that you'll want it to start prior to the time you want to awake. That is, if you have programmed the light to take 11 minutes to reach full brightness, start the event 11 minutes prior to your usual alarm time.

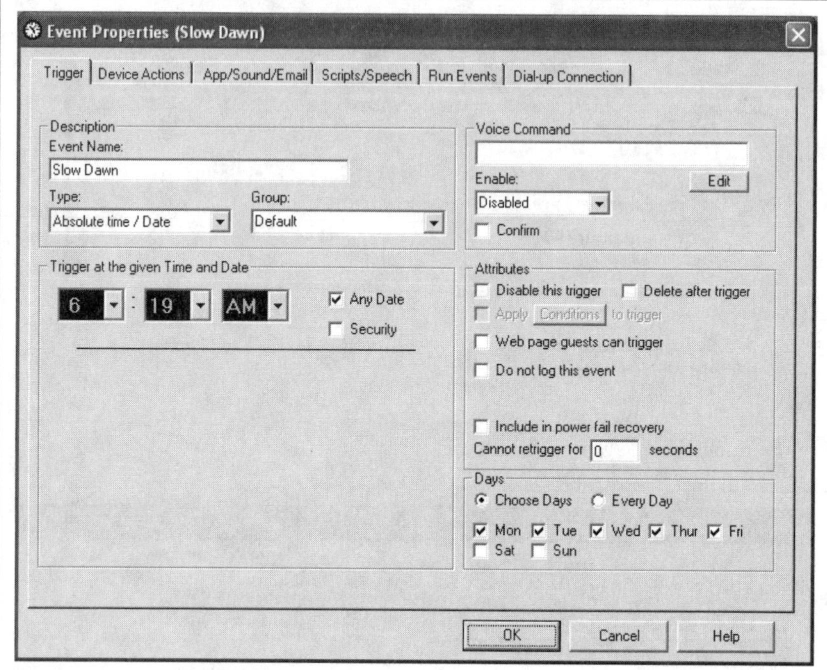


*Figure 4-4. A weekday-only event*

You will want to experiment with how early you begin the process of brightening the light, as well as how much it changes at each step. In addition to your own personal preferences, you'll need to take into account that some lamp modules might not visibly change brightness very much when told to increase by a small increment. That is, you might not be able to notice the difference when a lamp is set to 80% or 85% brightness.

## Hacking the Hack

A light that slowly dims, instead of brightens, is a nice way to end the day, too. It can also be a nice signal for your child's room; her lights turn down as bedtime approaches. You implement a simulated sunset using the same technique as a sunrise, of course: just turn down the lights with every iteration.

Combine this technique with "Educate Your Alarm Clock" **[Hack #47]** to create an alarm that goes off only when you're at home, along with other nice touches. If you keep a Palm Pad **[Hack #5]** near your bed, you even can add a snooze command that resets the brightening light. (See HomeSeer's `hs.RemoveDelayedEvent` command for an easy way to remove events that have been scheduled already.)

# HACK #52 Motorize Your Window Blinds

**An inexpensive motor and some clever pulley assemblies enable you to automate the opening and closing of window blinds.**

I designed and built a system that uses two motors to control the eight window blinds I have in my two-story family room. The system consists of pulleys placed under the floor, limit switches, gear motors, and the electronic circuitry needed to control the system.

Visit *http://www.edcheung.com/automa/blinds.htm* to view an animated picture of the blinds in action.

Here are the advantages of this design:

- You can remove the hardware without scarring the home.
- You still can operate the blinds manually, using the traditional method.
- The motors and machinery are hidden from view.

I accomplished this by hiding the hardware in the recreation room in the basement, directly below the family room, as shown in Figure 4-5.

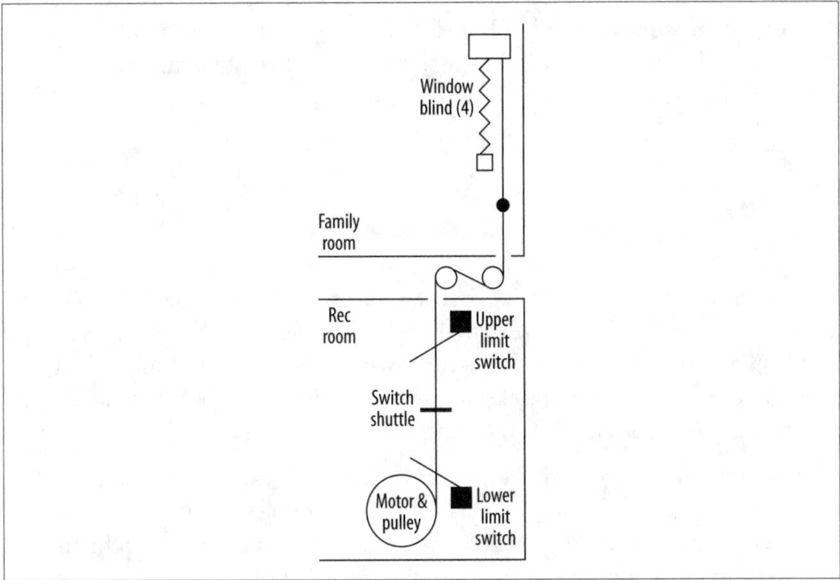


*Figure 4-5. The blind-pulling mechanism*

The only unusual thing you can see from the family room is a steel cable connected to the pull cords and running into the floor, through a hole in the carpet. An *S hook* connects the steel cable to the pull cords through an eyelet (as shown in Figure 4-6) so that you can disconnect the pull cords easily from the motor drive for manual use.

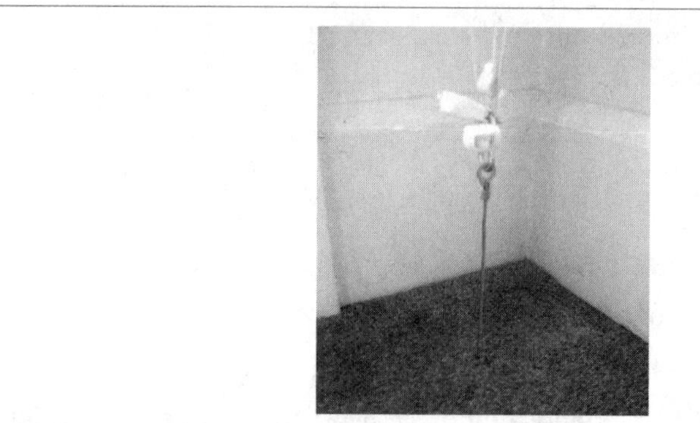

*Figure 4-6. Cable assembly*

I purchased two gear motors from Grainger Industrial Supply (*http://www.grainger.com*). The motors I use are similar to item #2L003 ($43). With a 6-inch-diameter pulley, they can provide 30 lbf of pull on the steel cable. I previously measured that about 21 lbf was needed to pull the blinds when they are in the almost fully open position (i.e., when the amount of weight on the pull cord is the highest). Each motor controls four blinds.

You also can use modified cordless drill motor assemblies. I've seen other designs use these, and they work well for blind control.

An important consideration is that we have HunterDouglas (*http://www.hunterdouglas.com*) dual-cell *Duette* blinds. The blinds do not lock into position when the cord is pulled straight down. Locking occurs only when you pull the cord sideways. Thanks to this, the motor can raise or lower the blinds freely, without having to disengage a cord lock.

Two limit switches govern the distance that each cable can travel. One is near the ceiling and the other is near the floor, in the recreation room in the basement. The motor stops when the shuttle hits the limit switch; then it can go only in the opposite direction. Figure 4-7 shows the motor and limit switch assembly.

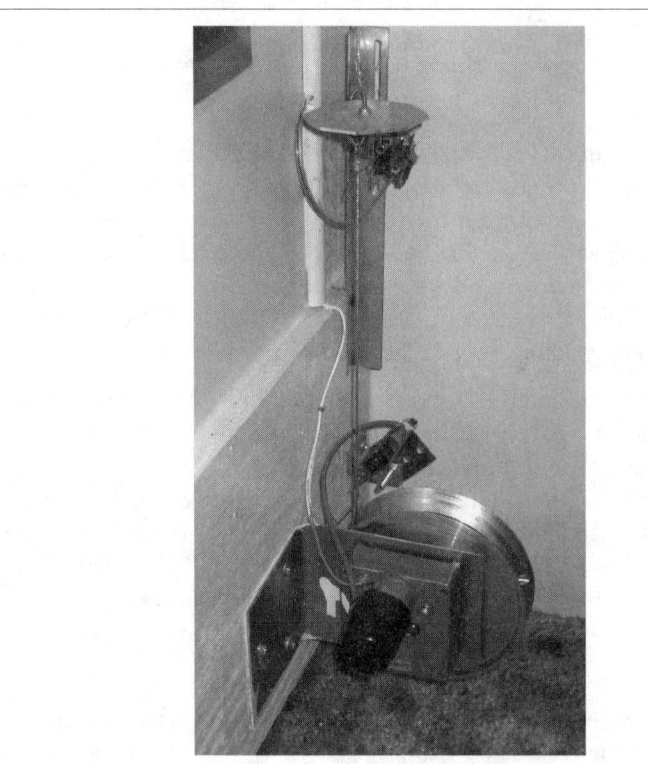

*Figure 4-7. Limit switch detail*

Note the limit switch at the top of Figure 4-7. It is mounted on a bracket with a long slot so that you can adjust the shuttle travel. Above the switch is the shuttle. A third limit switch is right next to the pulley.

I control the system using X10. Each motor is connected to an appliance module **[Hack #3]**. Sending an `On` command causes the motor to run, and the associated four blinds to open, until the shuttle hits the limit switch. Sending an `Off` command causes the motor to reverse direction. When an `On` command is sent to a third address, both motors halt, enabling me to set the blinds to any position.

We have been using this system for several years, and it performs reliably. Because the motors are one floor below, the moving blinds create little sound. The only problem we've experienced is that occasionally (about twice per year), one of the cord locks will stick while the system is trying to lower the blinds. In that case, we need to release and close the blind manually.

*—Edward Cheung*

# Outdo Big Ben

The classic Westminster chimes ring forth from your home automation system.

All of us crave it, but only a select few of us obtain it. I'm referring to acceptance of our home automation hobby by our spouses, of course—the elusive SAF **[Hack #14]**. We can wax quite eloquently of the benefits of computer-controlled lighting **[Hack #16]**, talking CID **[Hack #27]**, driveway vehicle detectors **[Hack #54]**, and automated sprinklers **[Hack #65]**. But seldom do we get more than a glazed look in return. Well, here's a simple hack that will add a touch of class to your home, and might get a smile from your better half.

All you need is a computer in a room where it can be heard easily (or, ideally, a whole-house audio system) and some quality chime recordings. In fact, the hardest part is finding a quality recording of the chimes—you'll need each hour's chime in a separate sound file. A search at the HomeSeer message board (*http://ubb.HomeSeer.com/*) will turn up many to choose from, or visit the *Smart Home Hacks* web site (*http://hacks.oreilly.com/smarthomehks*) for pointers to appropriate files.

## The Code

Before we set up HomeSeer **[Hack #19]** to play at the proper times, first let's cover the script that plays the chimes. We want the chimes to play only between the hours of 8:00 a.m. and 9:00 p.m. (Try that with a grandfather clock!) So, the first thing the script does is to make sure it's between those hours. If it's not, no chimes are played and no one will be disturbed:

```
sub main( )
dim h
dim prior_volume_level_left
dim prior_volume_level_right
dim chime_volume_level
dim start_time
dim end_time
chime_volume_level = 3

if now() > cdate(date() &  " 7:15:00 AM") and now() < cdate(date( ) &  " 9:
05:00 PM") then
```

Note the space in the code between the time and AM or PM; it's necessary for the comparison to work correctly.

Next, examine the hour of the current time (h) and reduce it to a 12-hour clock, if necessary. That is, if the hour is 18, change it to 6 so that the correct sound file will play:

```
        h = datepart("h", now( ))
        h = cint(h)
if h > 12 then h = h -12
```

The script saves the current sound volume settings so that they can be restored later, then calls the `hs.setvolume` function to adjust the volume of the right and left channels to a lower level so that the bells aren't too loud:

```
prior_volume_level_left = hs.getvolume(0,0)
prior_volume_level_right = hs.getvolume(1,0)
hs.setvolume chime_volume_level,chime_volume_level,0
```

Loud sound files result in low SAF.

A simple `select` statement plays the correct sound file, based on the hour value. HomeSeer's `hs.PlayWavFile` expects a complete path to the WAV file to play, so this script assumes the sound files are stored in the *HomeSeer\sounds* directory:

```
select case h
      case 1
        hs.PlayWavFile "sounds\One.wav"
      case 2
        hs.PlayWavFile "sounds\Two.wav"
      case 3
        hs.PlayWavFile "sounds\Three.wav"
      case 4
        hs.PlayWavFile "sounds\Four.wav"
      case 5
        hs.PlayWavFile "sounds\Five.wav"
      case 6
        hs.PlayWavFile "sounds\Six.wav"
      case 7
        hs.PlayWavFile "sounds\Seven.wav"
      case 8
        hs.PlayWavFile "sounds\Eight.wav"
      case 9
        hs.PlayWavFile "sounds\Nine.wav"
      case 10
        hs.PlayWavFile "sounds\Ten.wav"
      case 11
        hs.PlayWavFile "sounds\Eleven.wav"
      case 12
        hs.PlayWavFile "sounds\Twelve.wav"
    end select
```

Finally, the sound volume is restored to its previous setting:

```
   hs.setvolume prior_volume_level_left,prior_volume_level_right,0
  end if
end sub
```

Next, create a HomeSeer event [Hack #19] that executes this script. I have it trigger at the top of every hour; it exits without playing the chimes if the time is outside the time range specified in the script.

My family really enjoys our virtual Big Ben, and my sons actually learned to count (to 12) by listening to it. Most importantly, though, my wife actually likes this feature of our home automation system. Who could ask for more?

## Hacking the Hack

Instead of coding the time range that you want the bells to play in the script, you can set up a more sophisticated event to execute the script. Figure 4-8 shows a recurring event that triggers once an hour, at 00 minutes, between the hours of 8:00 a.m. and 9:00 p.m (the "Apply Conditions to trigger" option limits its actions).

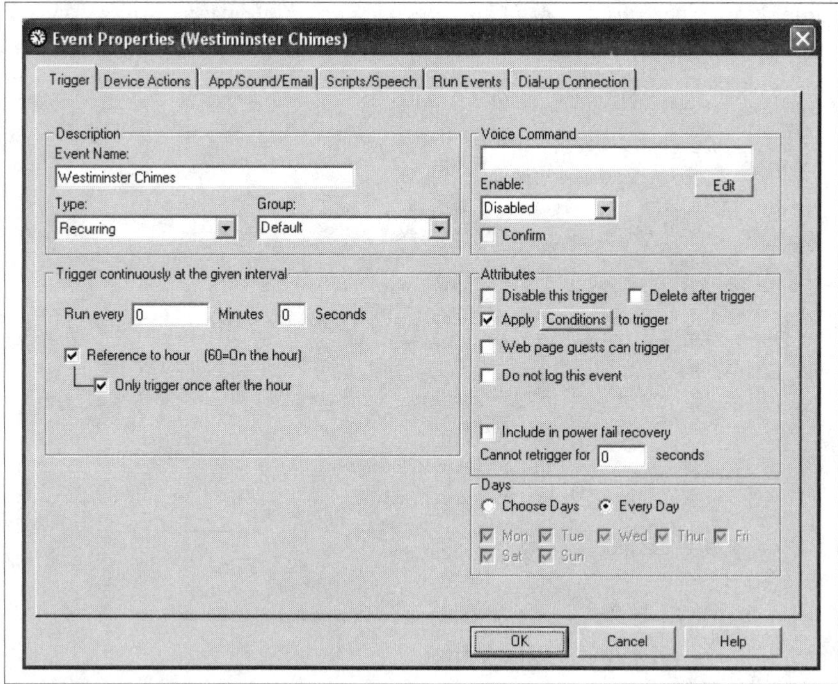


*Figure 4-8. An event to play the chimes*

Using this technique, you can create additional conditions under which to silence the bells, such as when you're not at home [Hack #70] or have gone to bed for the night [Hack #48].

—*David Kindred*

# CHAPTER FIVE

# Garage and Yard

## Hacks 54–68

Of all the areas of your home, the garage, garden, and yard provide the most enjoyment, and require the most work. Use the hacks in this chapter to increase your enjoyment and greatly decrease the hassles associated with these areas.

You'll learn how to dispense with mundane tasks, such as closing the garage door at night **[Hack #56]** and checking to see if the mail has been delivered **[Hack #62]**. Other hacks help you save money by being smarter about buying fuel oil only when you need it **[Hack #61]** and using the weather forecast **[Hack #64]** to run your automated lawn sprinklers **[Hack #66]** only when necessary **[Hack #67]**. Finally, speaking of the lawn, don't miss the outrageous automated lawn mower that really works **[Hack #63]**.

## Monitor Your Driveway

**Use a couple of sensors in your yard to greet you when you get home and alert you to visitors when they pull up to your driveway.**

This hack provides a simple but effective method for detecting and reacting to cars that enter and leave your driveway. It can do the following things for you:

- As you drive away from the house, your home automation system can light the driveway and then turn off the lights a few minutes after you're gone.
- When you arrive home, it can light the driveway and begin getting the house ready for your arrival (e.g., by turning on lights, adjusting the thermostat, and so on).
- If you're at home and a car enters the driveway, your house can alert you to the arrival of a visitor.

Once you have the system in place, you'll think of other things you can trigger based on knowing that someone is leaving or arriving. Let's begin by discussing the sensors that make all this possible.

## The Sensors

I use two X10 PR511 Motion Detector Floodlights (*http://www.smarthome.com/4080.HTML*; $50) to monitor and light the area in front of my garage and the area where my driveway meets the street. These devices have two floodlights and a passive infrared motion detector **[Hack #10]** integrated into one unit, as shown in Figure 5-1.

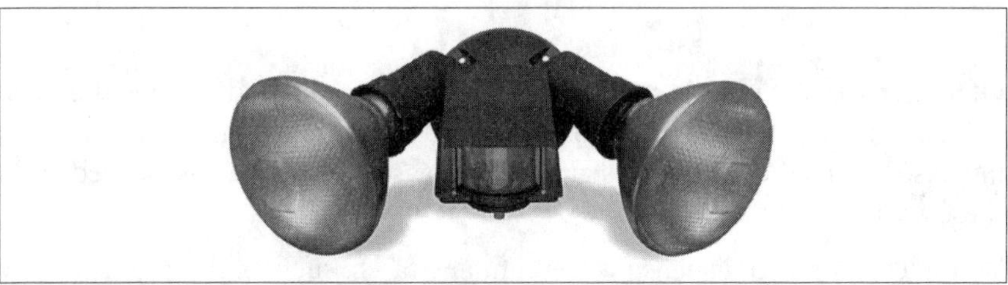

*Figure 5-1. X10 Motion Detector Floodlights*

You can set these units so that they will turn on the floodlights automatically when they sense movement at night, or you can control them with X10 commands **[Hack #4]**. For the purposes of this hack, let's set the unit so that, in addition to turning on the floodlights when motion is detected, it sends an `On` command to another address.

To set the address that the PR511 sends to when activated, first you must set its base address **[Hack #1]**. Let's say you set it to `B1`. Next, you use the switches inside the unit to set `Address +1` to `In`. Now, whenever motion is detected, the floodlights turn on and `B2 On` is sent to the power line. After a time delay that you've configured has expired, the unit sends `B2 Off`. Refer to the PR511's manual for details **[Hack #22]**.

Set the second PR511 unit to send its command to a different address, such as `B8`.

## Positioning the Sensors

Place one PR511 near the entry to the driveway, but far enough so that it won't be tripped by a bicycle on the sidewalk or a car that is just turning around. You might even aim it away from the street, back toward the house, but angled so that it monitors only the driveway area.

Place the other PR511 near the garage so that it is activated only when motion occurs within about 10 feet of the garage door, or approximately where you stop the car to open the garage door.

When you position the sensors, keep in mind that they can be falsely triggered by areas that reflect a lot of heat during the day or by trees that sway in the breeze.

See "Sense What's Happening" **[Hack #10]** for more information about how motion detectors work.

You can adjust the detectors' sensitivity using the controls in the PR511, which is probably necessary for most situations.

## Reacting to the Sensors

The sensor by your garage sensor automatically turns on the floodlights if someone leaves or approaches the house after dark. It also sends a `B2 On` command to indicate that it has been triggered. The `B2` unit is called `Front Drive Motion` in XTension **[Hack #17]**, and here is its On script:

```
if time delta of "Front Walk Motion" is less than 30 then
     turn off "Driveway Lights" in 20
 else
     turn on "Foyer Chime"
     turn on "Driveway Floods" for 2 * minutes
     write log "Someone entering driveway"
 end if
```

The `time delta` function is used to determine if the sensor by the driveway's entrance, `Front Walk Motion`, has been activated within the last 30 seconds. If it has, an event to turn off `Driveway Lights` is scheduled. These architectural lights line the driveway. Otherwise, an indoor chime module **[Hack #9]** is activated to alert anyone who's at home, and the lights are scheduled to turn off two minutes later.

Here's the Off script for `Front Drive Motion`:

```
if status of "Front Drive Motion" is false then
    if time delta of "Front Walk Motion" is less than 30 then
        turn off "Driveway Lights" in 20
     else
        turn on "Foyer Chime"
        turn on "Driveway Floods" for 2 * minutes
        write log "Someone entering driveway - missed the ON"
     end if
end if
```

The `Off` command indicates that the motion detector has reset itself after detecting motion. First this script checks to see if the unit is already turned on in XTension. If it's not, that means the `On` command was missed, due to line noise or a signal collision, so the actions to alert the house are repeated here. As a rule, it's a good idea to have scripts for motion detectors check to make sure an `On` event wasn't missed and, if it was, perform at least some of the tasks that should have been done.

Now, let's talk about the sensor installed at the driveway's entrance, near the sidewalk in front of the house: `Front Walk Motion`. Again, the floodlights on the sensor are not explicitly controlled; the PR511 turns those on and off automatically. The On script for `Front Walk Motion` does most of the work:

```
turn on "Front Walk Light"
if time delta of "Front Drive Motion" is greater than 30 then
    turn on "Foyer Chime"
     write log "Movement on the front walkway"
end if
```

Here's the unit's Off script:

```
if status of "Front Walk Motion" = false then
    turn on "Front Walk Light"
    if time delta of "Front Drive Motion" is greater than 30 then
         turn on "Foyer Chime"
         write log "Movement on front walk, missed the ON"
     else
         write log "Someone has arrived, missed the ON"
     end if
end if
```

The important point to notice here is that the status of the sensor in front of the garage, `Front Drive Motion`, is used to decide if someone is arriving or leaving. If a person is leaving, the chime module is not activated because there's no need to alert the house's occupants to this. You'll need to adjust the `time delta` value so that it works for the typical length of time it takes to drive between the two sensors. Don't expect it to be perfect, though; even humans sometimes have a hard time figuring out if they're coming or going!

## Hacking the Hack

If you don't want to turn on floodlights when motion is detected because it might disturb the neighbors, or if you just want a subtler response, you can remove the bulbs from the PR511 units. Or, use a photoelectric beam sensor instead (*http://www.rpelectronics.com/English/Content/Items/E-960-D290.asp*; $170), as shown in Figure 5-2.

*Figure 5-2. A photoelectric beam sensor*

Although much more expensive than the PR511, photoelectric sensors are more reliable and less subject to false alarms. They work by projecting an invisible laser beam across your driveway to a reflective sensor. When an entering car breaks the beam, a relay is triggered. Connect the unit's relay to a Powerflash module [Hack #10] so that an X10 command is sent to your home automation system. The price for these sensors varies based on their effective range, so, depending upon your property, you can get a long-range sensor to cover more than just your driveway, or a shorter sensor for less money.

Finally, note that the action programmed here (turning on a chime module) is just the beginning. If your house knows who is at home, it can send you an email [Hack #73] that someone is on your property and immediately start turning on lights to make the house look occupied [Hack #72], in case the visitor has bad intentions. If you've implemented a system to control your garage door [Hack #56], you can use that to open or close the door based on the time of day and whether it's likely to be you that's arriving or leaving.

As with most things in home automation, once the basics are in place, you can take your system in new directions by knitting together pieces with new logic and thinking.

—*Michael Ferguson*

## HACK #55 Know If the Garage Door Is Open

Use a magnetic switch and a Powerflash module to keep track of the status of your garage door.

One of the most common motivations for setting up your home automation system to keep track of whether the garage door is open is so that you don't forget to close it at night. It's a common thing to overlook and a common problem to want to solve. But in my case, I was motivated to add this capability for a different reason.

A keystone in my home automation system is that it knows when the house is unoccupied **[Hack #70]**. When the house is empty, the alarm system is activated, visitors coming to the front door are logged **[Hack #74]**, and the network camera that I use to check in on my dog **[Hack #82]** is turned on. Furthermore, the first person who comes home to the previously unoccupied house is greeted with appropriate lighting and thermostat setting, and important reminders and announcements **[Hack #28]**.

For all this to work, the first person home has to tell the system he has arrived, by pushing a button that sends an X10 command. One day, I had an epiphany that the button could be eliminated, simply by watching the garage door. Like most California suburbanites, I always drive where I'm going, which means, obviously, that when I return home, the first thing I do is open the garage door so that I can park my car. Therefore, if the garage door opens and the house is unoccupied, someone is returning home and the house can begin getting ready for my arrival—a much smarter approach, don't you think?

I use a simple method to keep track of the garage door's open or closed status. Here's how to set it up.

You need a sensor that will tell you when the door opens. I use a *normally open* (NO) magnetic contact switch (*http://www.smarthome.com/7113.HTML*; $6). As explained in Chapter 1, NO means that when the door is closed, the switch will be open. When the door is open, the switch closes the circuit. This allows the switch to work with a Powerflash module **[Hack #10]**, which is how the home automation system monitors the door's state.

Connect wires to the terminals on the magnetic switch and to the terminals on a nearby Powerflash module. Set the Powerflash to `Input B`, `Mode 3`. This will cause the Powerflash to send an `On` command when the door opens and an `Off` command when it closes.

The magnetic contact switch has two pieces. The switch portion, the one where you connected the wires, mounts on the garage wall. You attach the other piece, which contains a magnet, to the garage door. When the two pieces are next to each other, the magnet keeps the switch open. When you open the door, the magnet moves away and the switch closes.

You must mount the two pieces securely and ensure they are close enough for the magnet to work. I found the best location was at the top edge of the door. The switch is mounted to the garage door, and the magnet is glued to the top of the door with a bit of epoxy, as shown in Figure 5-3.

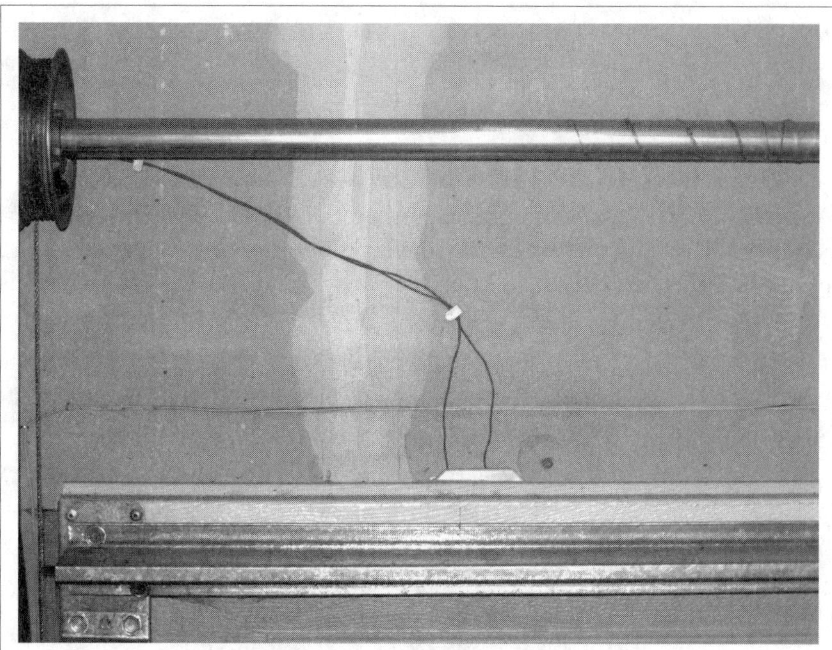

*Figure 5-3. The magnetic switch in place*

Use a multimeter to make sure the magnet is activating the switch before gluing everything in place. A surprisingly small gap can prevent it from working properly, as I learned the hard way when setting this up for the first time.

Instead of using a contact switch, a garage door contact sensor (*http://www.smarthome.com/7455.HTML*; $15) might work better. Designed specifically for garage doors, it's heavy-duty and will activate correctly even when the switch and magnet are far apart, which makes it work better with ill-closing garage doors.

Now that the switch is in place and you've configured and set the Powerflash to transmit the X10 address **[Hack #1]** you want to use, it's time to set up your home automation software.

Here's an On script for the Powerflash unit in XTension **[Hack #17]**:

```
If (status of "Nobody Home") is true then
    If (daylight) is false then
        Turnon "First Floor Lights"
    End if
    If (status of "summer mode") is true then
        Turnon "Thermostat 68 degrees"
    Else
        Turnon "Thermostat 73 degrees"
    End if
End if
```

First this script examines the `Nobody Home` pseudo unit. This unit is maintained by the scripts that are executed when house members indicate they're leaving for the day **[Hack #70]**. After the last person leaves, `Nobody Home` is turned on.

If `Nobody Home` is on (`true`) when the garage door opens, the group of units assigned to `First Floor Lights` is turned on, providing that it's after sunset. XTension automatically sets the `daylight` variable to `false` at sunset.

Next, another pseudo unit that tracks the current season **[Hack #24]** determines the temperature setting **[Hack #41]** for the heating and air conditioning system. Note that I live in California, where we have only two seasons: summer and not summer. You might need to account for more seasonal variance where you live.

It's a simple script, but don't overlook the logic. One key thing is that nothing happens if the house is occupied when the garage door opens. You almost certainly do not want to change lighting, and definitely not the thermostat setting, every time the garage door opens. Doing otherwise would not please your housemates **[Hack #14]**.

### Hacking the Hack

Although I didn't put this in place to remind me when I'd left the garage door open, you can use it for that purpose. For a simple approach, set up a lamp with a lamp module [Hack #2] that's set to the same address as the Powerflash connected to the garage door sensor. When the garage door is closed, the light will be off. When it's open, the light is on.

If you've implemented [Hack #48], it's easy to have that process check the status of the Powerflash unit and announce, or otherwise indicate, that the door needs to be closed.

HACK #56

## Control Your Garage Door

**By combining multiple modules and techniques, an experienced automator can create the ultimate garage door automation solution.**

If there's one thing in our homes that nearly screams out for automation, it's the garage door. After all, it's probably the only remote-controlled motorized door in the house! Here's a multifunction approach that uses several different components, each working together, to provide everything from basic to advanced garage door control. You can implement the whole hack or just pick out the parts you like.

Table 5-1 lists the parts you need and the functions they provide.

*Table 5-1. Parts required for an automated garage door*

| Part | Used for | Comments |
|---|---|---|
| Appliance module (*http://www.smarthome.com/2002.HTML*; $13) | Security lockout | Make sure the load rating is OK for your garage door opener. |
| Universal module (*http://www.smarthome.com/2010.HTML*; $22) | Auto close<br>Auto open | Two are required for this hack. |
| NO magnetic security switch (*http://www.smarthome.com/7113.HTML*; $6) | Auto close<br>Status monitor | The switch is open when it's next to the magnet. You might need a wide-gap model depending on your door. |
| Mini-timer (*http://www.smarthome.com/1100X.HTML*; $30) | Security lockout<br>Auto close | Not needed if you have a computer-controlled home. |
| Powerflash module (*http://www.smarthome.com/4060.HTML*; $24) | Status indicator | |
| Wall-mounted keypad (*http://www.smarthome.com/4210R.html*; from $45) | Status indicator | Requires base ($30) and keypad module (price varies by number of buttons). |

Connect everything as shown in Figure 5-4. The numbers next to the modules indicate in which function, such as auto close, the module is used (refer to the key for definitions). If you're not implementing a function, you can omit the module from the diagram. Use solid-core, light-duty electrical wiring to connect the modules to each other.

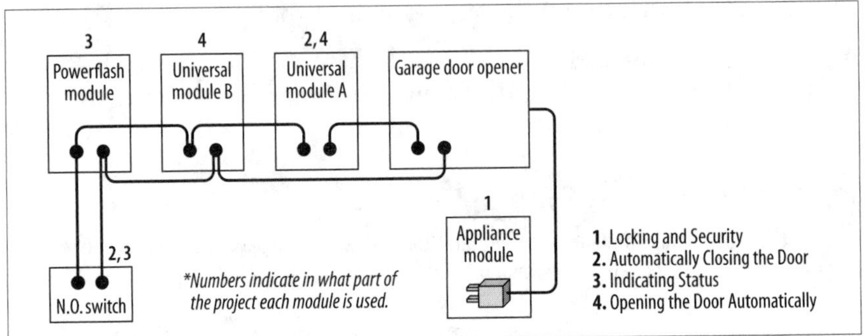


*Figure 5-4. Connection diagram*

Mount the security switch so that its magnet and switch are aligned when the garage door is closed. I like to put it at the top of the garage door to keep the wires out of the way. You can plug most of the modules into a single power strip, which you can fasten to the garage ceiling or rafters near the opener. You can install the wall-mounted keypad inside the house in any electrical box. Plug the mini-timer into any outlet.

Set the universal modules **[Hack #11]** to momentary and relay-only modes and set each to a different house and unit code **[Hack #1]**. Set the wall-mounted keypad to the same house code you set for the universal modules. Set the Powerflash module **[Hack #10]** to `input 2` (momentary contact closure), `mode 3` (single unit code on and off), and set the house and unit codes to match one of the buttons on the wall-mounted keypad.

## Locking for Security

Unless you have a newer-model garage door opener, it's not too hard for someone to find out your code and open your garage while you're not home. However, you can turn off the opener completely when you're away so that it will not work, even with your remote.

Plug the garage door opener into the appliance module. Schedule the appliance module, using the mini-timer or your home automation software **[Hack #16]**, to turn off any time you don't want someone to be able to open the door, such as when you're on vacation, at work, or sleeping.

Schedule an event that turns on the appliance module again shortly before you will arrive home. Be sure to have a way to get into your home if you decide to come home early, such as using the front door or controlling your home automation system with a telephone call [Hack #33].

## Automatically Closing the Door

I hate getting up in the morning and discovering that I left the garage door open all night. I've seen other solutions to this problem, but they sometimes don't work if the door is only partially open. This solution also has the advantage of being integrated with the other parts of this hack.

When the door is open, even partially, current will flow through the NO magnetic switch, which completes the circuit and allows universal module `A`, `connect to the opener`, to close the door (refer back to Figure 5-4). When the door is fully closed, the magnetic switch opens and no signal can reach the garage door opener, so nothing happens.

When you want to close the door, send an `On` command to universal module `A`. If the door is partially open, the command will open the door fully instead of closing it. Send two or more `On` commands with a short delay between them to make sure the door is closed. I send four commands at three-minute intervals just to be sure. Once the door is closed, the extra command will just be ignored.

Make sure your garage door has a safety beam feature that prevents it from closing if an obstacle is in the way, such as your car or the neighbor's cat.

## Indicating Status

I like to be able to tell at a glance if my garage door is open. The wall-mounted controller enables you to send X10 commands, but its buttons also have LEDs that light up when they receive a command from the power line. Whenever the garage door is opened, the magnetic switch activates the Powerflash module, which sends a signal and lights up the LED on the wall switch.

The LED will turn off only when the garage door is fully closed. You can use the extra wall-switch LEDs to monitor other doors or gates by adding additional magnetic switches and Powerflash modules. You also can use the extra buttons to close the garage door using the auto-close method described earlier.

### Opening the Door Automatically

There are times when it can be useful to open the garage door from your home automation system. Unfortunately, X10 just isn't reliable enough because power-line noise occasionally can activate a module. You wouldn't want that to open your garage door when you're not home. To prevent this problem, this hack uses two universal modules, requiring two different commands, to open the door.

Be sure the universal modules are set to different addresses [Hack #1]. To open the garage door, you must send `On` commands to both modules. Because they're set for momentary operation, you have about two seconds to activate both modules. This provides an extra layer of reliability and security.

One way to use this is with another Powerflash module that's connected to your front doorbell. Program your home automation software so that if you lock yourself out of the house, you can tap a code on the doorbell and have the garage door open to let you in.

### Hacking the Hack

For additional safety, consider installing a motion detector [Hack #6] in the garage. Check the status of the motion detector to determine if it's safe to close the door. If it has been triggered recently, someone might be nearby.

Another good idea is to install an X10 chime module [Hack #9], or set one of the universal modules to sounder mode, so that there's an audible warning before the door is opened or closed. Finally, when deciding how to respond to a request to open or close the door, have your home automation software check the house's status to see if the request makes sense. For example, if nobody is at home, a request to open the garage door that originates from the wall-mounted keypad is probably a spurious command.

*—Doug Smith*

## HACK #57 Control Your Home from Your Car

Use the HomeLink wireless transmitter as a remote control for your home.

Many recent-model cars and trucks have a HomeLink (*http://www.homelink.com*) transmitter. The HomeLink is a built-in wireless remote control that you can program to send commands to your garage door and, as luck would have it, to wireless X10 transceivers [Hack #5]. According to the product's web site, it is included, or available as optional equipment, in more than 150 models from nearly every manufacturer. If your car has a HomeLink transmitter, it will be designated with a house logo, as seen on the three leftmost buttons at the bottom of the Nissan rear-view mirror shown in Figure 5-5. If your car doesn't have a HomeLink, see the "Hacking the Hack" section for an alternative.

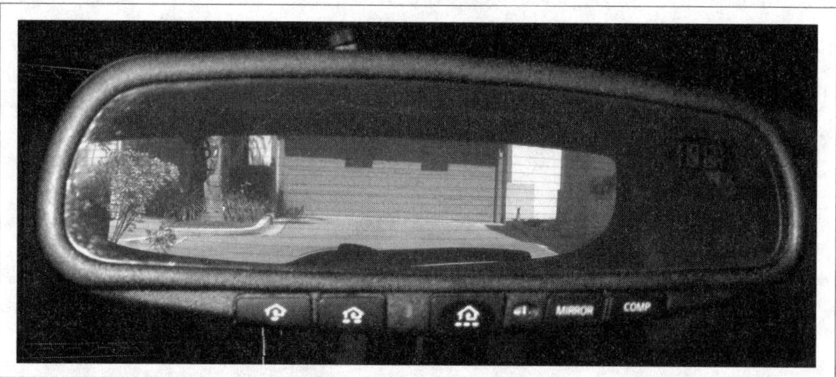


*Figure 5-5. A typical HomeLink transmitter*

To train the HomeLink to speak to an X10 transceiver, you'll need a Palm Pad that's set to transmit the X10 address **[Hack #1]** you want the HomeLink to be able to send. For example, if you want the HomeLink to send `C3 On`, you'll need a Palm Pad set to house code `C`. You'll also need a transceiver that's set to receive commands for house code `C` and is located within the range of where you park the car. I have my transceiver plugged into an outlet in the garage. After I open the garage door, the transceiver's range is excellent.

Here are the basic steps for programming a HomeLink, but be sure to check the owner's manual for specifics about your transmitter:

1. Turn the car's ignition to Accessory; then, press and hold the two outside HomeLink buttons until the indicator light begins to flash. This takes about 20 seconds.
2. Hold the Palm Pad about five to six inches away from the HomeLink.
3. Simultaneously press both the button on the Palm Pad (`C3 On`, for example) and the HomeLink button you want to program. Hold both buttons down until the HomeLink indicator light begins flashing rapidly; then release the buttons.
4. Now the HomeLink button is programmed to send the same command as the Palm Pad—that is, `C3 On`. If you want to program another button to send `C3 Off`, repeat the procedure.

If you're using a state machine to keep track of who is at home **[Hack #70]**, the HomeLink button is a handy way to signal your house when you're departing or arriving. Or, when you arrive home, use the HomeLink to activate a script that turns on the house lights (if it's dark), tells your computer to download your email, and turns up the thermostat.

Due to the limited number of HomeLink buttons you have available—most cars have just two or three—you'll probably want to script the button so that it acts as a toggle instead of a discrete command. This saves you from having to program one HomeLink button to send `On`, and another to send `Off`. This also saves you from remembering which button is which.

```
--Toggle for HomeLink button in car
--This same script assigned for both On and Off states of button
if (status of "Gordon Home") is true then  -- I must be leaving home
    turnoff "Gordon Home"
else -- I am not currently home, so I must be arriving!
    turnon "Gordon Home"
end if
```

## Hacking the Hack

If you don't have HomeLink in your car, and you're not willing to buy a new car to get it, try using an X10 Keychain Remote (*http://www.smarthome.com/4003.html*; $25) instead. Use a little Velcro to mount it on your sun visor or dashboard, and you've accomplished the same thing as a HomeLink, but without having it built-in.

# HACK #58 See Through Walls

**A portable wireless video camera, combined with a portable TV, provides remote viewing for times when you really could use an extra set of eyes.**

X10 Corporation's XCam2 (*http://www.x10.com/products/vk45a_how.htm*; $70) is a wireless color video camera that works by broadcasting its signal to a base station that's connected to a TV, VCR, or PC video digitizer. It's a handy device, but I've discovered a practical use that makes it almost indispensable for the home handyman.

I received an XCam2 battery pack for free (*http://www.x10.com/products/x10_zb10a.htm*; $20) when I bought the camera, but I hadn't put it to much use. The battery pack connects to the camera and turns it into a portable broadcaster because you don't have to find a power outlet for the transformer the camera normally requires.

The XCam2's broadcast range is about 100 feet, but the frequency it uses (2.4 GHz) is the same as microwave ovens, cordless phones, and WiFi equipment, and interference can reduce its range. Be prepared to do some channel-juggling (the XCam2 provides two channels to choose from) until you find the best selection for cohabitation with the rest of your gadgets.

I use the battery-powered XCam2 when I'm working alone and running wires for automation, such as a sensor wire for an alarm system, and I can't see what's happening on the other side of the wall or I don't know if my fish wire is going where I want it to. Sometimes, I mount the camera on a small tripod for easier positioning (*http://www.x10.com/products/x10_zt10a.htm*; $17).

At the receiving end, the XCam2 receiver is connected to a DC-powered, five-inch color LCD TV that I purchased online for about $130. The receiver and TV are in a large plastic project box in which I've drilled a hole to allow the television's antenna to protrude so that I can adjust it easily for best reception. A metal hanger on the box lets me hang it from a rafter for hands-free viewing.

Now, I have an extra pair of eyes I use to view anything in the house, from anywhere I want. It's the greatest tool I've ever had. When you need to adjust the dampers from your HVAC system, tape some strips of tissue paper over the air vents in your house. To adjust a particular room, place the camera so that it can see the vents, and then watch the tissue to judge the amount of air that's coming through, all without having to leave the furnace area to check your work.

The same technique is also useful to identify circuit breakers and outlets. Put a lamp on the circuit you want to turn off, and then place the camera so that it's pointing at the lamp. Take the monitor with you to the breaker box, and you will know when you have turned off the correct breaker.

In addition, I use the XCam2 when I need advice about how to best tackle a project or solve a problem. Instead of describing the situation, I just take my camera and transmit the images to my father, via the Internet, using X10's free XRay Vision software (*http://www.x10.com/products/xrayvision_software5.htm*; free) and a video-to-USB adapter (*http://www.x10.com/products/x10_va11a.htm*; $70). Now, I have both an expert available and a remote set of eyes!

—*Arthur J. Dustman IV*

# Use Indoor Modules in the Great Outdoors

Very few X10 modules are designed for outdoor use. Protecting the module from the weather is key to extending its life.

If you're using a lamp or appliance module to control outdoor lighting, fountains, or other equipment, you'll have to be clever about how you install and protect the module from the weather. Currently, no modules are rated for outdoor use, so the first thing to accept is the fact that any module you choose to use outdoors will succumb to the elements eventually. For that reason, use an inexpensive module that you don't mind replacing every once in a while. How often you need to replace it will depend on how severe your local climate is and how good a job you do shielding the module from the elements.

Using a module rated for indoor use in an outdoor environment also involves accepting the possibility of fire, electrocution, and other mishaps. Do so at your own risk. Consult an electrician to be sure you're proceeding safely.

## Temporary Installations

If you'll be using the module temporarily, or if you live in a mild climate, it might be sufficient to plug the module into the end of an extension cord, then seal it in a waterproof plastic bag and some duct tape, as shown in Figure 5-6. This is a handy technique for Halloween displays, which are in use for only a short period of time and often before the weather has turned very wet.

## Semipermanent Installations

If you need to use a module outdoors on a more regular basis, it's better to plug it into an outdoor outlet, then cover the outlet and module with a large, rain-tight cover. These are available in the electrical section of hardware stores. Make sure you get one large enough to allow the cover to close with the X10 module inside.

Another approach is to buy an extension ring for your existing outdoor outlet. These fit onto the face of the outlet and just about double its depth. It's a tight fit, but you can plug a lamp module into the top outlet, and then cover the whole thing with an outlet plate cover, sealing the module inside. This is often less expensive than buying a rain-tight cover, but the drawback is that the cord plugged into the lamp module blocks the second outlet, and the sealed cover prevents you from using the outlet for other purposes, such as your electric lawnmower.

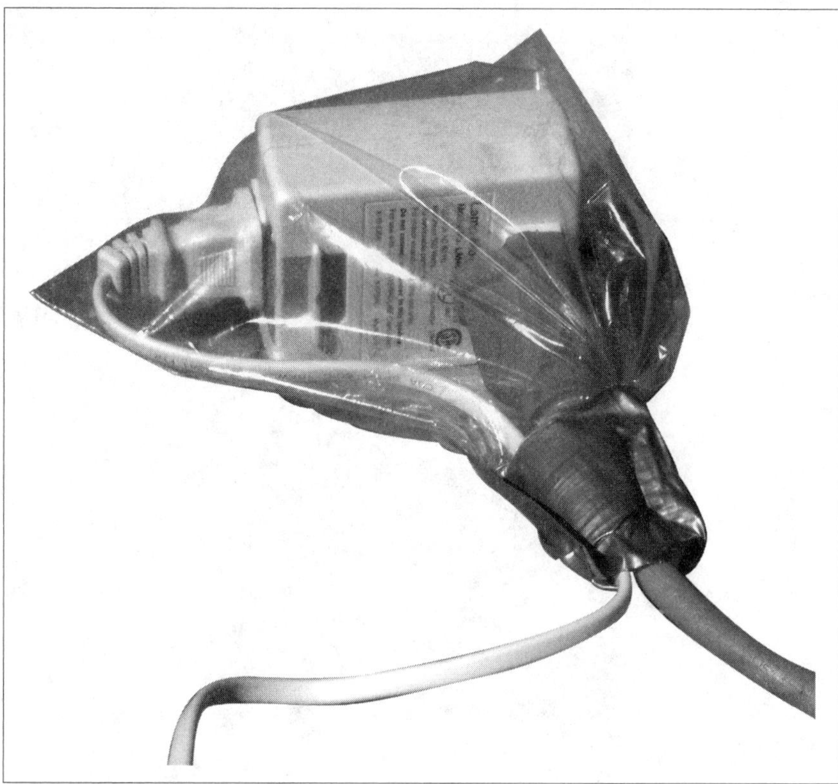

*Figure 5-6. A temporarily weatherized appliance module*

With either approach, you also should have an electrician install a Ground Fault Circuit Interrupter (GFCI) outlet, such as the one shown in Figure 5-7. The GFCI will shut down the outlet if an electrical short or other dangerous problem arises. You probably already have these types of outlets in the kitchen and bathroom; they're usually required by building code for any outlet that's near a water source.

## Permanent Installations

Instead of using an X10 appliance module, replace the regular power outlet with an X10 outlet **[Hack #14]**. An X10 outlet fits in the same space as a regular power outlet, but you can control one of the plugs with X10 `On` and `Off` commands. You'll still want GFCI protection, so have an electrician install a GFCI in your circuit box. Unfortunately, no manufacturer yet offers an outlet that includes both GFCI and X10 capability.

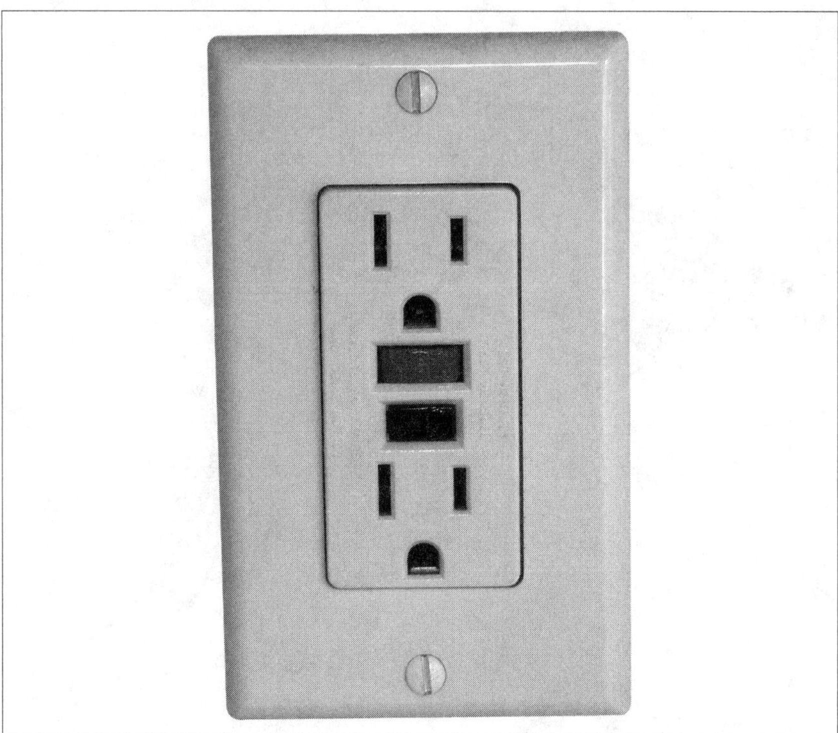

*Figure 5-7. A GFCI outlet*

Finally, if you're using a low-voltage device with its own DC power supply—an X10 XCam2 camera, for example—consider plugging the power supply into an *indoor* outlet, then running the lightweight connecting cord outdoors to its destination. If the cord won't reach, you can extend it using regular phone line (Category 3) wire. All it takes is the guts to cut the original cord and a little skill with a soldering iron to splice in the extension. The advantage of this technique is that it enables you to keep the sensitive and possibly hard-to-replace power supply indoors and away from the elements.

## HACK #60 Control Outdoor Lighting

**Controlling your outside architectural or garden lighting with a home automation system is much more flexible than using the simple timer that came with the lighting kit or was installed by your landscaper.**

Outdoor and garden lighting can be beautiful, but remembering to turn the lights on after sunset and off when you go to bed quickly gets annoying. That's why, as with sprinkler systems, they come with a simple, and dumb, timer. The built-in timer might be fine for some, but any self-respecting owner of a smart home definitely will want to do better.

## Making a Smart Light Timer

To begin, you need to stop your existing timer from controlling the lights. This control most often is built into the transformer, which you definitely need to keep, so the first task is to figure out how to prevent the timer from switching the lights on and off. Usually you can do this by removing the plastic nubs on the timer's rotating disk. As the disk rotates, making a complete trip every 24 hours, the nubs make contact with a switch that turns the lights on or off. Completely removing the nubs, as shown in Figure 5-8, means the transformer will remain on all the time. (You have to switch it on manually the first time, of course.)

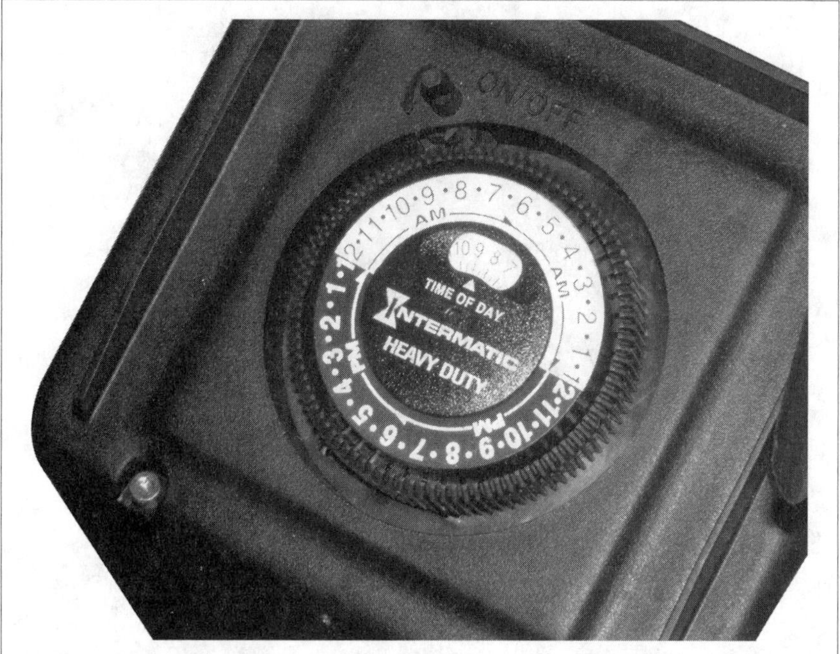


*Figure 5-8. A typical outdoor lighting timer*

If you can't remove the plastic nubs on your timer, simply put them as close together as possible, resulting in a quick on-off cycle. Position the nubs so that they trigger at a time when you won't mind the lights turning on and off, such as the middle of the day.

Instead of controlling the lights with the built-in timer, you'll be using X10. So, plug the transformer into an appliance module. Set the module's address [Hack #1], and add it to your home automation software. When the module receives an On command, all the lights connected to the transformer will turn on. If you have multiple transformers, connect each to their own appliance module [Hack #3] and consider creating an action group [Hack #18] for the transformers in your home automation software, as illustrated in Figure 5-9. This group is defined to toggle the state of each unit it contains. That is, if the backyard lights and front yard lights are on already, executing this action will turn them off.

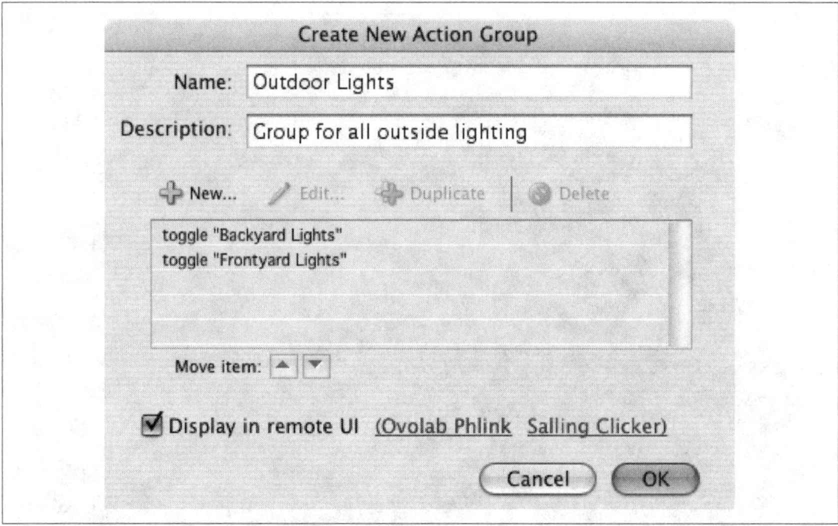


*Figure 5-9. An Indigo action group for outdoor lighting control*

Alternatively, instead of creating a group, you can set both appliance modules to the same X10 address. When you send a command, both modules will respond and turn their respective transformers on or off. The disadvantage to this approach, however, is that you won't be able to control each set of lights individually.

## Getting Flexible

Now that you have your home automation system in charge, it's time to consider your options for when the lights come on and go off. Your computer knows when sunset is, so that's a logical time to have the lights come on. However, often it's not dark enough to need them until several minutes after sunset, so you probably want them to come on several minutes after sunset. To accomplish this using Indigo, create a trigger action that occurs when the isDaylight variable "becomes false," as shown in Figure 5-10.

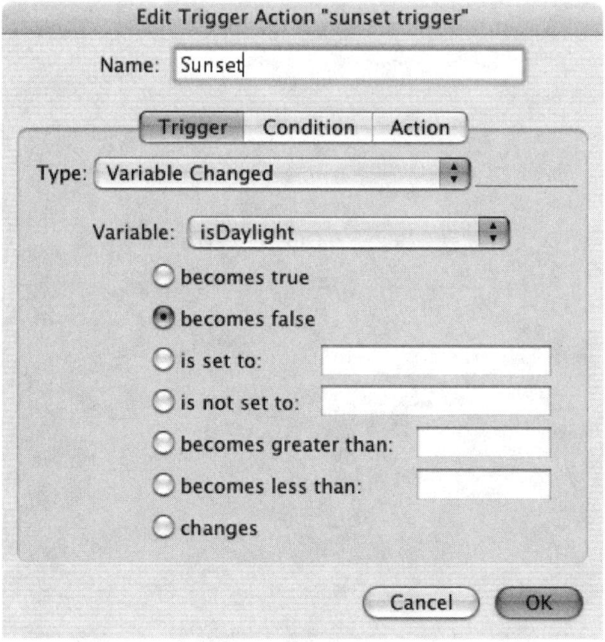

*Figure 5-10. Creating a trigger that runs at sunset*

In the Condition tab of the Trigger Action window, select Always, as shown in Figure 5-11. This ensures this action will be run every day at sunset.

Next, in the Action tab of the Trigger dialog, select Execute Action Group from the Type pop-up menu. Then, from the Group pop-up menu, choose the outdoor group that contains your lighting transformers.

Now let's set this action to turn on the lights 20 minutes after sunset. Check the "Delay action" checkbox and enter 20 in the text field. Because you specified that the Condition for this trigger is Always, uncheck the "Override previous delayed action" checkbox. This will enable you to set up additional triggers for use in other situations, as you'll see in a moment. Figure 5-12 shows how everything should appear at this point.

That's it. You're done. But what if you want to use the outdoor lights differently when nobody is home? Perhaps, for example, you want the lights in the front yard to come on earlier than those in the backyard.

To do this, add another trigger that is activated when `isDaylight` "becomes false," as you did before, but this time specify a different set of conditions, as shown in Figure 5-13.

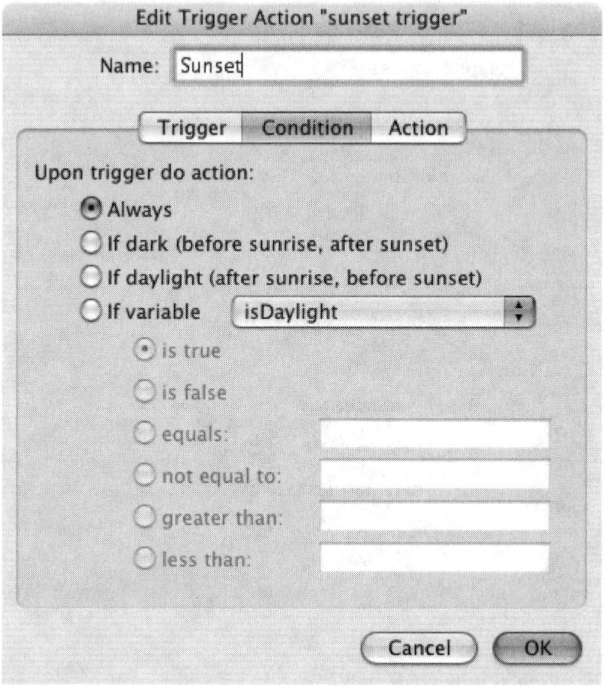

Edit Trigger Action "sunset trigger"
Name: Sunset
Trigger Condition Action
Upon trigger do action:
Always
If dark (before sunrise, after sunset)
If daylight (after sunrise, before sunset)
If variable isDaylight
is true
is false
equals:
not equal to:
greater than:
less than:
Cancel OK

*Figure 5-11. Using condition settings to determine when the trigger will run*

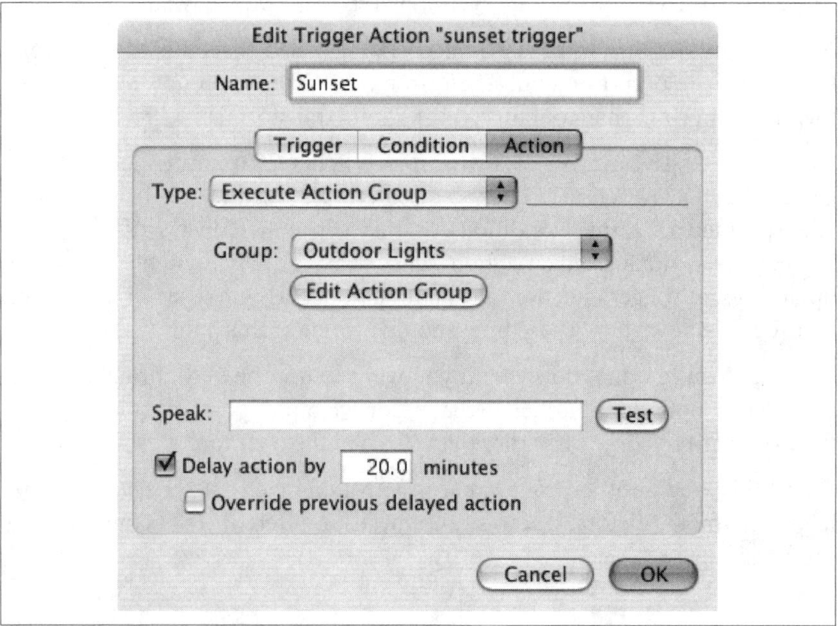


*Figure 5-12. Defining the actions for when this trigger executes*

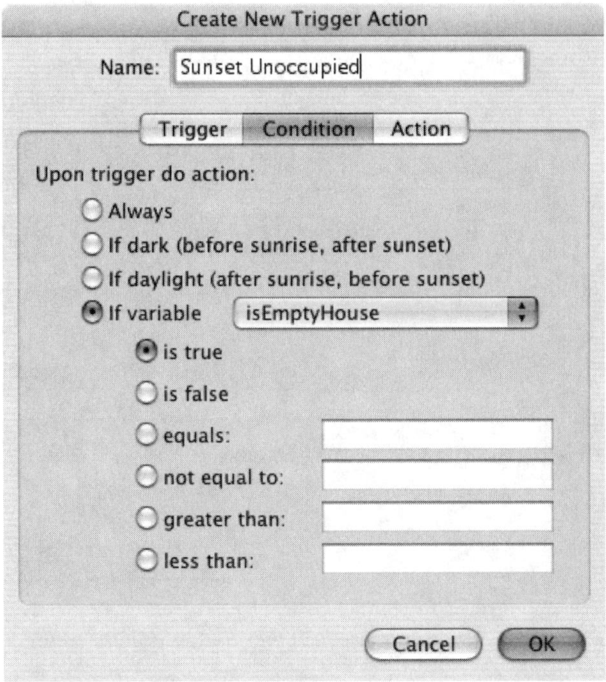

*Figure 5-13. A trigger that will execute only when the house is unoccupied*

We want this trigger to be executed only when nobody's home **[Hack #70]**, so the variable `isEmptyHouse` is examined. If it's true, the actions will be performed.

Next, define the actions for this trigger, as shown in Figure 5-14. Instead of using the `Outdoor Lights` action group, just the front yard lights are turned on, and the action is delayed to only 10 minutes after sunset. Also, the "Override previous delayed action" checkbox is checked, which will cause Indigo to remove any preexisting actions for this event. This is good practice for when you're creating conditional actions. It ensures that only the last action scheduled is active and helps avoid unexpected results.

Now, you have two triggers that will turn on the outdoor lights at sunset, but what about turning them off in the morning? Simply create another trigger that activates the `Outdoor Lights` action group when `isDaylight` becomes true. This is a good idea even if you have other events that might have turned off earlier; there's no harm in turning them off again, and it ensures they won't be burning all day long.

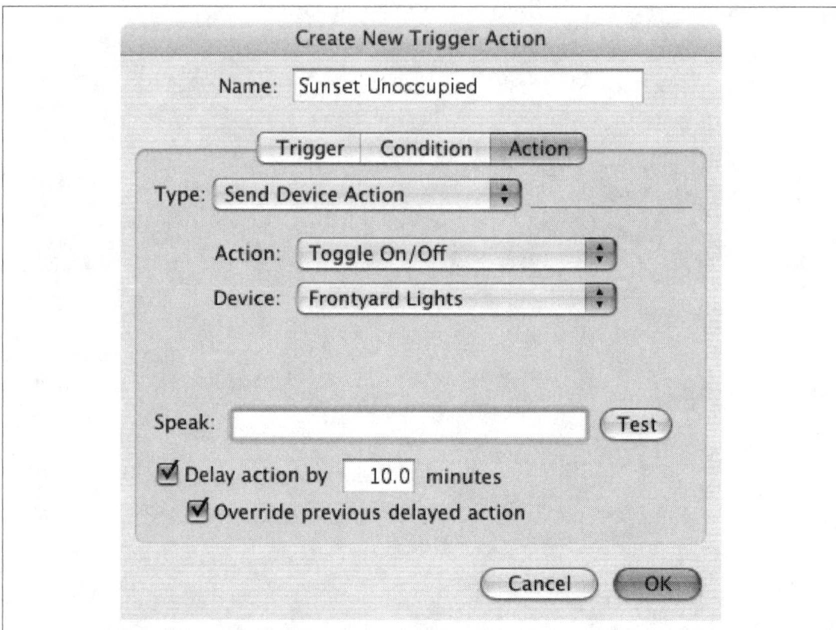


*Figure 5-14. Defining an action that overrides previously scheduled events*

## Final Thoughts

The techniques used in this hack are the building blocks of home automation techniques—scheduling events, taking action based on variables, and commanding groups of units. You can apply these ideas to many other aspects of your smart home, such as watering the lawn **[Hack #66]** and making your home unattractive to burglars **[Hack #72]**.

# Track Fuel Consumption

**Collecting information about fuel oil usage patterns can be straightforward, and when you know how much oil you're using, you can plan ahead and buy fuel when rates are lowest.**

My house has a 1,000-gallon underground tank to store fuel oil for the heating season. Instead of filling the tank on a scheduled basis, which means the amount you pay varies as the price of fuel oil changes, it's better to purchase oil during the summer when the demand is lowest and fuel oil is less expensive. During the more expensive winter months, it's best to purchase only the amount of addition fuel you need, if any. This system has saved me about $700 a year in fuel oil costs, which makes it well worth the effort.

To buy fuel *just in time*, I need to know how much fuel I have left and the rate at which I consume it. Unfortunately, due to the way the filler lines for my tank are installed, measuring the fuel level is unreliable. Instead, to figure out how much is remaining in the tank, I must keep track of how much fuel I've used since it was last filled.

The furnace's oil gun draws fuel oil from the tank and burns it to produce heat. The oil gun has a 1.25 tip on it, which means that it burns 1.25 gallons of fuel oil per hour. Said another way, it takes 22.5 seconds to burn one ounce of oil. To track the consumption of fuel oil, I need to track how long the gun is burning fuel on a daily basis. I hooked up a relay switch in parallel with the oil gun motor. The relay is connected to a Powerflash module [Hack #10]. When the oil gun turns on or off, the relay causes the module to send X10 `On` and `Off` commands to the power line. I use XTension [Hack #17] to measure the time between each on and off signal, calculate the accumulated time, and calculate the fuel consumed each day, week, month, and season.

Table 5-2 lists the units that are necessary for this hack.

*Table 5-2. Required units*

| Unit | Meaning |
|---|---|
| `Line Pump` | Universal module monitoring pump |
| `Wireless Pump` | Altered motion detector monitoring pump |
| `The Oil Pump` | Reflects state of the oil pump |
| `PavgOn` | Average pump on time |
| `PmissedOff` | Count of undetected `Off` states |
| `PmissedOn` | Count of undetected `On` states |
| `PmaxOn` | Longest running time so far |
| `PtimesOff` | Count of `Off` commands |
| `PtimesOn` | Count of `On` commands |

To ensure I don't miss any signals from the universal module, as can happen occasionally with X10, I also use a hacked motion detector to send a wireless signal directly to my computer [Hack #83] when the fuel gun starts and stops. The motion detector's dusk sensor is wired to the same relay that's connected to the universal module. When the relay is activated, XTension receives a wireless signal from the motion detector and a power-line signal from the universal module.

Visit Tom Laureanno's web site (*http://www.laureanno.com/x10-mod1.html*) for information about converting a motion detector's dusk sensor to a wired trigger.

Each module is a separate device in XTension. The motion detector is Wireless Pump, and the universal module is Line Pump. The On and Off scripts for each unit turn on a pseudo unit named The Oil Pump and use some logic to determine if an event was missed. Let's look at the On script for the Line Pump unit.

```
if (time delta of "The Oil Pump") > 10 then
    if (status of "The Oil Pump") is true then
        --we missed the last off...
        write log "Missed pump off event, assuming average."
        turnon "The Oil Pump" with  no script --keep the wireless event from
doing another missed event.
        set value of "PtimesOff" to (value of "PtimesOff") + 1
        if (time delta of "PmissedOff") > 20 then
            set value of "PMissedOff" to (value of "PMissedOff") + 1
            --If we're going to fudge for the missed event, we have to
            -- fudge the oil totals also.  22.5 seconds per oz...
            set pOil to (value of "PavgOn") / 22.5
        end if
    else
        turnon "The Oil Pump"
    end if
else
        turnon "The Oil Pump"
    end if
end if
```

The Wireless Pump unit receives its signal faster because it avoids the transmission delays of power line X10, so this script for Line Pump checks to see if the pump is already on and was turned on in the last 10 seconds. If both aren't true, the other unit missed the command. Pseudo units that track how many commands are missed are updated, as is PavgOn, which tracks consumption, to adjust for the missed event.

A similar method in the Off script adjusts for missed On commands:

```
if (time delta of "The Oil Pump") > 10 then
    if (status of "The Oil Pump") is true then
        turnoff "The Oil Pump"
    else
        --we missed the last on...assume average
        if (time delta of "PmissedOn") > 20 then --don't duplicate wireless
device
            write log "Missed pump on event"
            set value of "PtimesOn" to (value of "PtimesOn") + 1
            set value of "PMissedOn" to (value of "PMissedOn") + 1
            set secondsOn to value of "PavgOn"
        end if
    end if
```

```
else
    if (status of "The Oil Pump") is true then
        turnoff "The Oil Pump"
    end if
end if
```

The scripts that identify and adjust for missed commands are used in both the `Line Pump` and `Wireless Pump` units. Either could miss a command, and having the scripts in both places helps ensure accuracy.

The pseudo unit called `The Oil Pump` does all the calculation of elapsed time and fuel consumed. This is its On script:

```
write log "Pump on"
write log "the time delta of the pump is " & (time delta of (thisUnit))
set value of "PtimesOn" to (value of "PtimesOn") + 1
turnon "Attic fan" in 8 * minutes–this fan is turned on when the main
furnace air handling fan comes on.
write log "Attic fan schedules" color green
save database for unit thisUnit
write log "END OF ON SCRIPT"
```

This script increments the `PtimesOn` pseudo unit that's used for statistical tracking. It also schedules a fan in the attic to come on, thanks to an appliance module **[Hack #3]**, to help circulate the furnace's hot air.

The Off script for `The Oil Pump` does all the heavy lifting of calculations:

```
set secondsOn to time delta of (thisUnit)
write log "the time delta of the pump is " & (secondsOn)
set value of "PtimesOff" to (value of "PtimesOff") + 1
set tenthsOn to secondsOn / 6
set value of "PavgOn" to (((value of "PdayOn") * 6) / (value of "PtimesOn"))
```

The elapsed time since the unit was turned on, calculated using the `time delta` provided by XTension, is converted from seconds to minutes. The daily average time turned on (`PavgOn`) is updated using this information.

If the current elapsed time is the highest today so far, `PmaxOn` is updated to set a new high score:

```
if (value of "PmaxOn") < secondsOn then
    set value of "PmaxOn" to secondsOn
end if
```

Likewise, if this is the shorted elapsed time, `PminOn` is updated to reflect the new value:

```
if (value of "PminOn") > secondsOn then
    set value of "PminOn" to secondsOn
end if
```

Total usage for the season is stored as a *unit property*:

```
set t to (Get Unit Property "TimeOn")
Set Unit Property "TimeOn" to (t + secondsOn) -- in seconcds
set S to (Get Unit Property "SeasonUsage")
Set Unit Property "SeasonUsage" to round ((S + pOil) * 100) / 100 -- in
ounces of oil
```

This calculation determines the per-hour usage and updates a unit property:

```
set h to (Get Unit Property "HourUsage")
set h to round (h * 100) / 100
Set Unit Property "HourUsage" to round ((h + pOil) * 100) / 100
```

Daily, weekly, and monthly totals are also calculated and stored:

```
set D to (Get Unit Property "DayUsage")
Set Unit Property "DayUsage" to round ((D + pOil) * 100) / 100
set w to (Get Unit Property "WeekUsage")
Set Unit Property "WeekUsage" to round ((w + pOil) * 100) / 100
set M to (Get Unit Property "MonthUsage")
Set Unit Property "MonthUsage" to round ((M + pOil) * 100) / 100
```

Finally, `save database for unit` causes XTension to write all the unit properties to its database file:

```
save database for unit thisUnit
```

Usually, XTension saves to disk hourly; this statement ensures the latest info is saved in case a problem occurs.

XTension's unit properties provide a handy way to store the calculated values. Table 5-3 provides a comprehensive list, but you don't need to create these in advance. XTension will create them when you store a value.

*Table 5-3. XTension's unit properties*

| Property name | Value (example) |
|---|---|
| SeasonUsage | 94580 |
| DayUsage | 0 |
| MonthUsage | 657 |
| WeekUsage | 0 |
| TimeOn | 542221 |

You might wonder why these values are kept in unit properties, but some (`PavgOn`, for example) are stored using pseudo units. All values could be stored as unit properties, but properties aren't visible in the Master List and aren't as easily accessible when you access XTension using a web browser **[Hack #99]**. I like to keep some of the values more accessible so that I can check on them easily and ensure that the system is not missing too many signals or that the average consumptions aren't inordinately high.

Now that the data is being gathered, let's do some analysis. A scheduled script that executes hourly appends the information to a comma-delimited text file:

```
set MyDate to ((month of (current date)) & (day of (current date)) & ":" &
(round (time of (current date)) / hours) as string) as text
set t to (Get Unit Property "TimeOn" from unit "The Oil Pump")
set h to (Get Unit Property "HourUsage" from unit "The Oil Pump")
set D to (Get Unit Property "DayUsage" from unit "The Oil Pump")
set w to (Get Unit Property "WeekUsage" from unit "The Oil Pump")
set M to (Get Unit Property "MonthUsage" from unit "The Oil Pump")
set S to (Get Unit Property "SeasonUsage" from unit "The Oil Pump")
try
--in case the file was left open
    close access alias "G4 X HD:Applications:XTensionFolder:OilUsageGraph.
txt"
on error the error_message number the error_number
end try
set MyFileRef to open for access alias "G4 X HD:Applications:XTensionFolder:
OilUsageGraph.txt" with write permission
set MyData to read MyFileRef
set eof of MyFileRef to 0 - append to end of file
set MyData to ((MyData) & (MyDate & "," & h & "," & D & "," & w & "," & M &
"," & (t / 6) & "," & S) as text) & return
write MyData to MyFileRef
close access MyFileRef
```

You can import this file into a spreadsheet for further analysis or graphing. I use DeltaGraph (*http://www.redrocksw.com/deltagraph/mac/*; $300). Figure 5-15 shows a typical chart for slightly more than a week's usage.

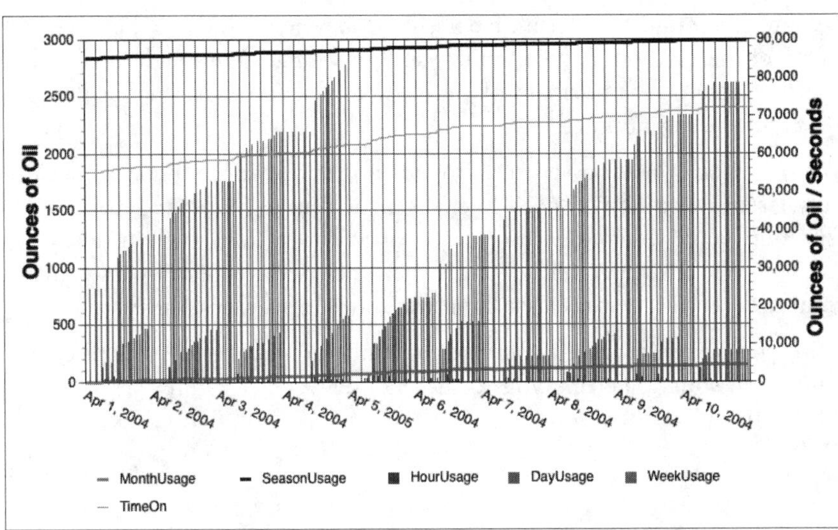


*Figure 5-15. Cumulative fuel usage chart*

If your system also tracks the current outside temperature **[Hack #64]**, plotting that value against daily usage would be an interesting addition. For more on plotting tools and methods, see "Chart Home Automation Data" **[Hack #92]**.

—*Henk van Eeden*

## Know When the Mail Arrives

**If you like to know when your mail has been delivered, to avoid needless trips to the curb or just to be sure you pick it up right away, here's an easy method to add a sensor to alert you to the delivery.**

To set up this mail-notification system, first you'll need to obtain a wireless doorbell kit from a hardware store. These typically are sold as replacement doorbell kits for use when you don't want to run new wires. An example is the Dimango Wireless Doorbell (*http://www.lamson-home.com*; $30). Wireless doorbells consist of a matched transmitter and receiver. The transmitter has the button you push to ring the bell, which is enclosed in the receiver unit.

Carefully open the transmitter and solder a lightweight, solid-core wire onto each button contact so that touching the ends of the wires together rings the doorbell. Then, solder the wires to a normally open (NO) magnetic reed switch or to a NO magnetic security switch (*http://www.smarthome.com/7113.HTML*; $6). Mount the switch so that it is aligned, thus keeping the circuit open when your mailbox is closed. You'll have to experiment a bit; if your mailbox doesn't close well at the top edge, mount the switch to the bottom, or vice versa.

Be sure to mount the transmitter itself *outside* the mailbox. Otherwise, the metal of the box will reduce the transmitter's range. You might mount the transmitter on the underside of the box, for example, or on the post on which the box is mounted.

Put the receiver unit somewhere in your house. If your mailbox is particularly far away, you might need to experiment a bit to make sure the receiver is within range. Also, keep in mind that as the battery in the transmitter loses power, the range will drop. If the unit suddenly stops working, the transmitter's battery is the first thing you should check, of course.

With everything in place, whenever your mailbox is opened, the receiver's chime will sound, alerting you to the mail delivery. If you're home, go see what arrived!

### Hacking the Hack

If you're not at home to hear the mailbox chime, you'll still have to check for mail when you get home. Let's improve on that. If you have a computer-controlled home automation system [Hack #16], consider connecting a Powerflash module [Hack #10] to the receiver. You can run wires from the speaker or chime module inside the receiver to the inputs on the Powerflash.

With this addition, your home automation system will receive an On command when the mailbox is opened, which in turn opens up a world of possible responses. You can announce that the mail has arrived when the first person arrives home to an empty house [Hack #70]. Or, if you're waiting for something important to arrive, send an email to your cell phone [Hack #73].

*—Reynold Leong*

## HACK #63 Mow the Lawn

**Although a robotic lawnmower might be nice, a little wire and an old spool (and a lot of guts and common sense) can get you partway there.**

Sometimes, a low-tech approach to automation is all that is needed to save a lot of work. Although I'm working on building a robotic lawnmower, it will be quite some time before it's perfected. Meanwhile, I've managed to cut my mowing time to about half of what it used to be. I'm not sure of the exact cost because I scrounged most of the parts. Even the mower was a $30 garage sale special. In total, I probably spent less than $50 on this project.

> What this hack proposes can be very dangerous. Damage to life, limb, and property can result. This is how I mow my lawn, but it's not necessarily how you should mow yours.

The basic idea behind this hack is that a self-propelled lawnmower can be tethered to a central spool so that it cuts the lawn as it winds itself around the spool, with each pass resulting in an ever-diminishing circle. You can see an animated picture of this in action at my web site (*http://www.smithsrus.com/HomeAuto/Mower/index.html*).

My lawn is nearly an acre, with most of the property residing in the backyard. The backyard is mostly a large rectangle with almost no trees. I calculated that cutting two large circles would cover most of the lawn. That leaves only the edges, outside the circle, which I need to cut by hand.

With a 21-inch cutting width, a wire spool with a circumference of 19 inches results in a small overlap for each pass around the circle. That's exactly what you want for this hack. I use an old wire spool, shown in Figure 5-16.

*Figure 5-16. Wire spool and pipe*

The spool has a pipe that runs through the center, which fits into another pipe that's set in place with a small pad of concrete, as shown in Figure 5-17.

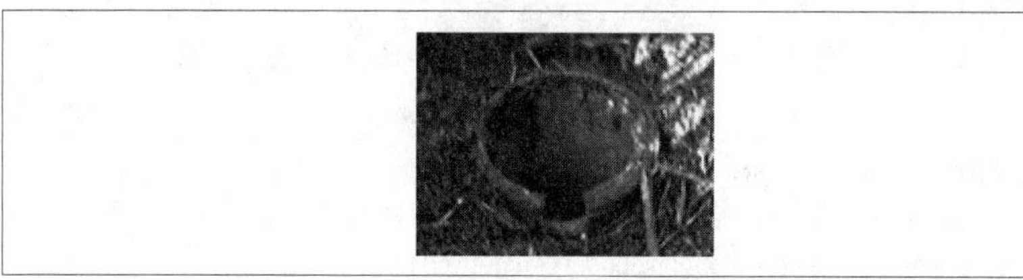

*Figure 5-17. The in-ground pipe*

The pad is sunk low and covered with a layer of dirt and grass to help hide it from sight. Notice the notch in the pipe. The spool pipe has a bolt that fits into this notch to keep the whole spool from spinning in place. I have two of these pipes in my yard: one at the exact center of each mowing circle that I calculated earlier.

For the cable, I use plastic-coated wire that I picked up at the local hardware store. I use one long piece and another shorter extension to make it easy to mow my two different-size circles. I use cable clamps to make loops in the end of the cable so that I easily can connect the two pieces to each other and to the lawnmower.

A screw eye and a spring latch clip quickly attach and release the cable to the mower, as shown in Figure 5-18. I have attachments on both sides of my mower, so I can change the mowing direction to help keep the grass from being matted down in one direction.

The self-propelled drive lever on the mower's handle is held in place by a metal clip that is usually used for hanging brooms on a wall. I keep it tied to the mower handle so that it doesn't get lost while the mower is stowed or fall off and get mowed over if it springs off in use.

*Figure 5-18. A quick-release clip on the mower*

As you can imagine, safety is a big issue here. Run this lawnmower in a fenced yard only. Thoroughly inspect the yard for toys or other objects that could get hit. Do not allow children or pets in the yard while the mower is running. And, most important, never leave the mower unattended.

That's about it. Put the spool in one end of the lawn, let out the cable, and position the mower so that it's somewhat taut. Then, let it run in circles until all the cable is wound up. When it finishes its first run, move it to the other end of the lawn and repeat the process using the second mounting point. All that's left is to trim the outside edges of the circles manually.

## Hacking the Hack

I am also working on a bumper system that will activate the engine kill switch if anything gets in the mower's path. This will have the added benefit of killing the engine when the cable is fully wound on the spool.

*—Doug Smith*

## HACK #64 Get the Weather

**If you have an Internet connection, or your own personal weather station, two Macintosh programs make it a snap to use the information with your home automation system.**

Weather information is a handy thing to integrate with your home automation system. Once you know the current outside temperature and conditions, you can use this information in a variety of ways. Here are just a few ways in which you can use it:

- Adjust your heater or air conditioning system [Hack #41].
- Announce the day's forecast every morning [Hack #47].
- Remind your wife to take a sweater with her when she leaves [Hack #70].
- Skip watering the lawn [Hack #67] if it's raining.

Your first consideration when deciding to use weather information is how precise and up-to-date you need the information to be. You can either rely on data from the Internet or use a personal weather station to measure conditions in your backyard.

Climate information from the National Weather Service typically is updated near the top of every hour. Stations from around the country (and the world) report their conditions to the National Weather Service. Usually, the reporting stations are located at airports, and if your local airport is close enough to you to have similar weather (and you don't mind getting updates only once an hour), this is the easiest route to go.

The National Weather Service data is available free on the Internet. To get the weather for your area, first you need to find the ICAO code for the nearest reporting station. You can look up a station to use at the National Weather Service web site (*http://www.nws.noaa.gov/tg/siteloc.shtml*). The closest station to my home in Muskegon, Michigan, is the Muskegon County Airport. Its code is `KMKG`. To retrieve the current report from this station, I can download a file from *http://weather.noaa.gov/pub/data/observations/metar/decoded/KMKG.TXT*. The results look something like this:

```
Muskegon, Muskegon County Airport, MI, United States (KMKG) 43-10-16N 086-
14-12W 191M
Aug 16, 2004 - 02:55 PM EDT / 2004.08.16 1855 UTC
Wind: from the SSW (210 degrees) at 10 MPH (9 KT):0
Visibility: 10 mile(s):0
Sky conditions: mostly clear
Temperature: 73.0 F (22.8 C)
Dew Point: 57.0 F (13.9 C)
Relative Humidity: 57%
Pressure (altimeter): 30.22 in. Hg (1023 hPa)
ob: KMKG 161855Z 21009KT 10SM FEW043 23/14 A3022 RMK AO2 SLP233 T02280139
cycle: 19
```

Although it would be simple to write an AppleScript that used URL access to download this file, the code necessary to parse out the individual bits of information, such as the current wind direction, is rather complex. It's also fragile because the format of this file isn't strictly consistent. You can get access to a few pieces of information, however.

Another option is to download and parse the station's XML feed. At the time of this writing, the XML feeds are experimental, but if they continue to be available, they'll be much easier to use programmatically. See *http://www.nws.noaa.gov/data/current_obs/* for information about the format of the XML files.

Instead of writing a script to download and parse the weather data, it's much easier to use the inexpensive Mac OS X program WeatherManX (*http://www.afterten.com/products/weatherman/index.shtml*; $10). WeatherManX accesses the National Weather Service for you and displays the result in an attractive manner, as shown in Figure 5-19.

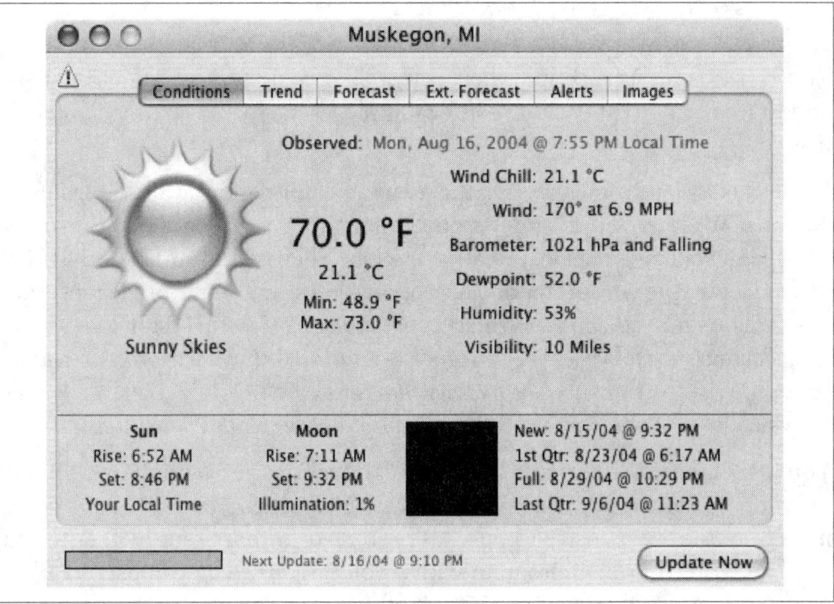


*Figure 5-19. WeatherManX*

Additionally, WeatherManX makes all the weather data available to other programs via AppleScript. This makes it easy to retrieve a 10-day forecast, severe weather alerts, and even satellite or radar imagery without having to spend hours trying to figure out how to parse this information out of raw National Weather Service feeds.

Here's a script for XTension that demonstrates the technique:

```
tell application "WeatherManX"
    set outsideTemp to Current Temperature of City "San Jose" as integer
    set outsideCond to Conditions of City "San Jose" as string
```

```
end tell
Brighten unit "WT Temp" to outSideTemp
Brighten unit "WT Temp" to outsideCond
```

Save this as a global script [Hack #17] and schedule a repeating event that executes it every hour. The script sets the dim level of `WT Temp` to the current outside temperature and sets the unit's description to the current sky conditions. For example, after executing this script, `WT Temp` would be set to `78.3` with a description of `partly cloudy`.

To use this information, you access the properties of the `WT Temp` unit:

```
set conditions to description of "WT Temp"
    if conditions is not "" then
        set sayString to sayString & conditions & " , "
    end if
    say "Outside it is; " & conditions
    say ((value of "WT Temp") as string) & " degrees"
```

You can execute this script when you wake up in the morning [Hack #47] to announce the current weather—for example, "Outside it is partly cloudy, 78 degrees."

If the quality and timeliness of the weather information available from the National Weather Service are not good enough for your location, you should consider installing a personal weather station. They cost from $200 to $1,500, depending on the type of station and options you select. Some of the most popular stations for the consumer market are from La Crosse Technology (*http://www.lacrossetechnology.com*), Oregon Scientific (*http://www.oregonscientific.com*), Davis Instruments (*http://www.davisnet.com/weather/*), and WeatherHawk, a division of Campbell Scientific (*http://www.weatherhawk.com*).

You install these stations outdoors and connect them to a serial or USB port on your computer. Some stations are wireless, which makes them easier to install but more expensive to purchase. To read the data sent by the station, you can write your own program with AppleScript, or you can use WeatherTracker for Mac OS X from AfterTen Software (*http://www.afterten.com*).

The Campbell Scientific WeatherHawk comes with a special version of WeatherTracker for Mac OS X, but the program is also available for use with stations from Oregon Scientific, Davis Instruments, and La Crosse Technology. In addition to displaying data from your weather station, as shown in Figure 5-20, the program creates charts and graphs that you can upload to a web site, and it can report data to Weather Underground (*http://www.weatherunderground.com*) so that others can benefit from your largesse.

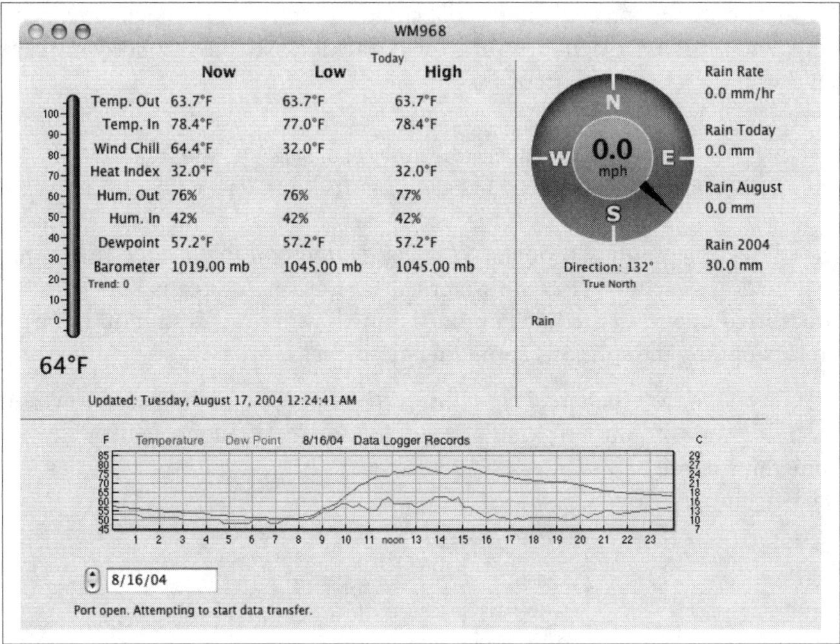


*Figure 5-20. WeatherTracker for Mac OS X*

To integrate WeatherTracker information with your home automation system, you use an approach similar to the one you used with WeatherManX:

```
tell application "WeatherTracker"
    set TheOutsideTemp to Outside Temperature of Station "Vantage Pro 1"
    set TheInsideTemp to Inside Temperature of Station "Vantage Pro 1"
end tell
```

This script retrieves the current outdoor and indoor temperature from a Davis Vantage Pro 1 weather station. This demonstrates a key advantage of having your own weather station. Not only are the outdoor measurements, quite literally, from your own backyard, but most stations come with indoor sensors, too. This can be helpful for determining when to turn on the air conditioning or heating.

*—Dean Davis*

## HACK #65 Safely Water the Garden

**A simple circuit helps prevent sprinklers from turning on accidentally.**

In-ground sprinklers are controlled with electric water valves (24V AC), so it seems like a sprinkler system would be a good candidate for home automation. However, because X10 is not 100% reliable, I was very leery of using it for this purpose. An accidental signal could turn on the sprinklers unexpectedly, and without a signal to turn them back off, serious property damage (and water waste) could occur.

Another way to mitigate the risks of mixing X10 with sprinkler control is to use a dedicated sprinkler controller [Hack #66].

To address these safety problems, I devised a method that notifies my home automation system whenever a water valve turns on, so it can check to see if it occurred due to a scheduled event. If not, all water valves shut off immediately, stopping the spurious sprinkling in its tracks.

Figure 5-21 shows the circuit to control the water valves. It uses a universal module [Hack #11] and a Powerflash module [Hack #10], along with a certain something extra.

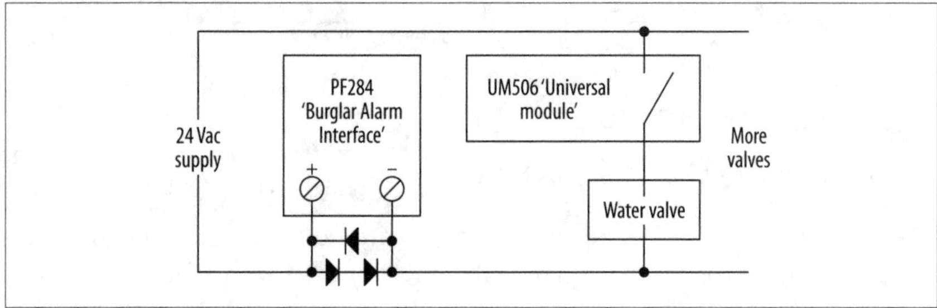


*Figure 5-21. A safer sprinkler circuit*

The secret is that, when the universal module closes its contacts, the diodes register a voltage drop that causes the Powerflash to send its X10 command. Use three 1N4004GICT-ND diodes (*http://www.digikey.com*; 60 cents each), wired as shown in Figure 5-21.

To turn on the sprinklers with your automation system, you need to keep track of whether the sprinklers are expected to currently be on. For example, in HomeSeer [Hack #19], set up a virtual device, named `Sprinklers OK`, that you turn on whenever you send an `On` command to the universal module in the sprinkler circuit. Shortly thereafter, HomeSeer will receive an `On` command from the Powerflash, which confirms that the sprinklers are now on. Set up an event that is activated whenever a signal from the Powerflash is received. This event checks the value of `Sprinklers OK` and, if it's `false`, immediately sends an `Off` command to the sprinkler's universal module, which stops the watering.

## Final Thoughts

After implementing this circuit, I programmed my home automation system to keep track of how many times an inappropriate sprinkler signal was detected. It happened once or twice a month, and each time, the sprinklers were shut down after just a few seconds. Since adding a whole-house X10 signal blocker **[Hack #12]**, however, I have not had any false triggers. So, although this system might be an unnecessary precaution now that I've improved my system's reliability, it certainly continues to provide peace of mind.

Now that your sprinkler system is automated, consider making it smart by incorporating weather data into your scheduling decisions **[Hack #67]**.

*—Edward Cheung*

# HACK #66 Foster Green Pastures with a Smart Sprinkler System

**For the ultimate in automated watering of your lawn and garden, use a dedicated sprinkler controller.**

The invention of zoned sprinkler systems certainly has made watering your lawn and gardens simpler; you get to avoid the hassle of hoses and running outside every 30 minutes to move your sprinkler from one part of the yard to the next. However, although modern sprinkler timers finally have become easier to program than first-generation VCRs, they still leave much to be desired for truly hands-off operation.

You can trace most of these shortcomings to the fact that you cannot specify the amount of water your lawn and gardens need using a simple time duration (e.g., 20 minutes) and frequency (e.g., three times a week). What if it has rained for the last three days in a row? Worse, what if it is raining when your sprinkler is scheduled to run **[Hack #67]**? What if it has been abnormally dry and hot the last several days? When it comes right down to it, you should use several variables, in addition to duration and frequency, when calculating an optimal watering schedule.

By using computer-based home automation control, you can consider these variables and avoid wasteful watering while still keeping your yard healthy. In addition, automating your sprinkler system makes it easier to make a change to its schedule. You'll be able to cancel a watering schedule using your computer or remote control without having to make a trip to the garage or the side of your house. You can use a Palm Pad [Hack #5] to turn the different zones on and off—a real timesaver for periodically checking the sprinkler heads.

## Choosing a Controller

If you have only a single-zone sprinkler system, or not very sophisticated needs, you can use an X10 universal module [Hack #11] to control the zone valve [Hack #65]. For more than one zone, you can use multiple X10 universal modules or an X10 multirelay controller (*http://www.smarthome.com/2310.html*; from $99). But the best approach is a controller that's designed for irrigation control, such as the IrrMaster (*http://www.homecontrols.com/cgi-bin/main/co_disp/displ/prrfnbr/669/sesent/00*; $200) or Rain8 (*http://www.wgldesigns.com/rain8.html*; $100).

Using an irrigation controller is ideal because it has safety features that prevent a valve (zone) from being on for an extended period, even if the X10 command to turn off the zone has not been received for some reason. Similarly, irrigation controllers prevent more than one valve from being turned on at once. This hack focuses on using the Rain8 with Indigo [Hack #18], but you can apply the principles to other controllers and software.

HomeSeer users should visit the HomeSeer message board (*http://ubb.HomeSeer.com*) to learn about the Rain8 plug-in for their systems.

The Rain8 comes in several different models, some with special features for operating pumps, controlling drip irrigation, or even allowing extended watering times for orchard irrigation. We'll use the basic Rain8B that includes the X10 power-line interface needed to receive X10 commands. If you have a spare power-line interface (such as a PSC05 or TW523), you can get the basic Rain8 model and save about $20. If you need to control more than eight zones, look at the bundles (Rain8B2, Rain8B3, Rain8B4) that support up to 24 zones. Macintosh users will want the Rain8MB, which allows easier configuration of the house code because Mac computers don't have the standard serial port necessary for configuring the Rain8 hardware.

## Setting Up the Hardware

The Rain8 is easy to connect to your sprinkler system. Simply use it to replace the conventional sprinkler timer you already have—connect the valve wires, common wire, and 24 AC power (the adapter is included with the Rain8). Additionally, you plug in a power-line interface (a PSC05 is included in the Rain8B), which passes the X10 commands received from the power line to the Rain8.

The Rain8 operates in two modes: the *automatic zone programmed* mode, which is useful when you are not using a computer-based home automation system, and the *manual computer-controlled* mode for when you have your computer handle the logic and scheduling of all the zone actions. Because we are using Indigo home control software, we will use the latter mode.

The Rain8 has a serial port connector, which is used to connect your computer to the Rain8 for configuration. In the manual computer-controlled mode, you use your computer to set the house code for the Rain8 and the maximum time the valve will be on for each zone. Mac users, who won't have a serial port connection unless they have a USB-to-serial adapter, can configure the Rain8MB using the Indigo Rain8M Set Address script (*http://www.perceptiveautomation.com/hackbook/rain8setaddr.dmg*; free).

Once configured, the Rain8 claims all the addresses in its house code **[Hack #1]**, depending on the number of zones you have. For example, if you set the Rain8 to house code `I`, you control zone `1` by sending `I1 On` or `I1 Off`. When you turn on a different zone—say, `I2 On`—zone 1 is turned off automatically.

## Scheduling Irrigation

To install the Indigo script for the Rain8 (*http://www.perceptiveautomation.com/hackbook/sprinkler.dmg*; free), copy the *sprinkler attachment.scpt* file into your *~/Documents/Indigo User Data/Scripts/Attachment/* folder and then select Reload Attachments from the Scripts menu in Indigo. This attachment script **[Hack #88]** defines two AppleScript functions, `DoSprinkler()` and `CancelSprinkler()`, which you call from a time/date action scheduled to water your yard at specific intervals.

`DoSprinkler()` takes two arguments, the X10 address of the Rain8 and a list of integers that define the zone watering durations in minutes. For example, the following command schedules the Rain8 that is set to address `F1` to water for 20 minutes on zone 1, 20 minutes on zone 2, 15 minutes on zone 3, and so on:

```
DoSprinkler("F1", {20,20,15,15,10,10})
```

The list of durations can contain from one to eight integers, and you can skip zones by specifying a `0` value. For example, this command waters zone 3 for 12 minutes and zone 5 for 5 minutes:

```
DoSprinkler("F1", {0,0,12,0,5 })
```

Multiple Rain8s are supported; just call the function with the base address assigned to the Rain8 you want to control.

When you call the function, the attachment script automatically creates the appropriate time/date actions in Indigo to handle turning each zone on and off. As each subsequent zone is turned on, the time/date actions delete themselves automatically.

If you want to cancel a watering schedule already in progress, call `CancelSprinkler()` with the Rain8's base address:

```
CancelSprinkler("F1")
```

Now, we have all the pieces to water the lawn automatically. All that's left is to create a schedule for the days on which we want to water. To do this, create a time/date action that executes a `DoSprinkler()` command at the desired times. Figure 5-22 illustrates an event that waters 30 minutes before sunrise every Monday, Thursday, and Saturday.

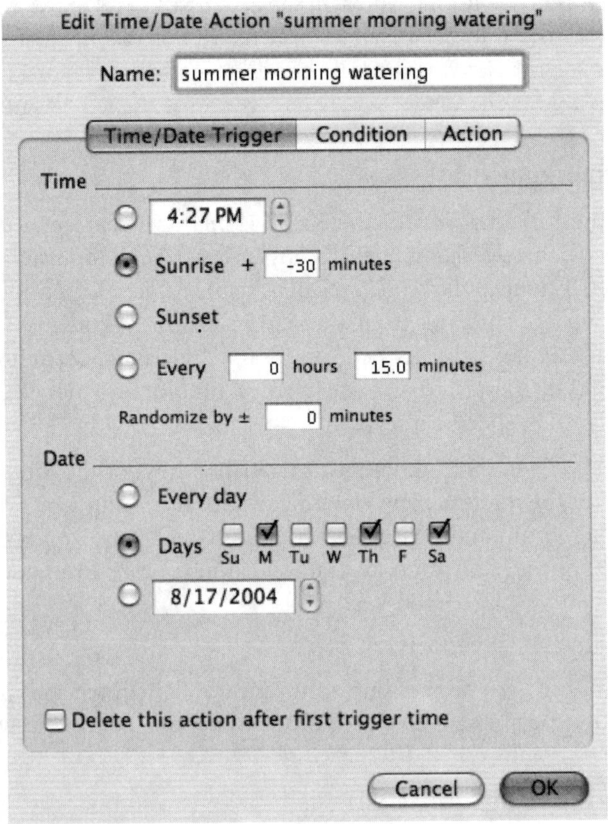

*Figure 5-22. A watering event*

The action for this event, shown in Figure 5-23, schedules watering times for all eight zones.

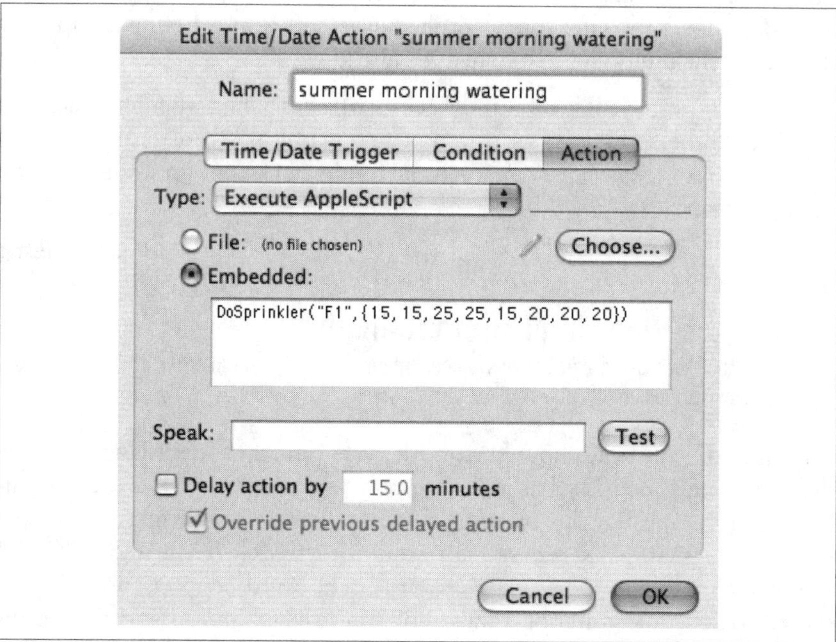


*Figure 5-23. Setting the watering durations*

From the main window in Indigo, you can use the enable checkbox to temporarily suspend time/date actions. You can create multiple watering schedules for different seasons and watering conditions and then suspend all time/date actions except the ones you want active; just use the Enable checkbox next to the action in the main Indigo window.

## Hacking the Hack

The attachment script automatically maintains two Indigo variables that can be useful for enhancing your watering routines. The `sprinklerF1_CurrentZone` variable (where `F1` is the address of the Rain8) indicates which zone is currently being watered. You might use this information in conjunction with window sensors **[Hack #78]** to make sure windows in the vicinity of the sprinklers are closed.

The `sprinklerDurationMulti` variable provides a way for you to adjust the watering durations dynamically. The value of this variable is used as a multiplier when the attachment script schedules the watering events. It's initially set to `1`, but if you change it to `2.5`, for example, the cycle for each zone will be 2.5 times longer than specified in the `DoSprinkler()` command. If you maintain this variable using weather information, you'll have a sprinkler system that knows to water longer on hot days and skip watering completely (by setting the multiplier to `0`) when it's already raining.

To consider the current weather conditions when deciding whether to activate or schedule the sprinklers, see "Stop Watering During Rainstorms" [Hack #67] and "Adapt Sprinkler Schedules and Solar Water Heating to Available Sunlight" [Hack #68].

*—Matt Bendiksen*

## HACK #67 Stop Watering During Rainstorms

**Macintosh users can utilize weather information from the Internet to adjust lawn sprinkler schedules.**

This hack demonstrates how Macintosh users can adjust their lawn and garden watering schedules based on current and forecasted weather conditions. It accomplishes this by using three AppleScripts. The first two scripts use the WeatherManX [Hack #64] application to track current and historical weather conditions with Indigo [Hack #16]. The third script is used within Indigo to schedule watering times. You'll also need an X10-capable sprinkler controller, such as WGL & Associates' Rain8 (*http://www.wgldesigns.com/rain8.html*; $100).

As you might imagine, anything dealing with the murky art of weather prediction can get somewhat complex. Even as full-featured as these scripts are, their effectiveness is limited by the timeliness and accuracy of your weather data. WeatherManX (*http://www.afterten.com/products/weatherman/index.shtml*; $10) makes it easy to download the hourly weather from your closest U.S. National Weather Service, but that information is less than perfect. It's updated only once an hour, and if the closest station to you isn't nearby, it might not give an accurate snapshot of the weather at your house. See the "Hacking the Hack" section for information about alternative approaches.

### Feeding Indigo the Weather

Let's begin with the script that gathers data from WeatherManX and stores it in Indigo variables [Hack #18]. Set up a scheduled event that executes this script every hour; there's no need to run it more frequently, because the weather data is updated only once an hour as well.

The first portion simply loads the current weather information into variables. You'll need to change the value of WMcityname to the station name from which you're retrieving data, and that station must be activated already in WeatherManX's City Manager (choose Show City Manager from the Cities menu in WeatherManX):

```
property WMcityname : "Albuquerque, NM"
set StartTime to (current date)
set OldDelims to AppleScript's text item delimiters
set AppleScript's text item delimiters to ","
tell application "WeatherManX"
    set the updatetime to the Update Time of City WMcityname
    set the CondtionsTime to the Conditions Time of City WMcityname
    set the Temperature to the Current Temperature of City WMcityname
    set the Barometer to the Current Pressure of City WMcityname
    set the PressureTrend to the Pressure Trend of City WMcityname
    set the WindSpeed to the Wind Speed of City WMcityname as real
    set the WindGust to the Wind Gust of City WMcityname
    set the WindDirection to the Wind Direction of City WMcityname
    set the theHumidity to the Humidity of City WMcityname
    set the windchill to the Wind Chill of City WMcityname
    set the heatindex to the Heat Index of City WMcityname
    set the theDewPoint to the Dewpoint of City WMcityname
    set the theConditions to the Conditions of City WMcityname
    set the theWeather to the Weather of City WMcityname
    set the theVisibility to the (Visibility of City WMcityname) as integer
    set the windlongdir to the Long Compass Wind Direction of City
WMcityname
    set the windshortdir to the Short Compass Wind Direction of City
WMcityname
    set the forecastlist to Forecast of City WMcityname
end tell
```

Next, the weather information is stored in Indigo variables. This is a lengthy portion of the script because of the extra steps necessary to create the variable in Indigo if it doesn't exist already—a good practice to ensure your scripts are reliable:

```
tell application "Indigo"
    --create variables if they don't exsit
    if not (variable "weather_last_update" exists) then make new variable
with properties {name:"weather_last_update", value:" "}
    if not (variable "weather_temperature" exists) then make new variable
with properties {name:"weather_temperature", value:"0"}
    if not (variable "weather_barometer" exists) then make new variable with
properties {name:"weather_barometer", value:"0"}
    if not (variable "weather_pressure_trend" exists) then make new variable
with properties {name:"weather_pressure_trend", value:"0"}
    if not (variable "weather_wind_speed" exists) then make new variable
with properties {name:"weather_wind_speed", value:"0"}
    if not (variable "weather_ave_wind_speed" exists) then make new variable
with properties {name:"weather_ave_wind_speed", value:"0,0,0,0,0,0"}
```

```
    if not (variable "weather_wind_gust" exists) then make new variable with
properties {name:"weather_wind_gust", value:"0"}
    if not (variable "weather_wind_direction" exists) then make new variable
with properties {name:"weather_wind_direction", value:"0"}
    if not (variable "weather_humidity" exists) then make new variable with
properties {name:"weather_humidity", value:"0"}
    if not (variable "weather_wind_chill" exists) then make new variable
with properties {name:"weather_wind_chill", value:"0"}
    if not (variable "weather_heat_index" exists) then make new variable
with properties {name:"weather_heat_index", value:"0"}
    if not (variable "weather_dewpoint" exists) then make new variable with
properties {name:"weather_dewpoint", value:"0"}
    if not (variable "weather_current_conditions" exists) then make new
variable with properties {name:"weather_current_conditions", value:" "}
    if not (variable "weather_visibility" exists) then make new variable
with properties {name:"weather_visibility", value:"0"}
    if not (variable "weather_long_compass_wind_direction" exists) then make
new variable with properties {name:"weather_long_compass_wind_direction",
value:"0"}
    if not (variable "weather_short_compass_wind_direction" exists) then
make new variable with properties {name:"weather_short_compass_wind_
direction", value:"0"}
    if not (variable "weather_rain_count" exists) then make new variable
with properties {name:"weather_rain_count", value:"0,0,0,0,0,0"}
    if not (variable "weather_max_temps" exists) then make new variable with
properties {name:"weather_max_temps", value:"0,0,0,0,0,0"}
    if not (variable "weather_min_temps" exists) then make new variable with
properties {name:"weather_min_temps", value:"800,800,800,800,800,800"}
    if not (variable "weather_count_of_daily_updates" exists) then make new
variable with properties {name:"weather_count_of_daily_updates", value:0}
    if SubDate(value of variable "weather_last_update") is
SubDate(CondtionsTime) or CondtionsTime is "" then
        --no updated needed
    else
        --set the weather variables
        set value of variable "weather_temperature" to temp
        set value of variable "weather_barometer" to baro
        set value of variable "weather_pressure_trend" to PressureTrend
        set value of variable "weather_wind_speed" to WindSpeed
        set value of variable "weather_wind_gust" to WindGust
        set value of variable "weather_wind_direction" to WindDirection
        set value of variable "weather_humidity" to theHumidity
        set value of variable "weather_wind_chill" to windchill
        set value of variable "weather_heat_index" to heatindex
        set value of variable "weather_dewpoint" to theDewPoint
        if theWeather = "" then
            set value of variable "weather_current_conditions" to
theConditions
        else
            set value of variable "weather_current_conditions" to
theConditions & " skies with " & theWeather
        end if
```

```
        --set value of variable "weather_current_conditions" to wmCond
        set value of variable "weather_visibility" to vis
        if windlongdir is not "" then
            set value of variable "weather_long_compass_wind_direction" to
windlongdir
            set value of variable "weather_short_compass_wind_direction" to
windshortdir
        end if
```

The text of the current conditions is examined to see if it includes any reference to rainy weather, and if it does, the time is saved in weather_last_rain. A running counter that keeps track of how many rain reports have been noted today, weather_rain_count, is incremented:

```
    if ((theConditions contains "Rain") or (theConditions contains "Mist")
or (theConditions contains "Drizzle") or (theConditions contains "Snow") or
(theConditions contains "Sleet") or (theConditions contains "Thunderstorm")
or (wether contains "Rain") or (wether contains "Mist") or (wether contains
"Drizzle") or (wether contains "Snow") or (wether contains "Sleet") or
(wether contains "Thunderstorm")) and (theConditions does not contain
"Clear") then
            set value of variable "weather_last_rain" to
(SubDate(conditionstime)) as string
            set x to (value of variable "weather_rain_count")
            set theList to text items of x
            set item 1 of theList to (item 1 of theList) + 1
            set (value of variable "weather_rain_count") to theList as
string
        end if
```

Other daily statistics are updated, too, such as the low and high temperatures and average wind speed:

```
set x to (value of variable "weather_max_temps") as string
        set theText to the text items of x
        if the Temperature > item 1 of theText then
            set item 1 of theText to the temp
            set value of variable "weather_max_temps" to theText as string
        end if

        set VarVal to value of variable "weather_min_temps"
        set theText to the text items of VarVal
        if the Temperature < item 1 of theText then
            set item 1 of theText to the temp
            set value of variable "weather_min_temps" to theText as string
        end if
        --Set average wind speeds
        set VarVal to value of variable "weather_ave_wind_speed"

        set theText to the text items of VarVal
        set TempVar to item 1 of theText
        set item 1 of theText to (wind_speed + TempVar) as string
        set value of variable "weather_ave_wind_speed" to theText as string
```

```
        set (value of variable "weather_count_of_daily_updates") to (value
of variable "weather_count_of_daily_updates") + 1
```

Instead of trying to decide if the weather forecast text has changed, the latest version is simply stored in the appropriate variables. This simple approach works well because there's no need to maintain a historical record of previous forecasts; you want only the latest one.

```
        --go through all variable and delete variables related to forecast
        set i to 1
        repeat (count of variables) times
            if name of variable i contains "weather_forecast" then
                delete variable i
            else
                set i to i + 1
            end if
        end repeat
        --create variable for the forecast
        repeat with dayforecast in forecastlist
            set dayStr to item 1 of dayforecast
            set forecastStr to item 2 of dayforecast
            set varname to searchReplace("weather forecast " & dayStr, " ",
"_")
            --log "varname: " & varname & ", forcastStr: " & forecastStr
            if not (variable varname exists) then make new variable with
properties {name:varname, value:forecastStr}
        end repeat
        set AppleScript's text item delimiters to OldDelims
        set value of variable "weather_last_update" to
SubDate(CondtionsTime) as string
        log "Weather script completed in " & (current date) - StartTime & "
seconds."
    end if
end tell
```

Finally, these subroutines provide functions used throughout the script. The first is for date comparisons, and the second is for parsing information from the text-based data:

```
on SubDate(TheString)
    return (date TheString) as date
end SubDate
on searchReplace(theText, SearchString, ReplaceString)
    set OldDelims to AppleScript's text item delimiters
    set AppleScript's text item delimiters to SearchString
    set newText to text items of theText
    set AppleScript's text item delimiters to ReplaceString
    set newText to newText as text
    set AppleScript's text item delimiters to OldDelims
    return newText
end searchReplace
```

## Cleaning Up Nightly

The second script runs every night at midnight, triggered by an Indigo scheduled event [Hack #18]. It takes the day's accumulated weather data—high temperatures, wind speeds, rain counts, and so on—and transfers them to the next day's variables. This gives longer-term data from which to calculate the figures used in deciding the sprinkler's schedule.

```
using terms from application "Indigo"
    on NightlyWeatherCalculations( )
        tell application "Indigo"

            set OldDelims to AppleScript's text item delimiters
            set AppleScript's text item delimiters to ","

            set VarVal to value of variable "weather_max_temps"
            set theText to the text items of VarVal
            repeat with i from 5 to 1 by -1
                set item (i + 1) of theText to item i of theText
            end repeat
            set item 1 of theText to 0
            set value of variable "weather_max_temps" to (theText as string)

                      set VarVal to value of variable "weather_min_temps"
            set theText to the text items of VarVal
            repeat with i from 5 to 1 by -1
                set item (i + 1) of theText to item i of theText
            end repeat
            set item 1 of theText to 800
            set value of variable "weather_min_temps" to (theText as string)

            --Set average wind speeds
            set Dividor to (value of variable "weather_count_of_daily_
updates") as integer

            set VarVal to value of variable "weather_ave_wind_speed"
            set theText to the text items of VarVal
            set TempVar to item 1 of theText
            if TempVar is not 0 and Dividor is not 0 then set item 1 of
theText to (TempVar / Dividor)
            set item 1 of theText to ((item 1 of theText) * 100) as integer
            set item 1 of theText to ((item 1 of theText) / 100)
            repeat with i from 5 to 1 by -1
                set item (i + 1) of theText to item i of theText
            end repeat
            set item 1 of theText to 0

            set value of variable "weather_ave_wind_speed" to (theText as
string)

            set (value of variable "weather_count_of_daily_updates") to 0
```

```
            set theText to value of variable "weather_rain_count"
            set theList to ParseList(theText)
            repeat with i from 5 to 1 by -1
                set item (i + 1) of theList to item i of theList
            end repeat
            set item 1 of theList to 0
            repeat with i from 1 to (length of theList)
                if i is 1 then
                    set VarVal2 to item i of theList
                else
                    set VarVal2 to VarVal2 & item i of theList
                end if
            end repeat
            set value of variable "weather_rain_count" to VarVal2 as string

            set AppleScript's text item delimiters to OldDelims
        end tell
    end NightlyWeatherCalculations
end using terms from
```

## Scheduling the Sprinklers

Now that the data is in Indigo, you can use it in a variety of ways when evaluating when to water the lawn or garden, or for how long the water should stay on. For example, you could increase the watering time by factoring in the day's high temperature. You could set a base watering time, `IrrWaterTime`, to 40 minutes a day, and then set a small factor of the daily high. Using the following script excerpt, if the day's high temperature is 72 degrees Fahrenheit, watering time is extended by 7.2 minutes:

```
set the ave_temp to (value of variable " weather_max_temps ") as integer
set IrrWaterTime to IrrWaterTime + (ave_temp * 0.1)
```

You also might avoid watering if the current temperature is too low:

```
if (value of variable "weather_temperature" as  integer) < 50 then
    set IrrWaterTime to 0
end if
```

Or if the `weather_rain_count` variable, set by the WeatherManX script described earlier, indicates that at least two current condition reports have indicated precipitation, you can skip watering, too:

```
If (value of variable weather_rain_count as integer) > 2 then
    set IrrWaterTime to 0
end if
```

For more examples of weather-related calculations, including how to calculate a rain factor based on the week's history, see the extensive Indigo sprinkler script at my web site (*http://homepage.mac.com/gregjsmith/sprinkler.html*).

Similar calculations can take into account wind speed or direction, as well as the week's average temperature. Once you have adjusted `IrrWaterTime` appropriately, use the attachment script described in "Foster Green Pastures with a Smart Sprinkler System" **[Hack #66]** to create the watering schedules. In this example, the Rain8 unit with the X10 address `F1` will water zones 1 and 2 for the value of `IrrWaterTime`, and zone 3 will not be watered:

```
DoSprinkler("F1", {IrrWaterTime,IrrWaterTime,0})
```

## Hacking the Hack

If you want to obtain more reliable weather data, consider getting your own weather station. See "Get the Weather" **[Hack #64]** for Mac-friendly options. If all you're interested in is rain, however, you might use a Mini-Click Rain Sensor (*http://www.smarthome.com/7194.HTML*; $25) with a Powerflash module **[Hack #10]** to keep track of the current day's rainfall.

If you don't have a Rain8 or another dedicated sprinkler controller, consider using X10 modules to control your sprinkler system **[Hack #65]**.

In addition to being useful for sprinkler automation, weather conditions can inform other automation decisions, such as when to supplement solar-powered heating **[Hack #68]**.

—*Greg Smith*

# HACK #68 Adapt Sprinkler Schedules and Solar Water Heating to Available Sunlight

**If you have a solar-assisted water heater, or if you want to adapt your garden watering schedule to the day's conditions, this hack will get you started in the right direction.**

This hack creates an irrigation and water-heating schedule based on the weekly forecast. Like most algorithm-based control programs, the most complicated issue is that of defining thresholds—that is, deciding how the values affect the decisions of when to irrigate, turn on the water heater, and so on. It took a few trials and errors, as well as constructive criticism from family members **[Hack #14]**, to arrive at the values that work for me.

The system incorporates the five-day weather forecasts into its algorithm for deciding on irrigation and solar water-heating control. You can manage irrigation schedules and solar water-heating control more efficiently by considering the five-day weather forecast. This makes sense because both are impacted by the weather conditions. For example, passive solar panels need to be supplemented with an electric heating element when it's cloudy, but when the sun is out, the water is heated naturally via the solar panel and the electrical heater can be shut down. Similarly, irrigation is redundant when rain is expected, and it needs to be turned on only during dry spells.

Other factors that affect irrigation and water heating are the length of the day and the time of sunrise and sunset. For example, to minimize evaporation, it's best to irrigate lawns just before sunrise. For the solar panel, longer days mean additional solar energy can be harvested for heating the water.

If you're primarily interested in knowing if it has rained recently so that you don't water unnecessarily, see "Stop Watering During Rainstorms" **[Hack #67]**, particularly if you're a Macintosh user.

Here's what you'll need:

*A water heater with a passive solar panel*
: These are common here in Israel (in fact, they're mandatory on new homes). The main components are a boiler with an electric heater element and a passive solar panel. Copper pipes inside the panel conduct water and heat up the water as light is absorbed by the dark sheeting on the panel. The hot water rises up the panel and into the boiler, and cold water enters from the bottom of the panel. The system is quite efficient and easily can heat up a 75-liter boiler in about four hours on a sunny day.

*An irrigation system*
: I use a five-zone system. Each zone is controlled with a 24V AC valve, which in turn are controlled using X10 **[Hack #65]**.

*A computer with a connection to the Internet*
: Once a day the computer connects to the Internet and downloads the five-day forecast. I use AccuWeather.com (*http://wwwa.accuweather.com*) to retrieve a forecast for my city, Tivon, Israel, and I've found its three-day forecast to be quite accurate. The forecast is used to calculate a hot water and sprinkler schedule for the next five days.

## The Hot Water Schedule

The heuristic for calculating the hot water heater schedule is relatively simple:

- If it is summer (by month), never turn on the electric hot water heater. Here in Israel, it's never cloudy, so the water heated by the solar panel is sufficient during these months.
- If it is winter, always turn on the hot water heater because even during warmer days, the sun's azimuth is too low for efficient heating of the solar panel. Also, low air temperature and wind cool the exposed panel

so much that I found electrical heating is always necessary during these months.

- Additionally, because I know it takes 90 minutes to bring the water to my preferred temperature, I have my automation system anticipate when hot water will be needed and schedule the start times accordingly. I also take into account that I'll need the hot water at different times on weekdays, weekends, and vacation days, so there's still plenty to consider when deciding on a hot water schedule, even in the winter, when temperature isn't a factor.
- If it is spring or autumn, the text of the forecast is checked for specific keywords, such as *cloudy*, *sunny*, *rain*, *showers*, and so on. Also, each day's high and low temperatures are extracted, and the length of each day is calculated using the sunrise and sunset times.

## The Irrigation Schedule

The heuristic of irrigation control is more complex because it involves additional parameters and threshold conditions. The algorithm looks at the current day and at the entire weekly forecast. Here's the approach:

- Calculate an irrigation coefficient based on the number of forecasted rainy days, cloudy days, and sunny days. This coefficient is used to speculate how many days of irrigation are needed and when.
- Calculate the length of each day (measured from the sunrise azimuth to the sunset azimuth) so that more water can be dispensed on longer (summer) days.
- Calculate the irrigation start time so that it occurs 90 minutes before sunrise to minimize evaporation from the heat.
- Calculate the watering time for each sprinkler zone, based on the plants in the area it covers. The lawn is watered for longer than, for example, the vegetable bed.

## The Code

This code demonstrates the heuristics discussed in the previous sections, but it is only an illustration of how to approach the problem and implement your own threshold and logic. You must tailor it to your site to get the results you want. It's written in Visual Basic for use with a home automation program that I've written for my own use (shown in Figure 5-24), so you can't just cut and paste the code. But it will get you started, and that's all it is intended to do.

| | MON | TUE | WED | THU | FRI |
|---|---|---|---|---|---|
| Irr: OFF | A good deal of sun | A moonlit sky and chilly | Warm with a good deal of sun | Partly sunny and warm | Abundant sunshine and hot |
| High | 23°C | 23°C | 25°C | 28°C | 31°C |
| Low | 7°C | 9°C | 13°C | 16°C | 14°C |
| Spr Top 0 | | | | | |
| Spr Mid 6 | | | | | |
| Spr Bot 6 | | | | | |
| Drippers | | | | | |
| Boiler | | | | | |

*Figure 5-24. The forecast and calculated events*

I'm not a professional programmer, so please don't contact me with questions about coding and style.

Here's the procedure that implements the algorithms discussed earlier. To begin, irrigation and boiler events that are already scheduled are removed. This is an important step to remember when you implement your own system, particularly if you'll be polling for new weather data more than once a day.

```
Public Sub IrrigationAndHotWater( )
RemoveEvent "Irrigation"
RemoveEvent "Boiler"
```

This code segment schedules irrigation based on the current month:

```
Select Case Month(Now)
Case 6, 7 'Summer: Irrigation every day, HW boiler off  During June, July -
irrigate every day:
For Wkday = 1 To 7
ScheduleDrippers (Wkday) 'Drip every day
Next
ScheduleBotSprinklers (vbMonday)  'sub to irrigate turn bottom sprinklers
zone on Monday, Wed, Sat
ScheduleBotSprinklers (vbWednesday)
ScheduleTopSprinklers (vbSaturday)
ScheduleMidSprinklers (vbSaturday)  'similarly, turn middle sprinklers zone
on Saturday etc.
ScheduleBotSprinklers (vbSaturday)
Case 8 'Summer: Irrigation every day, HW boiler off  'On August, the hottest
month here, do some extra irrigation
For Wkday = 1 To 7
ScheduleDrippers (Wkday) 'Drip every day
Next
ScheduleTopSprinklers (vbSaturday)
```

```
ScheduleTopSprinklers (vbWednesday)
ScheduleMidSprinklers (vbSaturday)
ScheduleMidSprinklers (vbWednesday)
For Wkday = 1 To 5
If DaysForecast(Wkday).Hi > 32 Or _
DaysForecast(Wkday).Day = "2" Or _
DaysForecast(Wkday).Day = "4" Or _
DaysForecast(Wkday).Day = "7" Then
ScheduleBotSprinklers (DaysForecast(Wkday).Day)
End If
Next
```

During December, January, and February, turn on the boiler every day for which the forecast contains the word *rain*, *shower*, *cold*, or *cloud*. If the forecast contains *hot*, *warm*, or *pleasant*, the boiler is not turned off, as the solar panel takes over on those days:

```
Case 12, 1, 2 'Winter: HW on every day, irrigation off

For Wkday = 1 To 5 'For each day forecasted:

If DaysForecast(Wkday).Hi < 10 Or _
CBool(InStr(1, DaysForecast(Wkday).Condition, "rain",
vbTextCompare)) Or _
CBool(InStr(1, DaysForecast(Wkday).Condition, "shower",
vbTextCompare)) Or _
CBool(InStr(1, DaysForecast(Wkday).Condition, "cold",
vbTextCompare)) Or _
CBool(InStr(1, DaysForecast(Wkday).Condition, "cloud",
vbTextCompare)) And Not _
CBool(InStr(1, DaysForecast(Wkday).Condition, "warm",
vbTextCompare)) And Not _
CBool(InStr(1, DaysForecast(Wkday).Condition, "hot",
vbTextCompare)) And Not _
DaysForecast(Wkday).Hi > 25 And Not _
CBool(InStr(1, DaysForecast(Wkday).Condition, "pleasant",
vbTextCompare)) Then 'If its cold, raining or cloudy:
ScheduleHotWater DaysForecast(Wkday).Day 'schedule the hot water for this
day
End If
Next
' For Wkday = 1 To 7
' ScheduleHotWater (Wkday)
' Next
Case Else 'Check weather. Base irrigation and HW on weather
'The following code creates a schedule for hot water boiler and lawn
irrigation
'based on the weather data

'First, the HW boiler:
NumOfRainDays = 0 'used for irrigation algorithm.
NumOfCloudyDays = 0
```

```
For Wkday = 1 To 5 'For each day forecasted:
If DaysForecast(Wkday).Hi < 10 Or _
CBool(InStr(1, DaysForecast(Wkday).Condition, "rain",
vbTextCompare)) Or _
CBool(InStr(1, DaysForecast(Wkday).Condition, "shower",
vbTextCompare)) Or _
CBool(InStr(1, DaysForecast(Wkday).Condition, "cold",
vbTextCompare)) Or _
CBool(InStr(1, DaysForecast(Wkday).Condition, "cloud",
vbTextCompare)) And Not _
CBool(InStr(1, DaysForecast(Wkday).Condition, "warm",
vbTextCompare)) And Not _
CBool(InStr(1, DaysForecast(Wkday).Condition, "hot",
vbTextCompare)) And Not _
DaysForecast(Wkday).Hi > 25 And Not _
CBool(InStr(1, DaysForecast(Wkday).Condition, "pleasant",
vbTextCompare)) Then 'If its cold, raining or cloudy:
ScheduleHotWater DaysForecast(Wkday).Day 'schedule the hot
water for this day
End If
If CBool(InStr(1, DaysForecast(Wkday).Condition, "rain",
vbTextCompare)) Or _
CBool(InStr(1, DaysForecast(Wkday).Condition, "shower",
vbTextCompare)) Then
NumOfRainDays = NumOfRainDays + 1
End If
If CBool(InStr(1, DaysForecast(Wkday).Condition, "cloud",
vbTextCompare)) Then
NumOfCloudyDays = NumOfCloudyDays + 1
End If
If DaysForecast(Wkday).Hi > 27 Then
NumOfCloudyDays = NumOfCloudyDays - 1
End If
Next
```

Here is a formula I found useful if it is expected to rain. The number of rain days in the five-day forecast, multiplied by 3, plus the number of cloudy days reduces the amount of irrigation needed:

```
IrrigationCoefficient = NumOfRainDays * 3 + NumOfCloudyDays
'Number ranges between 0 (no rain, no clouds) and 15 (five days
of rain).
'Inbetween we have a mix of rainy days and cloudy days.
'3 cloudy days are equivalent to 1 rain day.

Select Case IrrigationCoefficient
Case Is <= 0 'Clear, dry.
For Wkday = 1 To 5
ScheduleDrippers (DaysForecast(Wkday).Day)
If DaysForecast(Wkday).Day = vbSaturday Then 'Sprinklers
ON Saturday
ScheduleTopSprinklers (vbSaturday)
ScheduleMidSprinklers (vbSaturday)
End If
```

```
If DaysForecast(Wkday).Day = vbMonday Or
DaysForecast(Wkday).Day = vbSaturday _
Or DaysForecast(Wkday).Hi > 30 Then
ScheduleBotSprinklers (DaysForecast(Wkday).Day)
End If
Next
Case 1, 2, 3
' dry to cloudy
For Wkday = 1 To 5
If Not CBool(InStr(1, DaysForecast(Wkday).Condition,
"cloud", vbTextCompare)) Or _
CBool(InStr(1, DaysForecast(Wkday).Condition, "hot",
vbTextCompare)) Then 'no clouds
ScheduleDrippers (DaysForecast(Wkday).Day)
If DaysForecast(Wkday).Day = vbSaturday Then
'Sprinklers ON Saturday
ScheduleTopSprinklers (vbSaturday)
ScheduleMidSprinklers (vbSaturday)
ScheduleBotSprinklers (vbSaturday)
End If
End If
If Not CBool(InStr(1, DaysForecast(Wkday).Condition,
"cloud", vbTextCompare)) Or _
CBool(InStr(1, DaysForecast(Wkday).Condition,
"hot", vbTextCompare)) Or _
DaysForecast(Wkday).Hi > 30 Then
ScheduleBotSprinklers (DaysForecast(Wkday).Day)
End If
Next

Case 4, 5, 6, 7, 8
' 1 day rainy and others cloudy
For Wkday = 1 To 5
If Not CBool(InStr(1, DaysForecast(Wkday).Condition,
"Rain", vbTextCompare)) And _
Not CBool(InStr(1, DaysForecast(Wkday).Condition,
"Shower", vbTextCompare)) And _
Not CBool(InStr(1, DaysForecast(Wkday).Condition, "cold",
vbTextCompare)) And _
Not CBool(InStr(1, DaysForecast(Wkday).Condition,
"cloud", vbTextCompare)) Then
If DaysForecast(Wkday).Day = vbSaturday Then
'Sprinklers ON Saturday
ScheduleTopSprinklers (vbSaturday)
ScheduleMidSprinklers (vbSaturday)
ScheduleBotSprinklers (vbSaturday)
End If
If DaysForecast(Wkday).Day = vbMonday Then
'Sprinklers ON Saturday
ScheduleBotSprinklers (vbMonday)
End If
```

```
If DaysForecast(Wkday).Day = vbWednesday Then
'Sprinklers ON Saturday
ScheduleBotSprinklers (vbWednesday)
End If
End If
Next

'Other cases have enough rain
End Select
End Select
' Now add manual irrigation
For Wkday = vbSunday To vbSaturday
If CBool((Val(GetSetting("Homer", "Forecast", "Add" &
CWeekDay(Wkday))) And 1) = 1) Then ScheduleTopSprinklers (Wkday)
If CBool((Val(GetSetting("Homer", "Forecast", "Add" &
CWeekDay(Wkday))) And 2) = 2) Then ScheduleMidSprinklers (Wkday)
If CBool((Val(GetSetting("Homer", "Forecast", "Add" &
CWeekDay(Wkday))) And 4) = 4) Then ScheduleBotSprinklers (Wkday)
If CBool((Val(GetSetting("Homer", "Forecast", "Add" &
CWeekDay(Wkday))) And 8) = 8) Then ScheduleDrippers (Wkday)
If CBool((Val(GetSetting("Homer", "Forecast", "Add" &
CWeekDay(Wkday))) And 16) = 16) Then ScheduleHotWater (Wkday)
Next Wkday
End Sub
```

## Final Thoughts

I've been using this system since 1997 with few changes over the years. It's working surprisingly well, and at times, if I half-shut my eyes, it almost looks intelligent.

For example, one time I went out in the morning and had a quick look at the garden and thought it could use irrigation. I meant to look into it in the evening. At noon, a storm came in and the garden got soaked. Unlike me, the system saw this storm coming and held off irrigation.

—*Ido Bar-Tana*

CHAPTER SIX

# Security

## Hacks 69–82

Home security is the primary reason most people get started with home automation, and for good cause. A smart home can be much better than a simple alarm system when it comes to preventing burglaries. The hacks in this chapter will help you make your home look occupied while you're away [Hack #72], scare away prowlers with an electronic dog [Hack #79], and keep watch on your house, or pets, using network cameras [Hack #82].

## HACK #69 Check for an Empty Home

**If your home is unoccupied, your home automation system wants to know about it. But when you're in a hurry, it's easy to forget to set the alarm or otherwise tell the house it's on its own. This script uses your motion detectors to decide for itself.**

When you leave home for the day, sometimes it's hard to remember to push a button that tells the house you're leaving [Hack #24]. Besides, your home is supposed to be smart, so why can't it figure out you've left? It can, with a little clever scripting and knowledge about what an occupied home should look like, from your home automation system's perspective.

The key to detecting that the house is unoccupied is in identifying the normal activity patterns the motion detectors you have installed in your home will see. To help you do this, review a few days' worth of home automation event logs from typical days when you were home, compared to days when you were absent. Here are some things to look for:

- Are there periods of time that you're home, but none of your motion detectors is triggered? How long are these periods?
- Do certain motion detectors get triggered even when you're gone (by a pet, for example)?

I have several motion detectors in my two-story home. Here's where they are located:

- Garage
- Master bath
- Office
- Upstairs hallway
- Guest bedroom
- Front door
- Backyard

A careful examination of the logs from XTension **[Hack #17]** shows that the detectors in the guest bedroom and the front door can go for long periods, sometimes days, without being triggered. For the purposes of this hack, then, let's ignore those units.

The other four units are very active. The detectors in the master bath and upstairs hallway are triggered dozens of times a day when someone is home—especially the one in the hallway, which sees everyone traverse the stairway. But they are completely silent when the house is empty.

The detector in the garage is triggered when someone leaves the house by car, which is really the only way a long absence is likely to begin, and the backyard detector signals occasionally due to my dog, cloudy skies **[Hack #6]**, or the gardener coming to visit.

Given these patterns, I concluded that of the four most active motion detectors, `Master Bath`, `Upstairs Hallway`, and `Garage` were the most consistent indicators of activity in the house. That is, it seems very unlikely that someone could be at home, going about their everyday activities, and not trigger at least one of these detectors every four hours. In XTension, I created a group **[Hack #17]** called `Empty House Motion Detectors` and added these three units, as shown in Figure 6-1.

The next thing to consider is the consequences of your home automation system being unaware that the house is empty. For example, if you rely on a `Nobody Home` state to trigger scripts that scare away burglars **[Hack #74]**, it might be very important to you that the house quickly notices it is unoccupied. If, however, the only consequence is that you aren't notified of missed phone calls **[Hack #27]**, it might be less critical that your system know everyone has left.

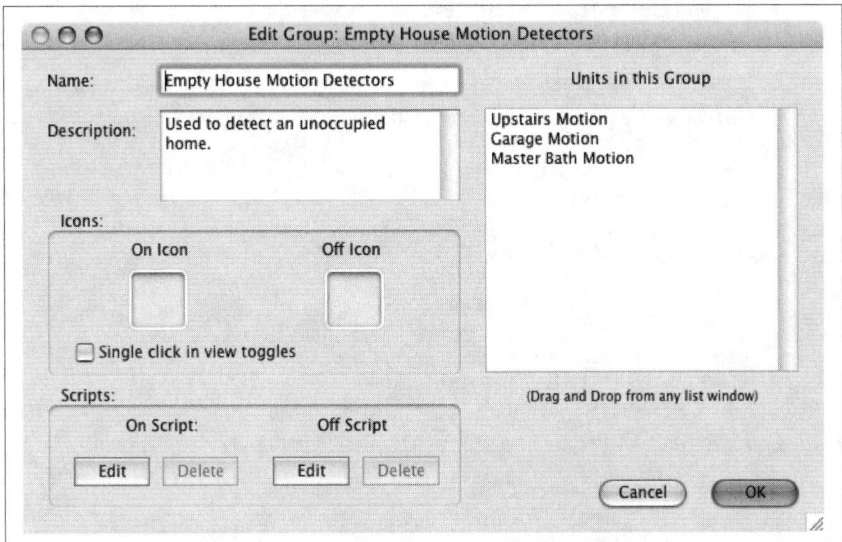


*Figure 6-1. A group of frequently used motion detectors*

Also, take into consideration what you'll do when the house decides it is empty. If you'll simply turn off all the lights, and the script accidentally decides the house is empty when it really isn't, it might be annoying but not troublesome. On the other hand, if you're going to have the script automatically arm the house and begin sounding alarms when it next detects inside motion, the psychological cost to the overlooked, but legitimate occupant, might be much higher.

Use your answers to these questions to decide on an acceptable period of inactivity and *slop-over* time. In my case, I use six hours. That's four hours for the basic motion pattern I identified using my logs, plus an extra couple of hours to allow for someone being home, but perhaps napping on the couch, or otherwise not moving around much. I've been using this script and logic for more than a year and it has never incorrectly identified an unoccupied home:

```
-- checks inside motion detectors for activity, if none, sets nobody home
-- uses group "empty house motion detectors" to determine detectors to
examine.
if (status of "Gone To Bed") is false then -- don't do this if we're
sleeping
    set recentMotion to false -- flag to decide if house is empty or not
    set mList to "" as list
    set mList to all of group "empty house motion detectors"
    set h to 6 -- how many hours of inactivity triggers decision
    repeat with x from 1 to count of items of mList
        set u to item x of mList
```

```
            if (time delta of u) is less than h * hours then
                set recentMotion to true
            end if
        end repeat
        --after above loop, if recentMotion is still false, then no motion
    detector has seen
        --action for at least H hours, so turn on nobody home, if it is off
        --and, suspend this script

        if (status of "nobody home") is false then
            if recentMotion is false then
                say "I'm getting lonely."
                say "Arming alarm system."
                write log "House seems empty, arming system."
                -- set all occupants to away
                if (status of "Gale Home") is true then
                    turnoff "Gale Home"
                end if
                if (status of "Guest Home") is true then
                    turnoff "Guest Home"
                end if
                if (status of "Gordon Home") is true then
                    turnoff "Gordon Home"
                end if
                set description of "Notify Gordon" to "House is now empty."
                turnon "Notify Gordon" in 3 * minutes
                suspend event "check for empty house"
            end if
        end if
    end if
```

The script is scheduled as a repeating event that runs every hour. It begins by getting the list of motion detectors in the `Empty House Motion Detectors` group and then examines each of them for their `time delta`, a property XTension maintains automatically that indicates when the unit was last active. If any one of the detectors has been active within the last six hours, a flag is set indicating that the house is occupied and the script exits without further action.

If none of the detectors has been active during the last six hours, text-to-speech is used to announce that it appears everyone has left. This is done in case someone is actually home, so the unseen occupant can correct the decision and check himself back in [Hack #70]. Then, the script sets the state of all occupants to away, writes to the log about what has happened, and sends a notification via email to my cell phone [Hack #73]. Finally, the script suspends execution of the repeating event that runs every hour because the status of the house is now set to empty, so there's no need to continue this verification procedure.

## Hacking the Hack

If you set the house to nighttime mode [Hack #48] when you go to sleep, and you're not using a motion detector in or near your bedroom, you'll probably want to add a step to that script that stops the empty house detection script, using the `suspend` command illustrated earlier. Otherwise, the lack of activity in your home while you're sleeping could cause the house to assume it's empty.

This script uses a short period of inactivity to detect that the house has become empty. You can use the same `time delta` and grouping techniques to take actions for longer absences. For example, if you have automated the curtains or blinds in the living room [Hack #52], perhaps you want to close them after you've been gone for longer than a day.

# HACK #70 Know Who's Home

**Knowing who is at home is an essential building block for smart home automation. You can easily signal your house that you're leaving, and when you've come home, so that your automations can react accordingly.**

Almost all the scripts in my home automation system rely on knowing either that the house is currently unoccupied, or which of the house members is away. For example, when my wife is traveling for work, her automated wake-up alarm [Hack #47] does not sound off. If someone comes to our front door when nobody is at home, the porch light is turned on for a few minutes [Hack #74] and a message is sent to my cell phone, alerting me to check for a package delivery when I return. Another example is when we have visitors staying overnight. The automation system also knows we have houseguests, so it stops controlling the lights in the guest bedroom so as not to perturb our friends who are used to good, old-fashioned manually controlled lighting.

The keystone in making all this possible is a simple in/out board system. I have a standard X10 minicontroller [Hack #4], mounted to the wall in the garage, that we use to tell the computer who is at home. I've added labels that indicate the purpose of each button, as shown in Figure 6-2.

The minicontroller is attached to the wall with double-sided tape and is located near the switch that opens and closes the garage door. These are the four buttons, from left to right:

- GORD, a.k.a. Gordon Home
- GALE, a.k.a. Gale Home
- GUEST, a.k.a. Guest Home
- EVERYONE, a.k.a. Everyone Home

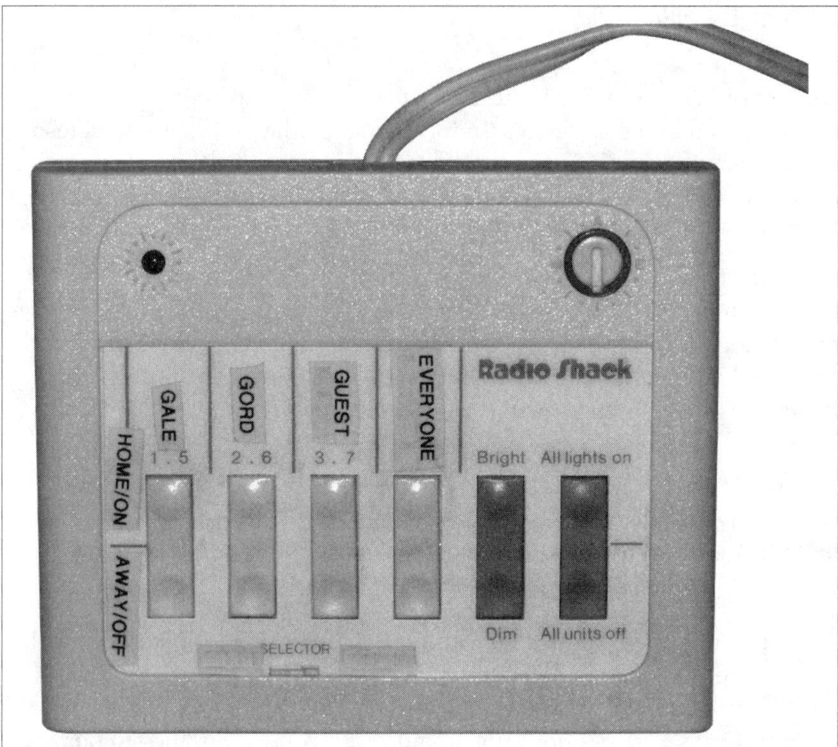


*Figure 6-2. The in/out controller*

The buttons labeled Gale and Gordon are used when my wife and I leave the house. A quick press of the Away/Off button tells the home automation system we're leaving, while Home/On indicates we've returned home. It's easy to become accustomed to pushing a button when you leave and arrive. Conceptually it's not any different from switching on an alarm system, and it's probably easier because only a single button press is needed.

If you have an alarm system, see the "Hacking the Hack" section for some ideas about how to integrate the in/out board with your system.

To set up the software side, add units to XTension [Hack #17] for each person whose status you want to keep track of. Figure 6-3 shows in/out board units that I use; the X10 addresses correspond to the buttons on the minicontroller (i.e., address `I1` for `Gale Home`, `I2` for `Gordon Home`, and so on). Figure 6-3 shows in/out board units in XTension for each of the four persons or person groups.

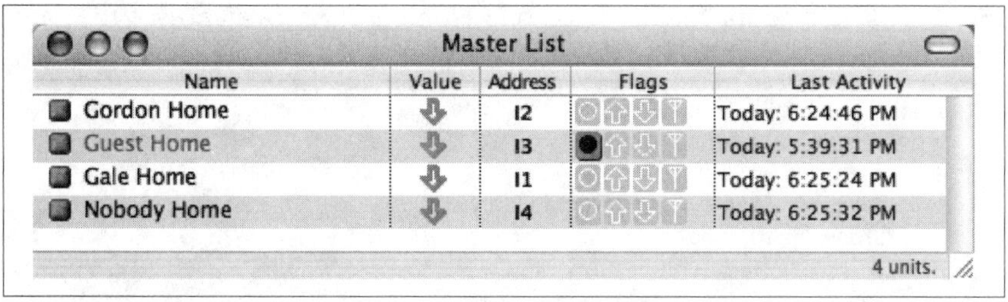

Master List

| Name | Value | Address | Flags | Last Activity |
|---|---|---|---|---|
| Gordon Home | | I2 | | Today: 6:24:46 PM |
| Guest Home | | I3 | | Today: 5:39:31 PM |
| Gale Home | | I1 | | Today: 6:25:24 PM |
| Nobody Home | | I4 | | Today: 6:25:32 PM |

4 units.

*Figure 6-3. In/out board units in XTension*

The buttons on the minicontroller have discrete on and off positions, but to simplify things, I think it's better to treat the switch as a toggle. That is, if Gale is currently home, pressing either on *or* off will set her status to Off—and vice versa. Little touches such as these make interacting with the home automation system less intrusive; just press a button and it does the right thing for you.

Add the following toggle code snippet for both the on and off actions **[Hack #17]** for the unit activated by the button on the minicontroller, modifying it so that it controls the correct unit for each individual:

```
If (status of "Gale Home") is true then
    Turnoff "Gale Home"
Else
    Turnon "Gale Home"
End if
```

The Everyone Home button is set up as a shortcut to save you from having to press individual buttons when the whole family leaves or arrives home. With the press of a single button, everyone who is currently home will be changed to Away status; each person's unit in XTension will be turned Off, and vice versa when the On button is pressed. To do this, you need separate scripts for the On and Off command for the Everyone Home unit. This is the Off script for the Everyone Home button (I4 in Figure 6-3):

```
if (status of "gale home") is true then
    turnoff "gale home"
end if
if (status of "gordon home") is true then
    turnoff "gordon home"
end if
if (status of "guest home") is true then
    turnoff "guest home"
end if
```

Notice how each person's status is checked to make sure it needs changing. This not only saves time, but it also provides an accurate accounting in the log file, should you care about such things.

Here is the On script for the Everyone Home button, which uses the same type of logic as the Off script discussed earlier:

```
if (status of "gale home") is false then
    turnon "gale home"
end if
if (status of "gordon home") is false then
    turnon "gordon home"
end if
if (status of "guest home") is false then
    turnon"guest home"
end if
```

## Houseguest Settings

The Guest Home unit is a special case. We don't have visitors all that often and because some of the other scripts, such as the one that turns off all the lights in the house at night **[Hack #48]**, use the status of Guest Home to change their behavior, it's important that it doesn't get turned on by accident or when the Everyone Home button is pressed. This is important because if Guest Home gets turned on when we don't have a guest, the house will not be considered unoccupied even after both my wife and I have left.

To avoid these problems, I block **[Hack #96]** the Guest Home unit in XTension when we don't actually have a houseguest. Once the unit is blocked, no controller or script can change its status. When we do have a guest, I unblock this unit and set Guest Home to On before the houseguest arrives.

To block or unblock a unit in XTension, click its button in the Master List to display the pop-up controller and then click Block, as shown in Figure 6-4. Or, open the Command window and enter `block unit "Guest Home"`.

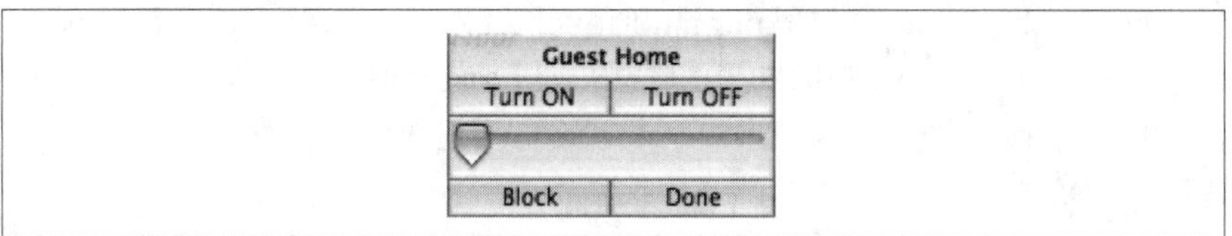


*Figure 6-4. Blocking a unit*

Instead of blocking a unit, set up a second variable, Ignore Guest Status. Whenever you receive an On or Off for Guest Home, check this variable to decide if you should toggle the status of Guest Home.

## Last One Out

The final piece of this solution is to keep track of whether the house is empty. You do this with a pseudo unit [Hack #17] called, appropriately enough, Nobody Home. You never set this unit's value manually. Rather, the scripts that are activated by each individual leaving or arriving handle this task. It is set to On when all the individual statuses—Gordon Home, Gale Home, and Guest Home—are Off. Its purpose is to provide a single place that other scripts can check to see if the house is unoccupied, instead of checking whether each individual is at home. For example, it's used by "Nobody Here but Us Ghosts" [Hack #72], a hack that discourages burglars by making the house look occupied.

The code to set the Nobody Home status needs to be part of the On and Off scripts for each individual unit because the logic differs slightly in each case. For example, when Gordon Home is set to Off, it has to examine that status for both Gale Home and Guest Home to decide if the house is now empty. Similarly, when Gale Home is turned Off, that script examines Gordon Home and Guest Home. This Off script for Gordon Home shows how it gets done:

```
say "Goodbye Gordon"
if (status of "gale home") = false and (status of "guest home") = false then
    turnon "nobody home"
end if
write log "-Gordon has left the building-"
```

When the last occupant leaves, Nobody Home is turned on. When someone comes home, the house is no longer empty, of course, so it's necessary to turn only Nobody Home off. There's no need to examine the other units. Here's the On script for Gordon Home:

```
say "Welcome back Gordon"
if (status of "nobody home") = true then
    turnoff "nobody home"
end if
write log "-Gordon has returned-"
```

Notice that the Everyone Home scripts, from earlier, don't need to set Nobody Home to on or off. That's because those scripts turn off the individual units, such as Gordon Home, which then sets Nobody Home appropriately, as we've just discussed. In other words, the last individual unit that gets turned off will turn on Nobody Home.

The On script for Nobody Home is where you can take care of things such as making sure all the lights are turned off, arming your alarm system, or other starting tasks that should happen whenever the house is left alone [Hack #72], such as activating your webcam [Hack #82].

## Welcome Back

Although it's nice to have everything turned off when you leave, and to have the house react intelligently while you're gone, it's what happens when you come home that you'll notice and appreciate the most.

As soon as the first person coming home changes his status to `On` by pushing his button on the minicontroller, the On script for his button turns off `Nobody Home`. This is the On script for `Gordon Home`:

```
Say "welcome back Gordon"
Turnoff "Nobody Home"
```

All the logic for what happens when the first person returns to an unoccupied home occurs in the Off script for `Nobody Home`. Several things happen in rapid succession.

First, if it's dark and the house was unoccupied until now, the script turns on the lights on the first floor and those closest to the door. Lights that I won't need right away, such as the lamp in the upstairs loft and those in the backyard, are scheduled to come on a bit later. (Because it takes a few seconds to transmit the X10 signals to the lights, minimize the number of actions that occur up front so that the script can continue without waiting to transmit those signals.)

```
if (daylight) is false then
    turnon "LR Lamp 1"
    turnon "Loft Lamp 1" in 20 -- this happens 20 seconds later
    turnoff "Outside Lights" in 10 * minutes
end if
```

Next, the script announces how long the house was unoccupied, determined by examining the timestamp of when `Nobody Home` was last changed. The elapsed time is calculated using an attachment script **[Hack #88]** function called `IntervalToString` **[Hack #97]**. At first I put in this spoken time announcement only for debugging purposes, but I decided to leave it because it's handy for knowing how long our dog, Scooter, has been at home alone so that I can guess if he'll need to go outside.

```
set goneTime to IntervalToString((time delta of "Nobody Home"))
say "Scooter was alone " & goneTime
```

Then, the number of missed phone calls is announced by examining the unit that gets incremented every time a phone call is received **[Hack #27]**:

```
say " calls  missed: "
set sayString to (value of "Absent Phone Calls") as string
say sayString
dim "absent phone calls" to 0 -- reset the phone calls counter
```

If anyone came to the front door, this is alsoannounced, in case a package has been left on the porch. This value is incremented every time the motion detector at the front door is activated and the house is empty **[Hack #74]**:

```
if (value of "Absent Front Door Motion") > 0 then
    say "Check the porch for packages!"
    dim "Absent Front Door Motion" to 0 - reset the counter
end if
```

Whew, that might seem like a lot of stuff happening, but it really just takes a few seconds when the computer is executing the code, and the amount of information it gives you while you're taking off your coat and setting down your briefcase is quite useful.

## Final Thoughts

If keeping track of whom is at home seems too Orwellian, or if you haven't implemented other hacks that depend on knowing exactly which individuals are home, you can simplify this concept greatly by only keeping track of when the house is empty. The last person to leave pushes a button that turns on `Nobody Home`.

Just knowing the house is empty will give you a lot of flexibility in creating a smart home. Later, as your needs and experience grow, you can implement statuses for individuals, if you need them. Don't feel like you're taking the easy way out if you do this; indeed, the important thing to remember about home automation is that it works best when it's kept simple and when it's tailored to do just what you need, and nothing more.

## Hacking the Hack

If you have an alarm system that can send an X10 command when it's armed, you probably don't need the minicontroller. Simply configure your alarm to turn on `Nobody Home` when the system is armed, and to turn it off when you set the alarm to standby.

# HACK #71 Avoid False Intrusion Alarms

**When your home automation software receives a signal from a motion detector, it's often useful to perform a few logical tests before deciding to sound an alarm.**

Motion detectors aren't perfect **[Hack #6]**, and they can be triggered by things such as warm air pockets, a moving shadow from a tree, or a pet. If you want to use a motion detector to help protect your home, it's best to add a little scripting that checks several conditions to make sure your detectors aren't just seeing phantom movement. This is particularly useful if you're going to sound an alarm or otherwise make a commotion when you think you have an intruder. Trust me, if you skip over this hack now, you'll be back the first time you get a false alarm such as this during dinner.

When it comes to deciding whether an intruder has set off a motion detector, the watchwords are *trust, but verify*. Don't rely on just a single signal from one detector; they're simply too prone to error. Here's a motion detector On script for XTension [Hack #17] that demonstrates a technique you can use with other systems, too:

```
    set recentMotion to false -- flag to decide if house is empty or not
    set mList to "" as list -- initialize list variable
    set mList to all of group "intruder motion" -- group of all motion
detectors
    set timeLimit to 8 -- eight minute threshold for recent activity
```

The basic idea here is that when a motion detector is triggered, we check to see if any other motion detectors in the `intruder motion` group also were triggered recently. The script begins by setting the `recentMotion` variable to `false`. This will be used later in the script to determine if anything in the group has been triggered.

```
repeat with x from 1 to count of items of mList
    set u to item x of mList
    if (time delta of u) is less than timeLimit * minutes then
        set recentMotion to true
    end if
end repeat
```

Next, the script sets up a list composed of the unit name for each member of the group. The `time delta` of each motion detector, which provides a timestamp for when the detector last sent a signal, is examined and the `recentMotion` variable is set when any detector has been active within the last eight minutes.

```
if recentMotion is true then -- another motion detector has fired
    speak "intruder alert! intruder alert!"
end if
```

Finally, after checking all the units, an intruder is announced based on the value of our flag variable. Let's take a closer look at the eight-minute threshold used to determine if any recent motion has been detected. I arrived at this value after thinking about where my motion detectors are located in my home and how long each one waits before sending an `Off` signal after motion has stopped. This time is programmed into each motion detector [Hack #6] and XTension's `time delta` property will be updated when the `Off` is received. In other words, `time delta` reflects the last time the unit sent *any* signal. The detectors in the `intruder motion` group are set to signal inactivity after five minutes. You might want to adjust the variable time appropriately for your units.

The intruder motion group consists of motion detectors that are located in the hallways of my home. I have motion detectors in other areas, but I use only the hallway detectors because they're sure to be triggered by a burglar moving through the house. Also, perhaps more importantly, they're less prone to false triggers because they're away from windows and are mounted high enough from the floor that my dog won't set them off.

This script's response to an intruder is innocuous—announcing "Intruder Alert!"—so it might not be a big deal if it goes off accidentally when you're at home. But that changes if you're taking more drastic measures such as setting off outside horns or flashing all the lights in the house.

Until home automation is common, a simple spoken announcement and a few lights turning on can be surprisingly effective in scaring away burglars. Several home automators have had success in this regard by having their system make a lot of racket and announce that authorities are being notified. What burglar is going to stick around to see if your seemingly living house is just bluffing?

To completely avoid false alarms such as this one, you can have the script check to see if anyone is at home [Hack #70] before it starts checking for recent activity:

```
if (status of "Nobody Home") is true then
    set recentMotion to false -- flag to decide if house is empty or not
    set mList to "" as list -- initialize list variable
    set mList to all of group "intruder motion" -- group of all motion
detectors
    set timeLimit to 8 -- eight minute threshold for recent activity

    repeat with x from 1 to count of items of mList
        set u to item x of mList
        if (time delta of u) is less than timeLimit * minutes then
            set recentMotion to true
        end if
    end repeat

    if recentMotion is true then -- another motion detector has fired
        speak "intruder alert! intruder alert!"
    end if
end if
```

You can also check other state variables [Hack #24], which might be helpful in determining if you have an intruder or deciding which actions to take if you determine that you do. For example, if you know your motion detectors are susceptible to being triggered by the afternoon sun, you might consider the time of day as a determining factor. Or, you might wait until after sunset if you figure the likelihood of a daytime intruder is too low to worry about.

All in all, this hack is an excellent example of how using a computer to run your home automation system **[Hack #17]** greatly expands the things you can do, especially when compared to typical alarm systems that react but don't *think*.

# Nobody Here but Us Ghosts

**McGruff the Crime Dog says that one of the best ways to prevent your home from being burgled is to prevent thieves from figuring out you're not at home. A little home automation can convince nearly anyone that your home is occupied.**

For years now, people have been using plug-in timers to turn lamps on and off while they're away on vacation. However, the timers typically offer only two or three on/off cycles per day, and once they're configured, they mindlessly repeat the same pattern over and over. Moreover, you have to buy one timer for each lamp, which means most people will have just one or two of them. The bottom line is that dumb timers are a pain to set up, are not very convincing, and are inconvenient enough that you're probably willing to use them only when you're gone for an extended period of time.

Luckily, you have an automated home and can do much better. By setting up a repeating event that makes your home look occupied whenever you're away at night, you can set it once and forget it—and enjoy the peace of mind knowing that your house is looking quite busy without any further thought from you.

## First Things First

For best results, you want this hack to start working automatically whenever your house is empty. To make this happen, your home automation system needs to know when the house is unoccupied and when it's dark out. See "Welcome to the State Machine" **[Hack #24]** for information about keeping track of statuses with your home automation software, and see "Know Who's Home" **[Hack #70]** for methods to signal who is at home. This hack uses techniques from both.

Next, carefully decide which of your automated lamps and other devices are safe to control when you're gone. A lamp that you know tends to get hot, or a fan that could tip over and overheat, are not good choices. In fact, a fan would be a darn silly choice because you also want to select devices that are going to make your home look occupied and are visible from the street. Mostly, then, you'll want to make a list of a few incandescent lights that you can control reliably. Be sure to select at least one light from each part of your home, such as upstairs and downstairs, so the simulated conditions this technique creates are consistent with how your house is lit when you really are at home.

In addition to the lights, consider setting up an old radio that you can dedicate for use by this hack, such as an old boombox that you can leave in the garage or out on the back porch. Turn it on, connect it to an appliance module, and tune it to a chatty news station. Set the volume control just loud enough to hear if you're at the front door or window and listening so that it sounds like someone might be home inside.

## Setting Up

Now that you know which lights you want to use, create a group [Hack #17] that contains all the units in your home automation software. Name the group `Security Lights`, as shown in Figure 6-5.

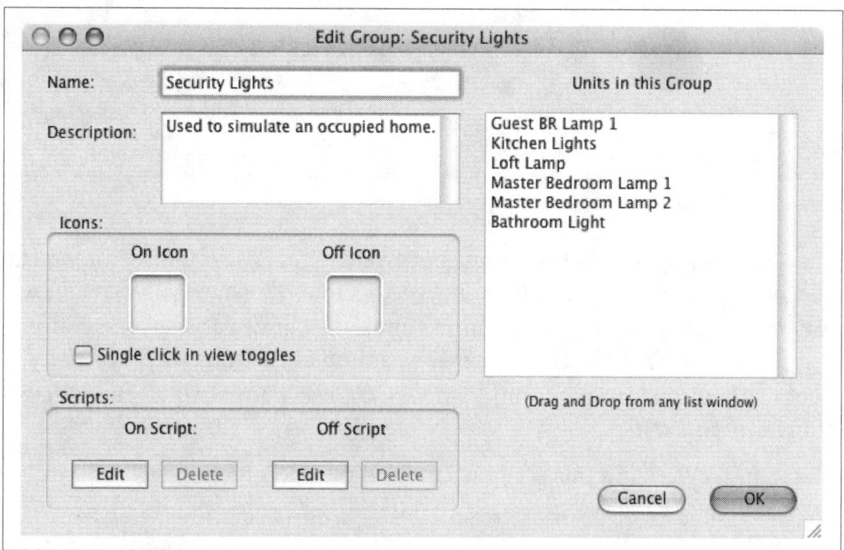


*Figure 6-5. Group list for security lights*

Don't add the radio to the group; you'll want to control that separately. If you have a two-story home, consider setting up two groups, one for upstairs lights and one for downstairs. This will let you make the home look more naturally lit when you're away.

## The Ghost Walker Script

The next step is to set up a repeating event that runs the script that makes the home look occupied while you're away, as shown in Figure 6-6. I named the script Ghost Walker because I like to imagine a ghostly watchman who wanders the home and turns the lights on and off for me. That's also the effect I want it to have for the outside world—as if a person is at home, occasionally moving from room to room.

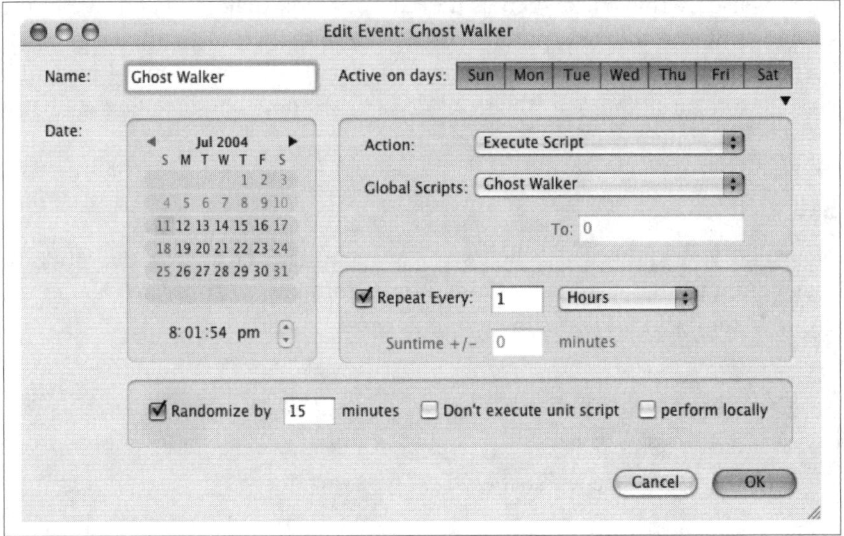


*Figure 6-6. Scheduling the Ghost Walker script*

The script is set to run every hour, but I've entered a 15-minute randomization parameter to make it a little less predictable. This means XTension will vary the script's schedule for plus or minus 15 minutes at every execution. I don't want this script to run while we're at home—it would drive us batty to have lights turned on randomly—so this event is paused when the house is not empty **[Hack #70]**.

A key strategy used in this script is randomness. Here's the first bit:

```
set seed to (random number from 1 to 2) -- a 50-50 chance if we'll do
anything
if seed is equal to 1 then ...
```

The first thing it does is to flip a virtual coin by randomly selecting between 1 and 2. Only if the chosen number is 1 does the script continue. This gives a 50/50 chance that the script will exit without taking any action.

```
if time of (current date) < (23 * hours + 59 * minutes) and time of (current
date) > (5 * hours + 59 * minutes) then ...
```

Next, the current system time is checked. If it's after midnight and before 6:00 a.m., I have the script exit without performing any further actions. My goal, after all, is to make it look like a normal person is at home, not an insomniac.

```
if (status of "Nobody Home") is true then
    if (daylight) is false then
        ...
    else
        write log "Ghost Radio On"
        turnon "Security Radio" in (random number from 1 to 20) * minutes
for (random number from 1 to 60) * minutes
    end if
```

The script then checks to make sure the house is empty and if it passes that test, it checks to see if it's still daylight. If it's not dark yet, only the security radio in the garage is considered for action; all the lights are left alone. The radio is turned on in 1 to 20 minutes, for a period of up to one hour.

However, if it is dark outside, the script tosses another virtual coin to choose between the first and second floors of the house. And things get even more interesting.

```
if (daylight) is false then -- make sure it is dark outside
    set seed2 to (random number from 1 to 2) -- select a floor
    if seed2 is equal to 1 then
        set randLites to all of group "First Floor Security"
    else
        set randLites to all of group "Second Floor Security"
    end if
    set uname to some item of randLites
    set EventNum to (random number from 1 to 1000) as string
    set InTime to (random number from 1 to 10)
    set OnTime to (random number from 7 to 50)
    set EventOnName to "Ghost Light On" & EventNum
    set EventOffName to "Ghost Light Off" & EventNum
    create event EventOnName that turnson unit uname in InTime * minutes
    create event EventOffName that turnsoff unit uname in OnTime * minutes
    write log "Ghost Walk Scheduled: " & uname

end if
```

Next, a light is chosen from either the `First Floor Security` or `Second Floor Security` groups. Then, it chooses to turn the light on in one to 10 minutes, for a period of seven to 50 minutes. Also, two events are created, each with a name containing the words `Ghost Light` and a random number that ensures the event names are unique, to avoid naming conflicts. Plus, it makes it easier to remove pending events programmatically. See the "Cleaning Up After the Ghost" sidebar for details.

After creating the two events, the script writes a note in the log to indicate which light it selected for action. That's not strictly necessary, but it provides a nice way to troubleshoot the script. After you implement it, you'll want to review your log files periodically to make sure the durations and selections the script is choosing are working out well for your home.

## Cleaning Up After the Ghost

When the script schedules its on and off events, it ensures that the event names include the words `Ghost Light`. This enables you to easily identify the events in both the log and the pending events window. But it's also handy for using a sweeper script to remove any events that are still pending events when you return home:

```
set EventList to all events
repeat with x from 1 to count of items in EventList
        set e to item x in EventList
        if e contains "Ghost Light" then
            remove event e
        end if
end repeat
```

This is a good, general-purpose technique for finding and removing any events you want to cancel when you've returned home **[Hack #70]**.

Here's the complete script, with comments, for your reference:

```
--Ghost Walk for XTension
--Randomly turns on lights and radio to make house seem occupied
set seed to (random number from 1 to 2) -- a 50-50 chance if we'll do
anything
if seed is equal to 1 then
    --only activate before midnight and after 6AM
    if time of (current date) < (23 * hours + 59 * minutes) and time of
(current date) > (5 * hours + 59 * minutes) then
        if (status of "Nobody Home") is true then -- make sure house is
empty
            if (daylight) is false then -- make sure it is dark outside
                set seed2 to (random number from 1 to 2) -- select a floor
                if seed2 is equal to 1 then
                    set randLites to all of group "First Floor Security"
                else
                    set randLites to all of group "Second Floor Security"
                end if
                set uname to some item of randLites
                set EventNum to (random number from 1 to 1000) as string
                set InTime to (random number from 1 to 10)
                set OnTime to (random number from 7 to 50)
                set EventOnName to "Ghost Light On" & EventNum
               set EventOffName to "Ghost Light Off" & EventNum
                create event EventOnName that turnson unit uname in InTime *
minutes
                create event EventOffName that turnsoff unit uname in OnTime
* minutes
                write log "Ghost Walk Scheduled: " & uname
            end if
```

```
            -- it's daylight, so only activate the old radio in the garage
            write log "Ghost Radio On"
            turnon "Security Radio" in (random number from 1 to 20) *
minutes for (random number from 1 to 60) * minutes
        end if
    end if
end if
```

## Hacking the Hack

For even more variability in the script's action, the last line of the script can reschedule itself to run again at a random interval:

```
execute script "Ghost Walker" in (random number from 35 to 50) * minutes
```

Make sure this line is outside the first `If` block so that the script is always rescheduled; otherwise, it will execute only once. And, of course, remove any repeating events you've scheduled manually, as described earlier in this hack.

For more about making your house seem occupied while you're away, see "Who's There?" [Hack #74].

# HACK #73 Send Notifications of Home Events

**When significant events occur at home, such as a missed call or package delivery, have a message sent to your cell phone or email at work.**

One of the benefits of having a smart home is that it can tell you about things that need your attention—sort of like an administrative assistant who handles most things and saves the important decisions for you. In fact, once you start having your house notify you, you'll find it so handy that you'll want to be able to add all sorts of routines in your home automation system. The best way to do this is to set up a general-purpose routine that any script or event can call. This centralizes the logic of how to send the notifications and enables you to quickly change all of them, perhaps switching one email address to another, by making the change in one place.

This script uses XTension's [Hack #17] ability to store text strings—up to 255 characters' worth—in a unit's description field. Set up a pseudo unit, which is a unit that does not have an X10 address. In my system, I have two units that I use for notifications, `Notify Gordon` and `Notify Gale`.

I have set up units for my wife and myself so that each of us can be notified separately of different messages. Her notification script sends messages to her cell phone only, while mine sends all messages to both my phone and an email address simultaneously. Having separate methods for each of us also makes it easy to block [Hack #96] one or the other, should we be traveling together and don't want to be overly pestered by automated messages.

The message to be sent is stored in the unit's `description` property. For example, a script that receives CID information from Phone Valet [Hack #27] might do this:

```
set description of "Notify Gordon" to ("Call Rcvd" & CIDdata)
turnon "Notify Gordon"
```

After putting the CID information into the `description` property, the `Gordon Home` unit is turned on. This activates the unit's On script, which has all the notification logic. This approach of having a single way to set and send notifications makes the system very easy to use and maintain.

The On script for `Gordon Home` starts out like this:

```
set theMessage to description of "Notify Gordon"
-- add current day and time stamp to message
set theMessage to theMessage & " " & " " & (weekday of (current date)) & " "
& (time string of (current date)) & " "
```

First the description text is pulled from the field, and a date and timestamp are appended to the end of the text:

```
if (status of "Gordon Home") is false then
    set theRecep to "gordon@example.org"
    set theRecep to theRecep & ",gordonwork@example.com"
    set theSubject to "Message From CoasterHaus"
    write log "Notifying Gordon: " & theMessage
    set theCommand to "echo" & theMessage & " | mail -s" & theSubject &
theRecep
    try
        do shell script theCommand
        on error msg number eNum
            write log "Sending Mail error:  " & eNum & " " & msg
    end try
end if
```

Next, the script checks to see if I am at home [Hack #70], based on the `Gordon Home` unit. If I'm not home, an email message is constructed and is sent using the Unix commands `echo` and `mail`.

I have my system set up so that when events are announced over the in-house speaker system, the same text that was spoken is used for the notification, as described in this hack. So, I have the script discard the notification if I am at home, under the assumption that I heard the audio announcement. In your system, you might want to have it logged, emailed to an account that you check frequently when you're at home, or sent to your cell phone.

Finally, before finishing, the script does a little housekeeping:

```
set description of "Notify Gordon" to ""
turnoff "Notify Gordon"
```

The description text is blanked out, and the `Notify Gordon` unit is turned back off. This isn't strictly necessary, but it provides some level of accounting in the XTension log file should a problem arise that you need to diagnose. Another reason to be fastidious about the unit's state is that in a system where messages are sent frequently, you could check to make sure the unit is off before changing the description text to another message to avoid overwriting one that hasn't been sent yet.

## Hacking the Hack

To implement this notification method in Indigo, set up a variable [Hack #18] to hold the notification message, then define a trigger action that is called when the contents of the variable change. The action runs an AppleScript similar to the one outlined for XTension in this hack.

## Who's There?

**It's said that burglary is often a crime of opportunity. This hack discusses methods for using your home automation system to welcome visitors and convince prowlers to look elsewhere for an easier target.**

I want my home to be both polite and defensive. That is, a visitor who comes to the door should feel welcome, but if I'm not home, I don't want it to be glaringly obvious that he's all alone. Here are two scenarios.

If the house is unoccupied and a visitor approaches the front door, he hears a radio playing softly inside. If it's after dark, the porch light brightens to full power, the visitor hears a dog barking, and a minute later a light inside the house turns on.

If I'm home when the visitor arrives, his presence is announced before he has a chance to ring the doorbell. If it's dark outside, the porch light is turned up, only to return automatically to its dimmed state a few minutes after I let the visitor in the door. Later, when I open the door to see him out, the porch light brightens again to light his way.

To be able to do this, your home automation system needs to know when someone has approached your house and whether anyone is home [Hack #70]. But even if you haven't implemented a method to keep track of who is at home, you still can adapt this technique so that it just politely welcomes and alerts you to visitors.

## Detecting Visitors

Let's focus on how to best detect when someone is approaching your home. To begin, assess how many different ways visitors can approach your property. In my case, my house sits on the property in such a way that visitors can reach my front door easily, but the side and back areas are surrounded by a wooden fence. In fact, to get to the gate to enter the backyard, they need to pass by the front porch.

Therefore, for my property, a single motion detector provides coverage for virtually all who would approach. I've mounted an X10 motion detector **[Hack #6]** in the eaves that overhang the front porch, as shown in Figure 6-7.

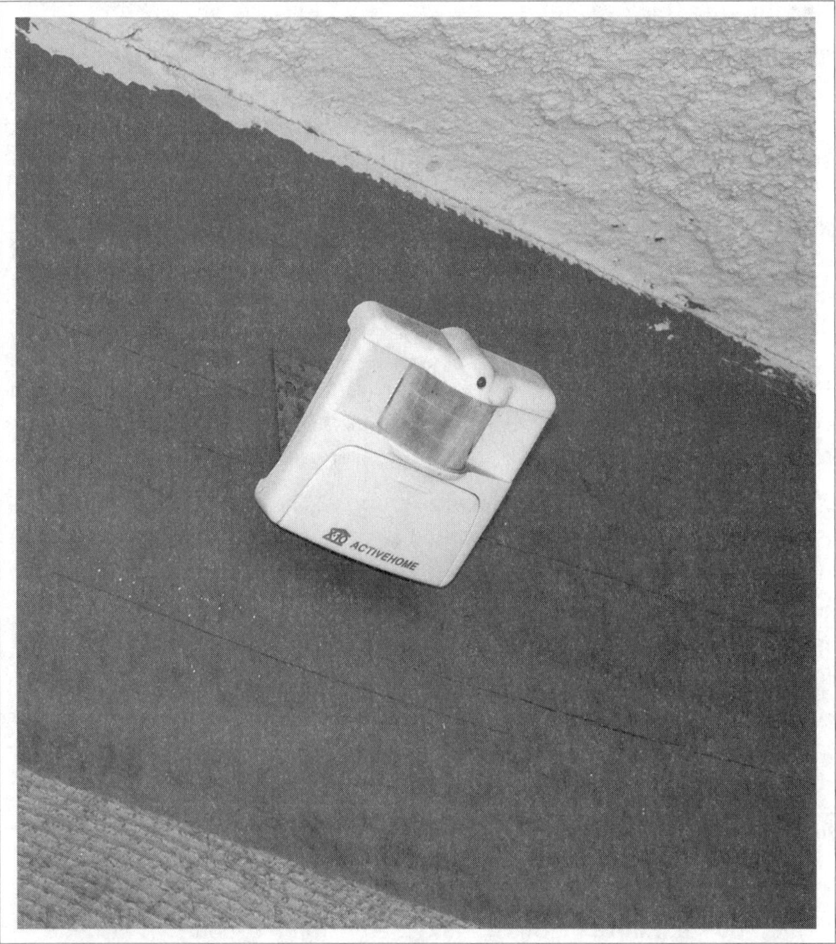


*Figure 6-7. Front door motion detector*

The motion detector is orientated so that it's pointing down and toward the door, not out toward the front sidewalk. I discovered this works best because it reduces false triggers caused by passersby. The motion detector's field of view is effectively limited to the door and the area just in front of it—exactly where you must stand to ring the doorbell. This positioning also makes it quick to detect when the front door is opened, which is handy for departing guests, too, as we'll see in a moment.

The motion detector is mounted to a metal bracket, made from a square piece of thin sheet metal from the hardware store. You can bend it easily by hand, so adjusting it to the proper angle to tune the detector's response zone is simple. The bracket is attached to eaves with two short screws. The motion detector is stuck to the bracket with double-sided tape. The assembly is very lightweight and is simple to assemble and install.

The motion detector is a low-power device, but I typically have to replace its batteries three or four times a year. That's more often than usual for a motion detector, but because the detector is outside, its signal needs to be strong enough to reach a transceiver inside the house.

That's all the hardware work you'll need to do for this hack; the rest is in programming your home automation controller.

## Reacting to Visitors

When the motion detector sends an `On` signal, meaning someone is standing on my porch, this script is executed because it's defined as the On script for the motion detector in XTension **[Hack #17]**:

```
If (status of "Daylight") is false then
    Dim "front porch light" to 100 for 10 * minutes
End if
```

First the script checks the global variable `Daylight` to determine if it's dark outside. XTension maintains the state of this variable automatically, based on the calculated sunrise and sunset times for your location.

The motion detector's dusk sensor also provides you with a way to know when it's dark out, if your home automation software doesn't calculate the sunrise/sunset time for you.

If it's dark, the front porch light is turned all the way on. I do this as the first thing in the script so that the light can react as quickly as possible. Even then, given the inherent X10 delay and the extra transceiver-induced delay **[Hack #5]**, it takes three or four seconds before the light comes up.

```
Say "Front Door Motion! Front Door! Front Door!"
```

After determining if the porch light should come on, the computer speaks a couple of times to announce the visitor over the household speakers **[Hack #28]**. Note that this happens even if the house is unoccupied. The voice does no harm and might add to the illusion that we're going to create in a moment—namely, that someone is around. (If the computer talks and there's no one to hear it, does it make a sound?)

```
If (status of "nobody home") is true then
    Set description of unit "Notify Gordon" to "Front Door Motion Detected"
    Turnon "Notify Gordon"
End if
```

After the announcement, the script determines if the house is empty. If it is, a text message is sent to my cell phone via email **[Hack #73]**. I do this so that I know to check the front door for notes or a package when I come home, in case it was a delivery person that I missed.

```
If (status of "nobody home") is true then
    Set description of unit "Notify Gordon" to "Front Door Motion Detected"
    Turnon "Notify Gordon"
    If (time delta of "Robodog") is greater than 10 * minutes then
        Turnon "Robodog"
    End if
End if
```

Next, the script turns on an X10 Robo-Dog module **[Hack #79]** that I keep in the back corner of the garage. The Robo-Dog (*http://www.x10.com/security/x10_dk9000.htm*) is a medium-size speaker and amplifier that plays a recording of a fierce dog bark. It's fairly realistic, and I make it even more convincing by locating it away from the door but still within earshot. The Robo-Dog will bark for 30 seconds when it receives an `On` command, so you don't want to trigger it every time the motion detector fires. It would be too obvious that it's not real. Instead, the script checks the last time the Robo-Dog was turned on, using XTension's `time delta` function, and activates the barking only once every 10 minutes.

Incidentally, I do have a real dog that barks at visitors, but if we're not at home, he might be out with us. So, the Robo-Dog provides a constant watchdog that supplements our real barker on occasion. It does not, however, fool my real dog for even an instant.

```
If (status of "nobody home") is true then
    Set description of unit "Notify Gordon" to "Front Door Motion Detected"
    Turnon "Notify Gordon"
    If (time delta of "Robodog") is greater than 10 * minutes then
        Turnon "Robodog"
    End if
    Turnon "Security Radio" for 10 * minutes
    Turnon "Office Light" in 2 * minutes for 15 * minutes
End if
```

Finally, the script turns on an old radio that I have dedicated to the security system [Hack #72]. It's tuned to a news-talk station and is set just loud enough to provide some voice noise, but not so loud as to be annoying to neighbors. Then, an upstairs light in the office is scheduled to turn on in a couple of minutes, providing a visible change inside the house. This might cause some visitors to think I'm ignoring them, but I'd prefer to let them think I'm at home and rude (or hard of hearing) than not at home.

That's it. Because most of the things that happen occur only when the house is empty, the resulting behavior when you are home is that the visitor is announced and the porch light is turned on for 10 minutes. Later in the evening, when the visitor again triggers the motion detector on his way out, the light will again be brightened, and the announcement will go off again. I've found that guests get a kick out that—it's not every day they encounter a talking house—but if it bothers you, adapt the script so that the `time delta` property of the motion detector triggers announcements only after several minutes of inactivity.

## Hacking the Hack

If you don't want to use a motion detector to know when someone has approached your home, you can use a Powerflash [Hack #10] and a reed switch attached to the front gate, or a driveway automobile sensor [Hack #54].

HACK #75

# Secure Your Construction Site

**With some battery-operated security equipment and a little ingenuity, you can help protect a construction site from vandalism and theft.**

Here's the scenario: you're constructing a home in a rural, wooded area that's far from your current residence. There's no electricity at the site yet, the sheriff's patrols are infrequent, and the closest city with a police force is 20 minutes away. The nearest neighbor is a quarter mile away and can't see the site easily. As a result, the site is a sitting duck for thieves. They come at night and help themselves to power tools, generators, and building supplies.

What you need is a way to discourage or prevent these thefts. The solution would have to last for a couple of days unattended. Subcontractors show up early every weekday, so it would have to cut off and on at certain times, then continue to work by itself on the occasional weekend when you're unable to visit the site. Here are some possible approaches.

## Driveway Warning

The site has a long driveway, so it's likely that thieves scout the area on foot and then drive their car into the site only if the coast is clear. A driveway monitor, such as The Reporter Wireless Driveway Alert System (*http://www.smarthome.com/7317S.HTML*; $80), can help convince thieves they're not alone. Each sensor ($44) is wireless and weatherproof and can transmit its signal up to 1,200 feet, provided that there are no obstacles between the monitor and sensors. I'm having success with them working 600 feet away from my house, although I have to put the base station near a window rather than behind the thick adobe and stucco walls of my home.

The Reporter base station runs off a 9-volt AC power supply, but it also has a backup battery that should keep it powered for a long time. If you find it runs out of power too frequently, try running it off a 12-volt car battery; many electronic circuits will do just fine with a little extra juice. When the base station detects a signal from a sensor (you can have up to four sensors), it activates a buzzer and triggers a relay switch. You can use this relay to activate other things, such as a few of the ideas that follow.

## Fake Headlights

Mount a pair of headlights about where they'd be if they were part of a car that is parked in front of the house, facing up the driveway. Enable them to turn on for a few minutes when the driveway sensor senses motion. Use a car battery to power the headlights, and connect them to The Reporter using a timed-relay switch (*http://www.smarthome.com/7279.HTML*; $23) that automatically turns them off after a few minutes. This sight of an apparent car, or at least just the lights coming on as it supposedly approaches, might scare the thieves away.

## Video Recording

If you want to try to record the perpetrators in action, consider using an Everfocus EDSR400H digital video recorder (DVR) (*http://cu1.com/evhire4chdvr.html*; $830). It records video and audio onto an 80 GB hard disk. Although it's expensive, it has several useful features:

- It can operate off a 12-volt car battery, as well as off a 120-volt battery.
- You can hook up to four cameras to it.
- You can record full-screen images of all four cameras, or a quad view where each quadrant of the screen shows a separate camera.
- It has built-in motion sensing. This means you can set it to start recording when the image coming from any camera changes enough to be considered motion.

- You also can trigger recording by using a switch or relay closure (such as from The Reporter driveway sensor).

The user interface is clumsy (especially the web interface), but for the price, there's nothing else I've found that has all these features, and it has good video quality. It uses about 40 watts when running off the power line; I haven't measured it when running on batteries, so I'm not sure how long it might keep running while in that mode.

For cameras, consider the KPC-EX230HL1 bullet camera (*http://cu1.com/ktckpcexullo.html*; $125). Buy it with the 3.6 mm lens for a 67-degree horizontal field of view or with the 2.97 mm lens for a 78-degree horizontal field of view.

These cameras are small and light, require only 12 volts, and have awesome low-light capability. They are rated at a sensitivity of 0.0003 Lux, without having to use infrared illuminators. The images these cameras deliver can look like dark twilight even when it's as black as coal outside to the naked eye.

If you want to record audio, you'll have to amplify the microphones to line level before the audio reaches the DVR. If you have only one microphone, a mic preamp kit (*http://www.partsexpress.com/pe/showdetl.cfm?&Product_ID=9563&DID=7*; $12) will do the trick. If you have multiple mics, you'll need a mixer to combine their signals for the DVR. I recommend the Behringer Eurorack MXB1002 (*http://www.behringer.com/MXB1002/index.cfm*; $100). This 10-channel mixer is a bargain—given all the useful features it includes. It runs for four hours using two 9-volt alkaline batteries. Like The Reporter, you probably can run it longer if you figure out how to attach higher-power batteries.

Some states require you to obtain the consent of anyone you record. Check your local laws. Perhaps you should put a sign on your driveway stating, "This property protected by video and audio surveillance. By entering this property you agree to be watched, listened to, and recorded." In fact, a sign such as that might be a bit of a deterrent by itself.

## Alarm System

Buy the X10 Protector Plus  wireless security system (*http://www.x10.com/security/x10_ds7000.htm*; $100). Install sensors on each window screen and on the door screens, if you don't have them installed already. This way, the alarm will sound when the prowlers remove a screen before getting to the door or window. Alternatively, place a piece of plywood in front of each door before you leave, with the sensor attached to the plywood so that moving the board trips the alarm.

You also can place motion sensors in key rooms inside the house. The motion detectors, as well as the door and window sensors, run off batteries with an expected life of more than two years, which should be more than enough to last the entire length of your construction project.

The Protector Plus base station has a replaceable 9-volt backup battery that is supposed to power the base station for 20 hours. But it's intended as a backup battery, so wiring it to an external 12-volt car battery should keep it running for much longer; its power consumption is only 2 watts when the alarm is not sounding off.

## Solar-Powered Security Lights

Mount solar-powered security lights (*http://www.smarthome.com/740122S.HTML*; $80) in the trees that line the driveway, as well as one or two near the house. As a person approaches the home, the lights will come on, producing a sense that he's being watched and followed. You also might find these lights are a nice permanent addition to your landscaping, to welcome legitimate guests [Hack #54] after you move in.

## A Driveway Gate

Depending on your driveway configuration, it might be relatively cheap (compared to the electronic equipment we're discussing) and effective to simply install a temporary metal gate. It doesn't have to be fancy; just one or more horizontal metal bars that prevent a car from entering the driveway. Each contractor can have a lock, and all the locks can be linked in a daisy chain so that any contractor can break the chain and open the gate.

## Final Thoughts

Combining all these elements gives you a nice system that runs off a car battery or two. Buy an extra set of batteries that you can swap out when you visit the site, taking the old ones home to recharge overnight. After construction is finished, you can keep the cameras and video recorder as a permanent part of your security system.

—*Bill Fernandez*

# Monitor Your Summer Home

**A couple of specialized sensors and an auto-dialer can keep a watchful eye on an unattended home and notify you when something needs your attention.**

Winter storms can cause all kinds of problems and concerns for homeowners, particularly if you're away from a vacation home and wondering how well it is faring in your absence. This hack describes an inexpensive system ($100–$200) you can use to monitor a remote vacation home. It can monitor some basic conditions and dial out for help if something happens, such as a change in temperature or if it detects a loud noise.

You can divide the tools necessary to do this into two categories: sensors and notification devices. The sensors will measure the conditions you want to monitor and signal the notification device to inform you of any problems.

Several general-purpose sensors are available. All the sensors described in this hack operate on the electrical concept of *contact closure*. Just as a light bulb comes on when you flip a wall switch, these sensors close or open a set of electrical contacts when the measured effect triggers the sensor. Using a temperature sensor as an example, if the room becomes too cold (you choose the set point), it will close a set of contacts. Once the room warms up and the temperature rises above the set point, the contacts open. This opening and closing is what signals your notification device to leap into action.

## Watching Weather Conditions

To watch for temperatures that are too high or too low, use a mechanical thermostat (*http://www.smarthome.com/7150.html*; $50) that triggers a relay when the set points you select are exceeded. No batteries or power supply is necessary, so you can place the device in the location you want to monitor. A freeze sensor (*http://www.smarthome.com/7193.html*; $25) is a usually closed switch that opens when the ambient temperature reaches about 37 degrees Fahrenheit. You can wire it inline with the switch that controls your sprinkler system to prevent the sprinklers from being turned on when it's cold, or connect its contacts to a notification device. A humidity sensor (*http://www.smarthome.com/7156.html*; $80) is useful for monitoring a greenhouse or for detecting when something is generally amiss in a home. Humidity sensors work by triggering a relay when the relative humidity exceeds the threshold you select.

## Monitoring for Water Problems

An undiscovered water leak is probably the single most troublesome thing homeowners worry about, particularly if the house has a history of that problem. The WaterBug Water Sensor (*http://www.smarthome.com/7160.html*; $60) can detect accumulated moisture and signal a notification device. You can attach up to five remote sensors ($18 each) to monitor additional areas.

## Monitoring for Power Failures

Next to a water leak, an electrical outage can really spell trouble for a vacation home, particularly if you're relying on power to run the air conditioning, security, and other systems you have in place. A power outage sensor (*http://www.smarthome.com/7154.html*; $50) plugs into an outlet and closes a relay switch after the power has been out for more than a few minutes. Of course, you'll need a battery-powered notification device to receive the signal, but that's a common feature.

## Watching for Outside Activity

If you want to be notified whenever a car enters your driveway, consider an in-ground magnetic sensor (*http://www.smarthome.com/7170.html*; $99) or laser-eye sensors [Hack #54] that can monitor a larger area for any activity.

## Watching for Inside Activity

If you want to be notified when something seems amiss inside your home, consider getting a glass breakage sensor (*http://www.smarthome.com/7496.html*; $40), a floor mat that sends a signal when it's stepped on (*http://www.smarthome.com/5195.html*; $70), or, of course, a motion detector (*http://www.smarthome.com/7481A.html*; $23). In addition to traditional smoke detectors, consider adding one that doesn't have a siren, but can signal your notification device (*http://www.smarthome.com/7497.html*; $60).

## Notification Devices

To keep the cost down, use an automated telephone dialer that can call out and play recorded messages to whomever answers the call. The Two-Channel Automatic Voice Dialer With Microphone (*http://www.smarthome.com/7437.html*; $120) can dial four phone numbers: two voice numbers and two pagers. You can connect up to two sensors to the device, and when one of the sensors is triggered, it will call each number in sequence and deliver a 20-second prerecorded message. You might have it say, when the first sensor is triggered, "This is Mike's house in the mountains. The temperature in the basement is now less than 35 degrees. Send someone immediately to fix the problem." You can set up a second message that is appropriate for the condition detected for the other sensor.

Never program an auto-dialer to call 911 or any other government agency. Most 911 operators will ignore the call and, in fact, might fine you for having an automated device contact them. Instead, auto-dialers should be programmed only to call someone who has the ability to fix the problem and is expecting the call.

Taking this one step further, you could use a Sensaphone (*http://www.smarthome.com/7005.html*; $380) instead. In addition to supporting more sensors, you can call the Sensaphone, enter an access code, and hear a synthesized voice report on the current inside temperature, status of the electrical system, and noise level in the home (useful for knowing if a smoke alarm is sounding).

—*Smarthome, Inc.*

## HACK #77 Protect Outdoor Cameras

**Build an inexpensive enclosure to protect your outdoor security cameras from the weather.**

A few years ago, my home burned down while I was out of town. As a longtime home automation enthusiast, I used the opportunity to rebuild my home with as much automation equipment and provisions for future expansion as I could. One of the additions was to have security video cameras scattered all around the property and outside the home. With that in mind, I installed more than two miles of RG-6 cable and Category 5 wire, one length of each to every potential camera location so that I can expand my system simply by adding cameras.

My current video setup consists of eight cameras connected to a 16-camera video controller, a 960-hour time-lapse VHS recorder, and a video-to-TV modulator for each camera. With this system, I can view the video from the cameras on my television sets, as shown in Figure 6-8.

Each camera is connected to an input on the video controller, and each output from the controller is connected to a TV channel modulator. Each modulator rebroadcasts the camera picture on a different TV channel, so to view a camera, I simply turn on a TV and tune in. The modulators are programmable and can broadcast on any channel from 14 through 64, so it's easy to set them to use unused channels on your cable system.

*Figure 6-8. The video command center*

ChannelPlus (*http://www.channelplus.com*) makes a wide variety of TV modulators and camera controllers.

The cameras I use don't have microphones, but the Category 5 wire that runs to each camera could carry sound back to the modulator if I decide to buy cameras that do have microphones. Currently, I use two wire pairs, of the four available in the Category 5 cable, to run electrical power to the cameras. The camera power supplies are plugged in near the video command center, as described earlier.

So far, everything I've described is standard equipment for video monitoring. The biggest challenge was the outside enclosures for the cameras. Instead of spending $50 to $100 each for commercial enclosures, I decided to build my own. These enclosures cost about $5 each and work just as well.

Here's what you'll need for this hack:

- One length of Schedule 40 PVC pipe cut to 4 inches longer than the camera and lens
- Two end caps that fit the PVC pipe
- One piece of standard window glass cut to fit inside the test cap

I used 4-inch diameter pipe, which should be big enough for almost any camera, but check to make sure your cameras will fit inside the pipe you select, as shown in Figure 6-9. You can cut the PVC pipe with just about any saw, but it might require two passes with the blade, given its size. Be careful, and make the cuts nice and even. Go slowly and make sure the pipe doesn't bind and shatter.

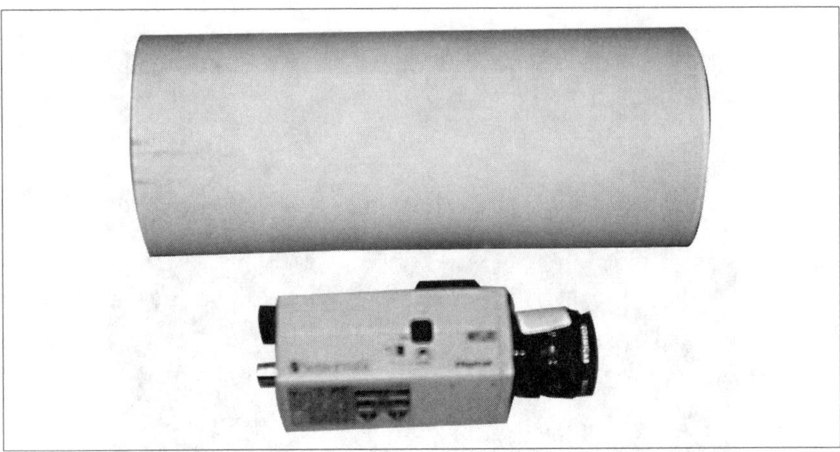

*Figure 6-9. Appropriately sized PVC pipe*

Carefully measure the location of the camera's mounting hole, and drill a corresponding hole in the pipe that will result in the camera's lens being set about 1 inch inside one end of the pipe. You'll use this mounting hole with a bolt to securely mount the camera inside the pipe during final assembly.

Drill another hole, this one just slightly larger than the coax cable you are using, 1 inch from the back end of the pipe. Add just one more hole for the Category 5 wire that carries power to the camera.

Mount the camera inside the pipe and connect all the wires before you cap the end of the pipe. For the front cap, use a drill hole saw to cut a 2-inch hole in the center of the cap. Put a small bead of tub caulk around the test cap, place the glass inside, and press tightly. The caulk should spread around so that it is waterproof. Remove any caulk that has leaked onto the glass.

While the caulk is setting, which takes about an hour, connect the camera and adjust its iris and focus. A small piece of duct tape will hold the settings in place. After the lens cap has dried, slide it onto the end of the pipe. It helps to coat the edge of the cap with petroleum jelly so that it slides more easily.

Mount the case on an L-bracket, and you are in business, as shown in Figure 6-10.

*—Don Marquardt*

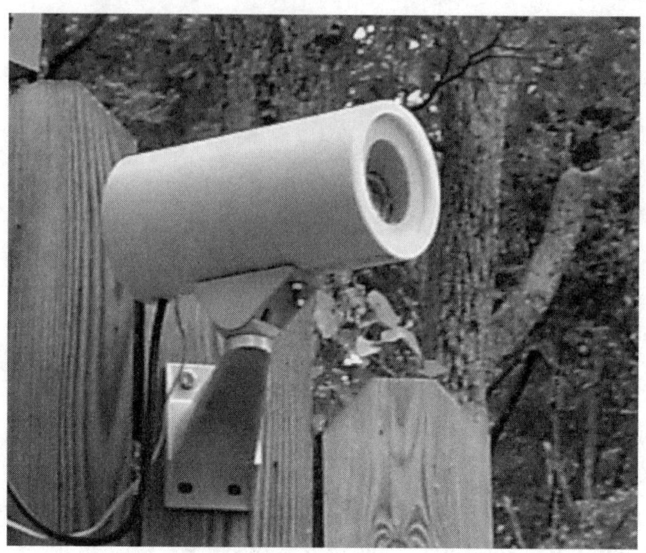

*Figure 6-10. The finished enclosure*

## Know When Windows and Doors Are Open

HACK #78

The security modules from X10 Corporation, such as the window sensor, can't talk to X10 home automation equipment unless you use a wireless receiver to bridge the communications gap.

For a long time now, home automation enthusiasts have looked forlornly at the nifty security modules offered by X10 Corporation. Despite being from the same company, and working in very similar ways, the modules just won't communicate with X10 home automation equipment, except via an X10 security console. Even then, it's not a direct or very useful route. It's a shame because the wireless security sensors, which indicate when a window or door has been opened or left open, could be very handy for home automation.

Thankfully, after years of bemoaning the inherent incompatibility, salvation finally arrived in the form of the W800RF32 from WGL & Associates (*http://www.wgldesigns.com/w800.html*; $75). This handy device connects to your PC using a serial cable and listens for signals from both regular X10 wireless devices, such as Palm Pads and motion detectors, *and* security modules. The signals are fed directly to your home automation system via a serial connection, which means they arrive more quickly and reliably [Hack #83] than those that have to go through a transceiver [Hack #6] to reach your system.

> Don't confuse the 32-bit W800RF32 with the 16-bit W800 receiver. The latter can't decode the signals sent by security modules and won't work for the purposes of this hack. For a hack in which you can use the W800, see "Improve the Response Time of Motion Detectors" **[Hack #83]**.

X10 Corporation's DS10A Powerhouse Door/Window Sensor (*http://www.x10.com/security/x10_ds10a.htm*; $20) is the most commonly used module for home automation purposes. It has a two-piece magnetic sensor that is attached to a transmitter with short wires. You mount each piece of the sensor on the door or window frame so that the pieces are next to each other when the door or window is closed. When it's opened, the pieces move apart, which causes a signal to be sent by the transmitter. It's the same idea as using a magnetic reed switch with a Powerflash unit **[Hack #62]**, but the DS10A is less expensive and a lot simpler to set up. Also, it's not terribly ugly, which helps its SAF **[Hack #14]** tremendously.

HomeSeer **[Hack #1]** enables you to integrate security devices by using a free plug-in that knows how to interpret the information the W800RF32 sends to your computer. To get the plug-in, use HomeSeer's built-in update function. Choose Updates from the Help menu, then select the plug-in, as shown in Figure 6-11. Follow the onscreen instructions to install the plug-in.

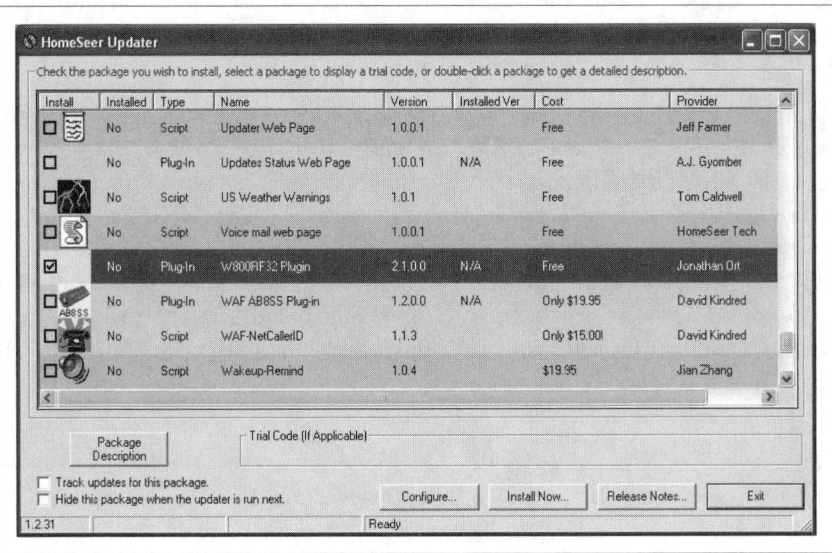


*Figure 6-11. Downloading and installing the W800RF32 plug-in*

After you've installed the plug-in, be sure to read the documentation installed in the *Docs* folder, inside the *HomeSeer* folder, on your hard drive. It contains all the step-by-step instructions for setting up the W800RF32, and a lot of useful information about the plug-in's advanced settings.

> The HomeSeer plug-in also enables you to use the X10 MR26A wireless receiver. It's a less-expensive alternative to the W800RF32, but it doesn't receive signals from security devices. If all you want to do is wirelessly receive signals from motion detectors and Palm Pads, however, it's an alternative to consider **[Hack #83]**.

After you've set up HomeSeer to use the W800RF32, you need to add your security devices to HomeSeer's device list. Choose Options from the View menu, then click the Interfaces tab. Select W800 RF (32 bit) Interface in the Installed Interfaces list, then click the Setup button. In the W800 32 Bit RF Receiver Setup dialog that appears, click the Security tab. Then, to add your new device, click the Add button.

The Add Security Device dialog, shown in Figure 6-12, is where you teach HomeSeer how to interpret signals received from a security device. Select the type of device you're adding from the Device Type list. As noted in the dialog, select Supervised Door/Win/Glass Sensor if you're using a DS10A. Also enter a Device Name and Device Location; just as when you add X10 devices, these names are used in scripting and for identifying devices throughout HomeSeer, so enter something meaningful.

Because the DS10A isn't directly X10-savvy, it doesn't have an X10 address **[Hack #1]**, and it doesn't send X10 commands such as `On`, `Off`, or `Dim`. Instead, it sends unique identifiers known as *security codes*. To teach HomeSeer how to recognize incoming signals from this device, you need to tell the device to transmit its identifying signal. Make sure you have fresh batteries in the DS10A, then press and hold its Test button for several seconds. This causes it to randomly select a new security code. Then press and release the Test button again so that it transmits its new code. HomeSeer automatically will sense that a previously unknown code has been received, and it will fill in the Device ID/Security Code appropriately.

Next, select the Operating Mode. This determines the amount of information that HomeSeer keeps track of for the device. In most cases, X10 On/Off Only is a good choice because it enables you to treat the device, when setting up HomeSeer actions, as if it were just another X10 module. In other words, although the device doesn't actually send X10 `On`, `Off`, or `Dim` commands as discussed earlier, selecting this option causes HomeSeer to translate signals it does send into these commands.

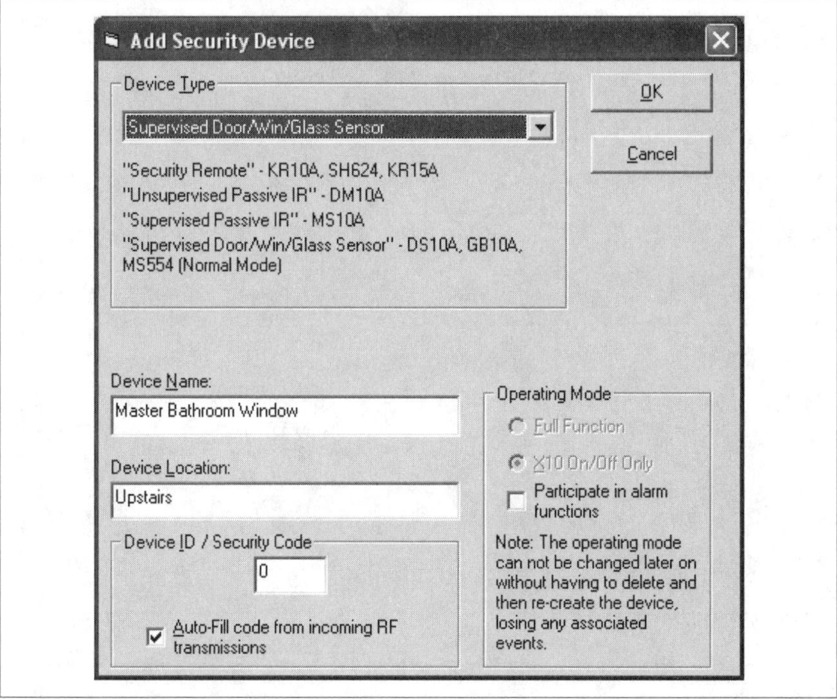


*Figure 6-12. Adding a security device to HomeSeer*

If you have an alarm console that uses these devices, consider selecting the Participate in Alarm Functions option. You'll want to read the documentation file described earlier for the details, but essentially this allows the device to do double duty; working with HomeSeer and your alarm console at the same time. Most alarm consoles, by the way, support only 16 security devices. This limitation doesn't apply with HomeSeer and W800RF32—you can use as many security devices as you need.

After you've added the security device to HomeSeer, you can set up trigger actions [Hack #19] to react when the sensor transmits a change of state. For example, you may want a trigger that activates when a window is opened, which will be seen as an On command in HomeSeer, as shown in Figure 6-13 (select the security sensor from the Device list).

If the house is supposedly unoccupied [Hack #70] when this occurs, you may have a burglar. Alternatively, if someone is home but it's cold outside [Hack #64], you may want to announce a reminder to adjust the thermostat accordingly.

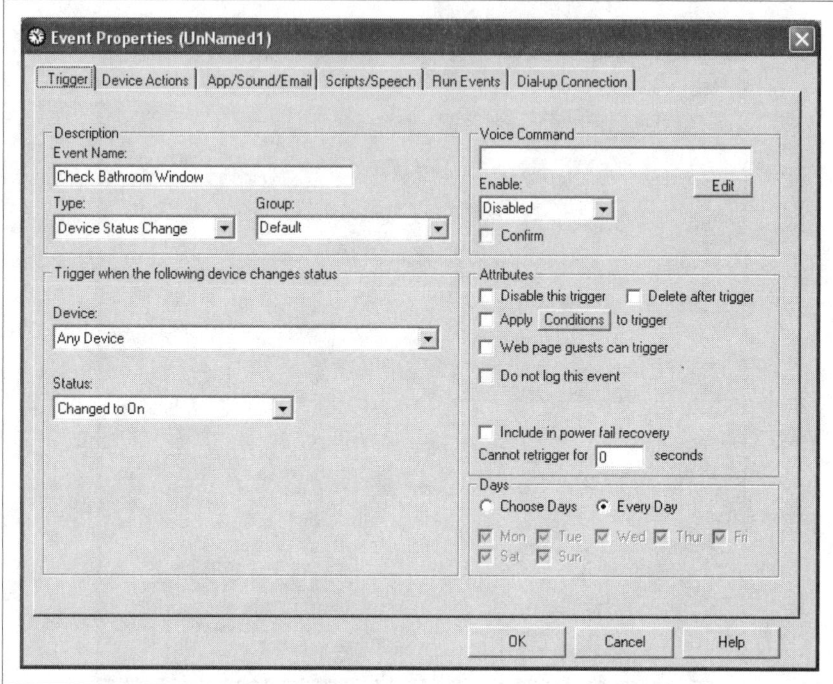


*Figure 6-13. An event trigger for a DS10A security sensor*

Technically speaking, the W80032RF supports regular and *extended* X10 commands. So, other devices that send extended commands work with it, too. For example, the X10 Entertainment Everywhere Remote Control, with its unusual mix of regular infrared and wireless X10 capabilities, works with it. You use the PC Remote tab in HomeSeer's 32 bit RF Receiver Setup dialog to enable support for the remote.

## Bark like a Dog

**You can protect your home with a real dog, or you can use the Robo-Dog to simulate one. And like any good dog, this module performs more than one trick.**

The X10 Robo-Dog module (*http://www.x10.com/security/x10_dk9000.htm*; $60) is truly a strange beast. Its purpose in life is to convince visitors that a vicious, barking dog is waiting to confront anyone who enters the home. It does this by playing a recording through the built-in amplifier and 4-inch speaker, which dominates the unit, as you can see in Figure 6-14.

The dog, affectionately named ReX-10, sounds like a big dog with a serious attitude problem, provided that you're not standing too close to the unit. If

*Figure 6-14. The X10 Robo-Dog*

you are, it sounds like a recording and might even induce a giggle. For this reason, carefully consider where you place the unit. You'll want it far enough away to mask its artificial qualities, yet loud enough to still be heard. Plan to spend a few minutes experimenting with different locations and volume settings when you set it up.

To use the Robo-Dog, you need only set its house code **[Hack #1]**. The module automatically assumes the unit code of `1`. In other words, if the module's house dial is set to `N`, its X10 address is `N1`. When the module receives `N1 On`, it will start barking madly. It will stop barking after about 30 seconds, or it can be silenced immediately by `N1 Off`.

The Robo-Dog is designed to work in partnership with other modules to provide a complete security setup. When you buy it, it comes with a wireless remote control that can start and stop the barking on demand. It works with security remote controls, too, so if you have an X10 alarm console, you can use the remote controls you already have. To do that, however, you'll need to follow the steps in the Robo-Dog manual to train it to respond to your system.

Additionally, the Robo-Dog can be triggered by a DM10A outdoor wireless motion detector (*http://www.x10.com/security/x10_dm10a.htm*; $30). The DM10A is a hybrid motion detector: it communicates with both home automation and security equipment by sending two different signals when it's triggered.

Instead of having the Robo-Dog controlled directly by a motion detector, one of the best ways to use a Robo-Dog is to make it part of what happens when a visitor approaches your unoccupied home. Anyone with bad intentions probably will be persuaded to look elsewhere when they hear what sounds like a fierce dog inside, see a few lights come on, and hear a distant radio playing **[Hack #74]**.

The Robo-Dog can run off batteries—a whopping nine C cells—but this is intended primarily for backup power so that it still can bark when the power goes out. However, don't be tempted to run it solely off batteries. If it's not plugged into the wall, it can respond only to wireless signals, which means you'll be unable to control it from your home automation computer.

There's something else that's unique about the Robo-Dog, and I bet most people who have one don't even realize it. The module isn't commonly used, and its documentation is incomplete, so I had one for a few years before a message from Michael Ferguson, on the XTension discussion list (*http://lists.shed.com/mailman/listinfo/xtensionlist*), tipped me off that the Robo-Dog is also a wireless *transceiver*. That is, any wireless commands it receives for its house code are echoed to the power line. This means that all nearby motion detectors and Palm Pads **[Hack #5]** can use it as their bridge to the power line, and thus to your home automation system, without you having to install a transceiver just for them. That's a nice bonus that not only saves a little money by pressing the module into double duty, but also enables you to eliminate yet another unsightly device from your home **[Hack #14]**.

It's a very good dog, indeed.

## HACK #80 Unite Your Alarm and Home Automation Systems

**If you have both an alarm and a home automation system that aren't currently on speaking terms, you might be able to bridge the gap with minimal effort.**

Although you can use your home automation system to help keep watch over your house, nothing beats a dedicated alarm system for doing the job right. As a result, many people have two separate systems in their homes: one for automation and the other for security. If your security alarm console has an unused relay output and your home automation controller [Hack #21] has ports where you can directly connect switches and input lines—such as the LynX-10, HomeVision, or, with extra modules, the Ocelot—you can get the two systems talking to each other.

If your home automation controller doesn't have input ports, you can use a Weeder Board (*http://www.weedtech.com*) or LabJack (*http://www.labjack.com*) instead.

First, let's cover how to allow your home automation controller to arm or disarm your alarm system. Most alarm systems have a feature called *key switch arming*. Key switch arming programs any unused zone on the alarm panel to act as a toggle that changes the alarm system from armed to disarmed, and back again, whenever the zone switch is tripped. By simply running a wire from the output relay switch on your home automation controller to the zone input switch you've defined as a key switch, you can get the two systems talking. Once connected, anytime your home automation controller closes the relay, the alarm system will turn off. When the relay is opened, the alarm system will be armed. It's the same as if you pushed the arm or disarm button on a keypad, but now the computer is doing it for you. For details on how you program your home automation software to send a command to the relay on your controller, see the software's onscreen help for how to access the special features of your controller.

Now, let's get the alarm system sending its status to the home automation system. This is necessary because the first connection we described only allows your home automation system to tell the alarm system to arm or disarm; there's no way for it to confirm what state it was already in, or if it actually obeyed the request.

To do this, program one of the relay output switches (often labeled AUX or Auxiliary) on your alarm panel or your alarm keypad if it has relay outputs, to toggle when the system is armed or disarmed. Run a wire from this relay to an input line on your home automation controller, and set up your home automation software so that it keeps track of the alarm's status, toggling between armed and disarmed whenever a signal on this line is received. You will gain a reliable way to know the status of your alarm system and, thanks to the first part of this hack, a method of changing its status programmatically.

In addition, most alarm panels have at least one relay output available for signaling when the alarm has been triggered. Run a wire from this relay to another input on your home automation controller, and now you can send an email [Hack #73], announce that the police have been notified [Hack #28], or perform whatever actions are otherwise required when you receive this notification. If you happen to have an alarm system that has additional per-zone relay outputs, and you have enough inputs available on your home automation controller, you can even extend this idea to report, and respond to, the status of each individual zone.

—*Jon Welfringer*

## Instill Peace of Mind for the Elderly

**If you have elderly relatives who you're concerned about, a few simple home automation techniques can bring you (and them) peace of mind.**

Instead of using home automation equipment to assist, you also can use it to monitor activity. Instead of controlling lights, it simply logs the activity of motion detectors and two-way X10 devices that send their status to the power line when they're activated manually. This can be useful for the young and old, who might not want to bother with home automation and lights that seem to have a life of their own, but would be willing to let the system notify a loved one if something seems amiss.

Set up an old computer—this is a great opportunity to reuse an old system you're no longer using—with just a few units that you can monitor. If you want to keep things unobtrusive, just one or two motion detectors [Hack #6] carefully placed can tell you a lot. Place them where they'll be triggered occasionally by everyday activities, such as in the kitchen or bathroom. Then, adapt the script described in "Check for an Empty Home" [Hack #69] to monitor for the motion you expect to see.

If the detectors haven't seen any activity lately, and it's during the day when the person at home should be active, have the script make an alert sound that asks the resident to press a button on a Palm Pad [Hack #5] to acknowledge that everything is OK. If there's no response after a while, have the computer dial into your ISP and send you an email, or have it turn on a lamp in the window that a neighbor has been enlisted to keep an eye out for.

You also might consider getting equipment that is designed for panic button or emergency use and combining it with home automation techniques for more flexibility and smarter behavior. The Telemergency Dialer (*http://www.smarthome.com/7962.HTML*; $130), for example, works with an external window sensor. Replace the sensor's contacts with a universal module [Hack #11], and you can activate the emergency dialer with an X10 command. Program it with your home and work phone numbers, and you'll know right away if something seems amiss.

Other modules can be useful for monitoring, too, such as the Powerflash [Hack #10]. Connect one to a pressure-sensitive mat (*http://www.smarthome.com/5195.HTML*; $70) that's activated when it's stepped on or to a reed switch that will be activated by opening a bedroom door. You also can use these techniques for someone who lives with you, but who shouldn't usually be leaving her room during the middle of the night. Set up the action that's triggered by these modules to check whether it's after sunset and, if it is, to sound a chime module [Hack #9]. If it's not during a time that you care to monitor, simply ignore the signal received from the module. Another idea is to adapt the techniques described in "Detect the Beer Thief" [Hack #38] to provide monitoring of medicine cabinets or refrigerators.

If none of these ideas seems to fit your needs, try a Google search for "elderly monitoring." You'll find a number of dedicated, and expensive, systems that are designed just for this task. However, at their heart, they're built on the same fundamental principles of home automation that are present throughout this book. With a little creativity, you probably can adapt the ideas to your existing setup and end up spending less to get a more flexible system.

## HACK #82 Monitor Your Home with a Network Camera

**If you have broadband network access at home, a standalone network camera is a great way to keep an eye on things. Your home automation system can manage the camera and save an occasional snapshot for later review.**

Network cameras are, in my opinion, one of the best examples of embedded computing. No larger than a video camera, they connect directly to the network and include a built-in web server for viewing the image and configuring the camera. The D-Link DCS-1000 network camera (*http://www.dlink.com/products/?pid=143*; $125) I use to keep an eye on my dog, Scooter, is a vital part of my smart home. Scooter stays in the kitchen while we're gone, and the camera is positioned for the best view of the area where he lounges out most of the time. We have an Ethernet jack nearby, so I use the wired version of the camera, but WiFi cameras are also available for the ultimate in portability and placement flexibility.

I've configured the camera's built-in web server to listen to port 8080, and I have opened that port on the firewall, which also reroutes the request to the camera's internal IP address. This enables my wife and me to see live pictures from the camera using any web browser, including those on our cell phones. Figure 6-15 is typical of what we see.

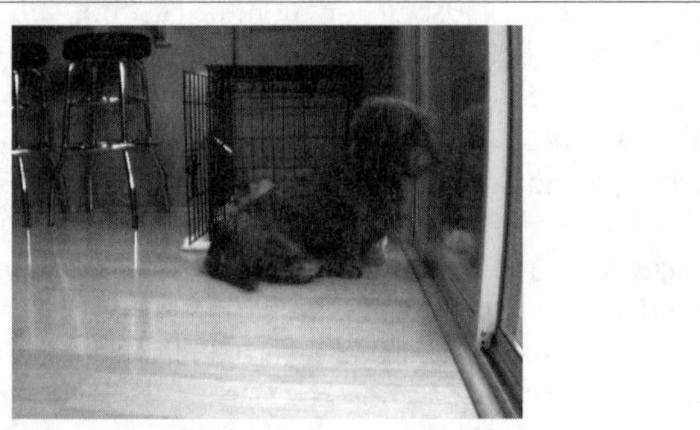

*Figure 6-15. Scooter, waiting patiently for our return*

When someone is at home with the dog, the camera is turned off. This protects our privacy by ensuring that the friends and family to whom we've given access to the camera can't peek in at other times. To accomplish this, the home automation system turns off the camera when the house is occupied **[Hack #70]**. The camera doesn't have a switch, so to accomplish this, I've plugged its power into an appliance module. When the last person leaves the house, the appliance module is turned on, putting the camera back online.

## The Code

In addition to the web-based access, a scheduled event in the home automation software saves the current camera picture every few minutes. The pictures are saved in a directory where we can review them later, both for fun and for investigative purposes if we suspect Scooter might have been misbehaving during our absence.

```
#!/usr/bin/python
import urllib2,random,time

# ############
# Configuration Variables
camurl = "http://192.168.0.19/image.jpg"
fpath = "/Users/gordon/Desktop/ScootWatch/"
# make sure folder and path actually exists!
# ############
def getpict(camurl):
        random.seed( )
        # seed random number with system time
        rn = random.randint(1,999999)
        # get a random number, used to trick camera into delivering new
photo for every request
        camurl = camurl + "?" + str(rn)
```

```
        # construct the url we'll use for this fetch using random int as
string
        # initialize our url object and fetch the picture image
        req = urllib2.Request(camurl)
        pict = urllib2.urlopen(req)
        return pict
def savepict(pict):
        tStamp=(time.strftime('%m%d%Y%H%M%S'))
        fname = fpath + tStamp + ".jpeg"
        # save to file
        fh = open(fname,'w')
        fh.writelines(pict)
        fh.flush( )
        fh.close( )
# ########################
# MAIN
pict = getpict(camurl)
savepict(pict)
# ########################
```

## Running the Hack

The script accesses the camera over the home network and saves the current picture, creating a filename that includes the current date and time, then exits. To save a picture at regular intervals, set up a repeating event that executes this script, as shown in Figure 6-16.

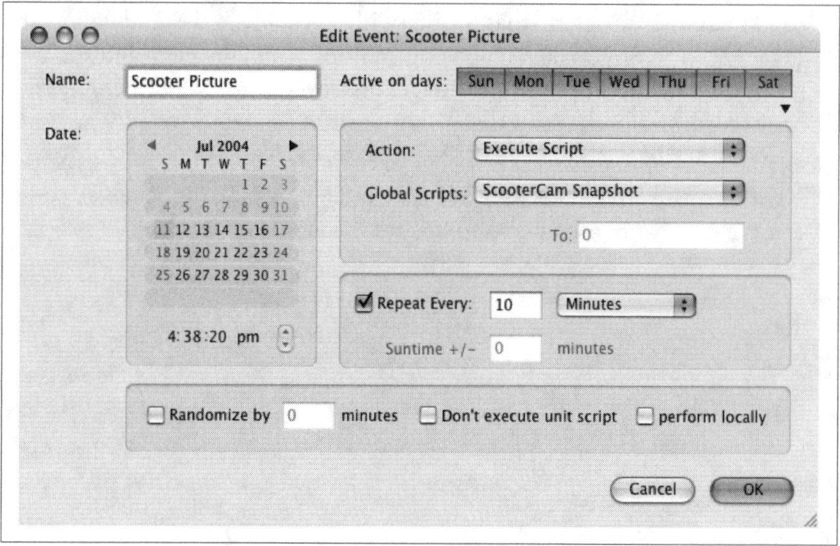


*Figure 6-16. An event to capture a photo every 10 minutes*

XTension can't execute Python scripts directly, so the event runs an AppleScript that, in turn, calls the Python script to snap the picture:

```
if (status of "nobody home") is true then
        do shell script "/Users/gordon/Documents/OneScootPict.py"
end if
```

To prevent the script from attempting to reach the camera and save pictures when the camera is turned off, the script checks the variable that indicates whether the house is empty. An alternative approach [Hack #70] is to suspend the script when the house is occupied, then resume it when the last person leaves the house.

If you have QuickTime Pro, it's easy to import the sequence of saved image files and create a time-lapse movie of the day's events. Choose Open Image Sequence from the File menu, and then select the first picture in the series that you want to import. Because the script names the files in sequence, they'll import in the correct order automatically.

## Other Approaches

These applications offer more ways to manage and view images from a webcam:

*SecuritySpy (http://www.bensoftware.com/ss/; $50)*
: A Macintosh program that works with network and standalone cameras. SecuritySpy makes managing multiple cameras easy and includes motion detection options, so images are saved only when there's action to record.

*GeekTool (http://projects.tynsoe.org/en/geektool/; free)*
: This is a versatile Mac OS X System Preference pane that enables you to put regularly updated webcam images in a desktop window.

# CHAPTER SEVEN

# Advanced Techniques

## Hacks 83–100

The hacks in this chapter represent the most creative, innovative, and often useful techniques for improving your home automation system. If you're interested in seeing what makes home automators tick, just dive into hacks that enable you to access your system from other computers on your home network [Hack #99], view information about your home using a calendar [Hack #91] or a chart [Hack #92], and get the most out of your motion detectors [Hack #85]. But lest you think this all sounds too serious, be sure to learn how you can have your kid's hamster earn his keep [Hack #89].

## HACK #83 Improve the Response Time of Motion Detectors

**Replacing your transceivers with a direct-to-computer wireless receiver greatly improves the response time from motion detectors.**

The standard X10 transceivers used to translate wireless commands [Hack #5] to X10 power-line commands are inexpensive and reliable, but they're not very convenient. Not only do you need to have a separate transceiver for each house code you use, but they're also slow. Their design introduces a brief delay as it translates the wireless command to its X10 equivalent, and that's in addition to the delay already inherent in the X10 protocol. Also, if you have many wireless devices sending signals at the same time, such as multiple motion detectors, the resulting X10 commands can flood the power line and cause some signals to be lost or further delayed. Fortunately, you can avoid these problems by replacing all your transceivers with a single receiver that connects directly to your home automation computer.

Currently, you can choose from two models of X10-compatible, direct-to-computer receivers: the MR26A PC RF Receiver from X10 Corporation (*http://www.x10.com/products/x10_mr26a.htm*; $30) and the W800RF32 Receiver from WGL & Associates (*http://www.wgldesigns.com/w800.html*; $75), shown in Figure 7-1.

*Figure 7-1. WGL & Associates' W800RF32 Receiver*

Both units have a DB9 serial port that connects to your computer (via a USB serial adapter if you have a Macintosh). This allows your wireless devices—motion detectors, Palm Pads, and remote controls—to send their signals directly to your computer. This completely bypasses the process of transmitting the command on the power line, as must be done when you're using transceivers. This saves time, and it generally improves X10 reliability by reducing the number of X10 commands that are competing for space on the power line. Additionally, unlike transceivers, the MR26A and W800RF32 listen for commands on all house codes once, so you need only one receiver for all your wireless devices.

> WGL & Associates sells two models: the W800 ($60) and the W800RF32 ($75). The W800 is comparable to the MR26A, but the W800RF32 has additional, and very useful, capabilities. It can receive extended X10 commands, which allows it to be used with security sensors **[Hack #78]** and multimedia remote controls. You can use the less-expensive W800 for this hack, but the W800RF32 might be a better buy if you want to expand your system's capabilities later.

## The Need for Speed

After you've set up your system for direct-wireless control, you'll be amazed at how much faster your motion detectors seem to work. You no longer have to wait three seconds between the time the motion detector sees you and when the lights that it controls turn on. With Palm Pads, modules seem to react before you've let up on the button.

Despite the expense, it's really well worth the effort to use a direct-to-computer receiver, and it's sure to increase your family's satisfaction with your home automation system. In addition, eliminating the ugly transceivers and freeing up those plug sockets for other devices are sure to tickle your interior decorator **[Hack #14]**.

## Improving Wireless Reliability

Sounds great so far? Well, just as power-line commands have occasional glitches, wireless commands aren't problem-free, either. There are so many wireless devices today that an MR26 or W800RF32 is certain to occasionally receive a spurious *phantom* command from an unknown source. When the phantom command matches the house and unit code of one of your X10 devices, your home automation software will react accordingly. For example, if your neighbor's cheap wireless doorbell happens to occasionally transmit a signal that your receiver interprets as `D14 On`, the light over your kitchen sink might come on in response. I suppose that's fine if you want to know when your neighbor has visitors, but diagnosing a problem such as that can be quite challenging.

To reduce the opportunity for phantom signals to wreak havoc with your system, you need to restrict the devices for which your home automation software will accept wireless commands. In Indigo, you do this by limiting wireless commands to certain house codes. For example, if all your motion detectors are set to house code `C`, and your Palm Pads are set to house code `D`, set Indigo to reject commands for house codes other than those two, and you'll eliminate the possibility of receiving phantom signals for other addresses.

To configure Indigo, choose Preferences from the Indigo menu. Then, click the Interface Options button in the Enable RF Receiver section of the dialog. The RF Interface Options window, shown in Figure 7-2, enables you to select the house codes that will accept wireless commands. Additionally, you can remap commands from one house code to another. This is useful when you've run out of addresses in a house code because you have more than 16 wireless devices.

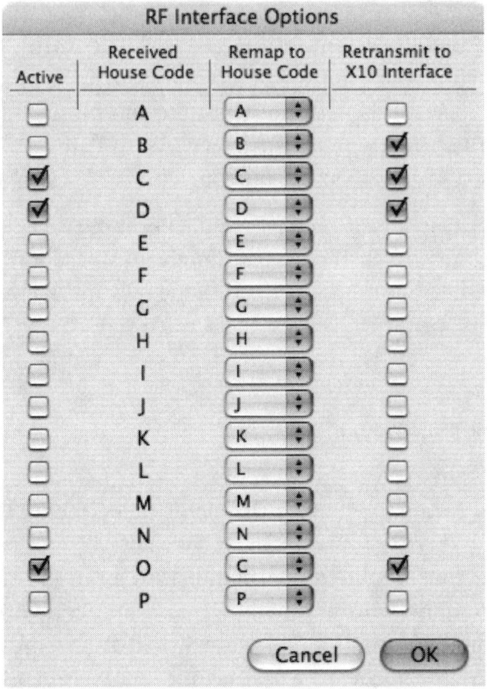

RF Interface Options

| Active | Received House Code | Remap to House Code | Retransmit to X10 Interface |
|---|---|---|---|
| ☐ | A | A | ☐ |
| ☐ | B | B | ☑ |
| ☑ | C | C | ☑ |
| ☑ | D | D | ☑ |
| ☐ | E | E | ☐ |
| ☐ | F | F | ☐ |
| ☐ | G | G | ☐ |
| ☐ | H | H | ☐ |
| ☐ | I | I | ☐ |
| ☐ | J | J | ☐ |
| ☐ | K | K | ☐ |
| ☐ | L | L | ☐ |
| ☐ | M | M | ☐ |
| ☐ | N | N | ☐ |
| ☑ | O | C | ☑ |
| ☐ | P | P | ☐ |

Cancel OK

*Figure 7-2. Setting wireless options in Indigo*

Finally, check the "Retransmit to X10 Interface" checkbox to have Indigo echo each wireless command it receives onto the power line. This is necessary if you want a motion detector to be able to directly control, for example, a lamp module. Remember that the whole point of a direct-to-computer receiver is that the signals aren't sent on the power lines. This means other X10 modules won't see the resulting commands, unless you choose this option.

If you're using XTension, the basic ideas are the same. To use a receiver, choose Preferences from the XTension menu and then click Communications. Check the Enable Wireless X10 checkbox and then select the device and serial port from the pop-up menus.

To allow XTension to receive wireless commands for a unit, select the unit in the Master List and then press Command-I. Check the "Wireless OK?" checkbox, and if you want commands for this unit echoed to the power line, check the "Pass-Thru" checkbox, as shown in Figure 7-3. XTension does not provide a wireless unit remapping feature similar to what we just described for Indigo, but as this book was going to press, a beta version of XTension added support for allowing duplicate house codes between wireless and standard addresses.

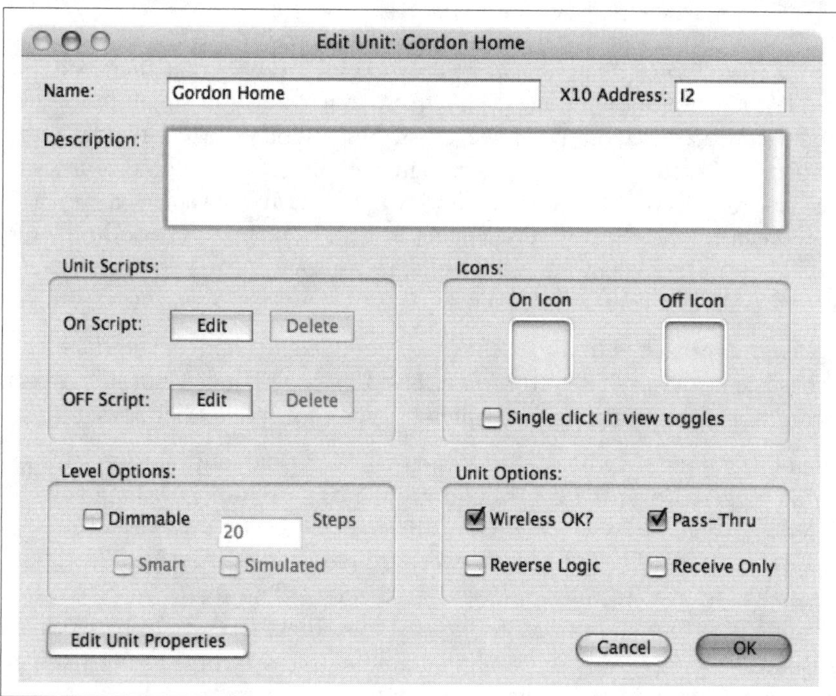


*Figure 7-3. Setting wireless options in XTension*

## Improving Reception

You need only one receiver. In fact, that's all you can have with most home automation software. But for best results, you want to it receive signals from devices located all over your home. The built-in antenna on the MR26A probably won't provide enough coverage, but you can hack an external antenna onto it. You'll find several sites that offer tips for this if you try a little Googling, but one of the best is Sand Hill Engineering (*http://www.shed.com/tutor/mr26ant.html*).

With either the W800RF32 or MR26A, you also can improve reception by using a long cable to position them far away from the computer and other sources of interference. Some automators are so enthusiastic about the benefits of this technology that they've installed their receivers in the attic to make sure every corner of the house is within range.

# Check for Dead Motion Detector Batteries

**Instead of scurrying about the house checking the batteries in your various motion detectors, have a script do the checking for you.**

The X10 motion detectors [Hack #6] draw so little power that their batteries can last for months at a time, even when the detector is in a high-traffic area and is activated frequently. However, as the batteries weaken, the detector's transmission range drops off and its signal might not be reaching your transceivers. Also, if you have a lot of detectors, it's more convenient to get notified when the batteries need changing instead of having to periodically test and confirm that each one is still working—provided that you can remember where they all are in the first place.

A better approach is to use a script that periodically checks to see how long it has been since each detector has received a signal. If a detector hasn't been heard from for a while, it's likely that it needs a fresh set of batteries.

First off, you need to decide how often you'd expect your motion detectors to be triggered over the course of a normal day, and for how long you can tolerate a detector whose batteries are dead. I expect most of my motion detectors to be triggered at least every couple of days. And because I have several motion detectors, one of them being offline for three days or so wouldn't be that great of a problem, as the others still would be working. Also, as much as I rely on my home automation system, I'm not likely to replace the dead batteries the same day they're discovered. I'll likely put it off until I have some spare time.

Given these parameters, I concluded that each detector should be triggered (under normal conditions) every other day, but I'm willing to have a nonworking detector for up to four days. So, I use a script that runs every other day and checks to see that every motion detector has been heard from within the last 72 hours.

## The Code

This is a global script that is executed every other day by a repeating scheduled event [Hack #17]:

```
if (status of "Nobody Home") is false then -- don't do this if house is
empty
    set mList to "" as list
    set mList to all of group "all motion detectors"
    set h to 72 -- how many hours of inactivity triggers decision

    repeat with x from 1 to count of items of mList
        set u to item x of mList
```

```
        if (time delta of u) is greater than h * hours then
            write log "Check batteries in motion detector: " & u
        end if
    end repeat
end if
```

The first thing the script does is to check if anyone is at home [Hack #70]. If the house is empty, and we've been away for a few days, the motion detectors will not have had a reason to fire lately and this might cause the script to falsely conclude there is a battery problem. (However, the detector's dusk detectors still should be sending a signal, so that's not an entirely sound assumption.)

If a motion detector hasn't been heard from in more than 72 hours, a message is written to the log with its name. Other possibilities would be to send an email, or to set the description of the unit so that it will stand out for you later.

```
if (time delta of u) is greater than h * hours then
    set description of unit u to "Check my batteries!"
end if
```

If you want to have different silent periods for some motion detectors, assign the detectors to different groups [Hack #17]. For example, create a `Daily Motion Detectors` group and a `Weekly Motion Detectors` group. The latter might include motion detectors you're using to alert you to activity in seldom-used guest rooms or your liquor cabinet [Hack #38].

```
set mList to all of group "daily motion detectors"
repeat with x from 1 to count of items of mList
    set u to item x of mList
    if (time delta of u) is greater than 72 * hours then
        write log "Check batteries in motion detector: " & u
        speak "Check batteries in motion detector"
    end if
end repeat
set mList to all of group "weekly motion detectors"
repeat with x from 1 to count of items of mList
    set u to item x of mList
    if (time delta of u) is greater than 168 * hours then
        write log "Check batteries in motion detector: " & u
        speak "Check batteries in motion detector"
    end if
end repeat
```

## Hacking the Hack

This technique, using a periodic repeating script to check up on usage patterns in your home, can also be applied to finding lights that have been left on at night accidentally, or a weather station [Hack #64] that has gone offline.

# HACK #85 Outsmart Motion Detectors

Motion detectors are an essential part of a smart home, but they have some limitations that make them less than perfect. However, with some clever positioning and scripting, you can overcome most of their quirks.

To use motion detectors effectively, you need to see the world as the motion detector does. The type used for home automation usually works on the passive infrared (PIR) principle. The physics of PIR are best left to scientists; all you need to know is that motion detectors work by noticing the movement of hot or cold objects **[Hack #6]**. For example, when a person enters the motion detector's field of view, the motion detector sees the person as a moving object that is radiating a different temperature pattern (higher, in most cases) than the surrounding background.

Because motion detectors are sensing patterns of heat and movement, it can be easy to fool their sensors. For example, if the motion detector is pointing toward a blank wall and it's a sunny day, a passing cloud that temporarily casts a cool shadow can trigger the detector. Indoors, a detector that's too close to an air duct can be fooled by a warm or cold air blast when the blower first comes on.

Aside from triggering accidentally, motion detectors sometimes fail to see legitimate motion. This can occur when a person's body heat is too close to the surrounding air temperature (such as a hot garage in the summertime) or the movement is too small in comparison to the overall area the detector is monitoring. To help avoid both kinds of problems, you should carefully consider where you mount the detector and change its field of view.

## Location Matters

The active sensing area of Hawk Eye-style X10 motion detectors is fan-shaped and is wider than it is tall, similar to that shown in Figure 7-4. As a result, it's more sensitive to motion that crosses the field of view than it is to motion coming toward the detector. Consider that a body walking across the detector's path will move two to three feet in a single step, but if the same body is moving straight toward the detector, it has gotten larger but hasn't moved much relative to the rest of the field of view.

If you're trying to detect motion in a confined space, such as a hallway, you often can make the detector trigger more quickly by mounting it on its side so that its view is taller than it is wide. Depending on the height at which you've mounted the detector, this will place your entire body into its sampling zone, which improves its ability to see the limited range of motion possible in the small space. Or, try mounting the detector on the ceiling, so

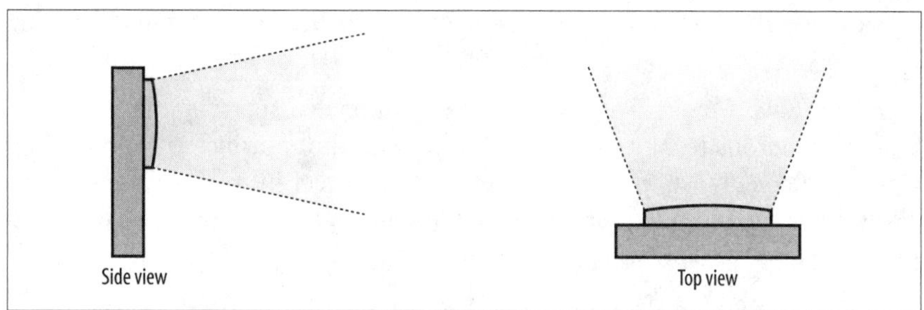


*Figure 7-4. How a motion detector sees the world*

it's looking straight down at the area you want to monitor. Mounting a detector on the ceiling, just inside the doorway, usually results in a very reliable method of knowing when someone has entered a room and not just passed by its open door.

Another handy trick is to attach cardboard flaps that block off portions of the sensor and limit its field of view; similar to the blinders placed on horses to settle them down in crowds. This technique is useful for avoiding false triggers—caused by passing traffic, moving tree branches, and the like—from the detector's vision.

Combine all of these methods to resolve other challenging situations. For example, a detector turned sideways and mounted four feet off the ground, with the bottom half of the sensor covered by opaque tape, is probably not going to be set off by pets or small children.

Although clever placement is important for helping the detector work well, you also can use clever programming of your home automation software to ensure the correct reaction when the detector is triggered.

## React Logically

Instead of always reacting to a motion detector, it can be useful to check on the status of other things to determine the best way to react. For example, if you want to use a motion detector to turn on the lights when someone enters a room, you can use an On script to turn on the lights only when it's after sunset:

```
if (status of "sunlight") is false then
    turnon unit "stairway light"
end if
```

This is a good first step to creating a smart home, but because of the inherent delay involved with wireless devices and X10 signaling **[Hack #83]**, you might find that your light comes on too late for a person who is just passing through. To avoid this problem, turn on the lights in the area where the motion was detected *and* the lights where you believe the person is heading next.

Here's how you make it work. Let's say you have a stairway that is potentially treacherous to navigate in the dark. Instead of putting a motion detector over the staircase, place the detector several feet in front of the stairs to allow time for the detector to trigger, for its signals to be sent and acted upon, and for the stairwell lights to turn on:

```
if (status of "sunlight") is false then
    turnon unit "stairway light" for 2 * minutes
end if
```

Turn on the stairway lights for just a minute or two—long enough for them to be useful, but not so long that if the person didn't use the stairs they won't be on, and unused, for very long, as shown in the previous script example.

If it bothers you that you might be turning on lights when they're not needed, use more than one motion detector. For example, when someone is detected in a hallway, turn on the lights in both of the bedrooms where the person might be heading. Then, when a motion detector in the bedroom where the person actually entered is triggered, turn off the lights in the room he didn't choose:

```
if (status of "guest bedroom lamp") is true then
    turnoff "guest bedroom lamp"
end if
```

In this On script for the detector in the master bedroom, the light in the guest bedroom is turned off. Of course, this works reliably only for homes with one occupant because it assumes only one room can be occupied at a time. If you have houseguests only occasionally, you can implement this so that it won't turn off your guests' lights, providing that your home automation system knows when you have visitors **[Hack #70]**:

```
If (status of "Guest Home") is false then
    if (status of "guest bedroom lamp") is true then
        turnoff "guest bedroom lamp"
    end if
end if
```

By carefully placing your motion detectors, and using the collective information they can provide about activity in your home, you begin to take your automation system to new heights. For example, you can determine if the house is empty **[Hack #69]** and reliably detect when you might have an intruder **[Hack #71]**.

# Improve X10 Reliability

**The electrical environment in your home can be a pretty harsh place; it's filled with noise and connections that aren't conducive to the propagation of X10 commands. But with a little work, you can improve it greatly.**

The most common criticism of X10 is that it isn't 100% reliable. And that's true; as a protocol, it's quite simplistic. There's no error correction and signals aren't acknowledged, so, in most cases, your modules simply send their commands and hope they get through to their intended recipient. To make matters worse, the medium that X10 uses (your home's electrical system) isn't designed for carrying data signals. In many ways, instead of being overly critical of X10's shortcomings, we should be pleased that it manages to work at all.

XTension's author, Michael Ferguson, says that as with any other technology, X10 works best when you establish a level playing field. For example, few would say that WiFi is a poor technology, but it, too, has its own environmental quirks and sensitivities. In my stucco home, my WiFi range is very limited, particularly when the microwave is in use. But the enjoyment I get from my AirPort Base Station far exceeds the occasional signal problems.

That's how I feel about X10, too. But I'm not immune from wanting the most reliable system possible, so let's take a look at the three most common sources of trouble that can cause home automation headaches.

## Electrical Noise

Electrical noise is the most common source of problems with X10. It usually occurs because something that's plugged into your electrical system is blocking or absorbing X10 signals. In an ideal world, all appliances would be designed and built so that they don't leak noisy signals onto the power line, but most just aren't designed that way. The most common culprits are items that have power supplies, such as electric toothbrushes, laptop computers, and halogen lamps.

The problems caused by devices that interfere with X10, commonly referred to as *signal suckers*, are easy to fix, if you can identify them [Hack #98]. Once you have located the source, you can use an X10 noise filter, such as a FilterLinc (*http://www.smarthome.com/1626.HTML*; $25), to solve the problem. To use a noise filter, you simply plug it into the wall, then plug the noise-causing device into the outlet that's on the filter. The filter will block any noise from entering the power line and interfering with X10 signals. It also blocks, by the way, any X10 signals from reaching what you've plugged into the filter. This makes it a handy way to set up an isolated X10 system for special purposes [Hack #13].

You'll need one noise filter for every device that interferes with X10. To save a little money, you can plug a power strip into the filter to isolate everything that's plugged into it, which is a fine approach for a workshop or office where you might have a handful of devices that could cause problems for your system. In fact, some power strips, particularly those with surge suppression, can block X10 signals as an accident of their design. This can work to your advantage if the suppression circuitry acts as a noise filter, but if you're unlucky, the power strip could generate noise and require a noise filter of its own. The only way to know for sure is to try your power strip and see the results.

## Weak Signals

If you've put noise filters on troublesome appliances, or otherwise eliminated the sources of electrical noise in your home, and your problems remain, you might need an X10 signal booster. The most common symptom of needing a signal booster is that you can send X10 commands to most of the devices in your home, but commands intermittently fail to reach some areas. Perhaps a lamp in the second bedroom will, on occasion, refuse to turn off at night.

In this situation, the first thing you should do is to make sure there isn't an occasional-use signal sucker that happens to coincide with the problem. For example, perhaps your daughter is charging her cell phone every other night, and on those nights, the lamp can't be controlled. If that's the case, put a noise filter on the charger and you're back in business.

Intermittent misbehavior from an X10 module is also a warning sign that the module might be failing. Try swapping it with another module that you know works reliably and see if the problem goes away.

Otherwise, consider getting an X10 signal booster or repeater. When X10 signals are sent, they go everywhere in your electrical system, and if there is noise, there are weak connections, or you have a large home, the signals can grow too weak to be understood. A signal booster, such as the BoosterLinc (*http://www.smarthome.com/4827.html*; $100), is a special-purpose device that listens for, then amplifies, X10 signals to increase their reach. Unless you have a very large home, you'll need only one. Simply plug it into the wall and hopefully your problems will be resolved. If not, try plugging it into a different outlet that is closer to the troublesome area of your home.

## Phase Coupling and Repeating

If your X10 transmission problems aren't intermittent and you can't reach some areas of your home with commands, it's likely that you need an X10 coupler. Most homes are wired with two-phase power systems. That is, the electrical feed coming into the home from the outside consists of two 120-volt lines, with each phase supplying approximately half of the power used in the home. Depending on how your home is wired, X10 signals that originate on one phase might be unable to cross over to the other phase, and vice versa. If you suspect this might be the situation in your home, turn on a 220-volt electrical appliance (an electric stove or clothes dryer) and see if commands can reach everywhere. If they do, you definitely need an X10 coupler, sometimes called a *signal bridge*. Turning on your dryer solves the problem because it draws power from both electrical phases, and the X10 signals are able to pass between the phases by going through the dryer when it's switched on.

Most couplers must be installed in your home's circuit panel, where an electrician can unite the two phases easily. Several different couplers are available, each with the ability to bridge, and sometimes amplify, X10 signals. Typical to Smarthome's approach, however, its SignaLinc coupler (*http://www.smarthome.com/4826.html*; $100) simply plugs into a 220-volt outlet and bridges the phases without requiring any work in your circuit panel. If you go this route, be sure to get one that matches the plug connector used by your dryer or oven because it comes in both three- and four-wire configurations.

An *uncommon* problem with X10 transmission is signals that originate outside your home. If your neighbor has an X10-based home automation system, and you happen to share the same electrical transformer on the power pole, it's possible for X10 commands to leave your neighbor's house and enter yours. This is probably unlikely, but keep it in mind if you run into a problem you can't explain in any other way. You'll need to either install a whole-house blocking coupler (*http://www.smarthome.com/4850.HTML*; $75) or agree with your neighbor not to use the same X10 addresses for your devices.

## Final Thoughts

Frustration over X10 signal problems can spoil your home automation experience if you let it get to you. By methodical use of the tools described in this hack, and by using higher-quality modules that are less sensitive to noise, it is entirely possible to create a reliable home automation system. But nothing is perfect, so it's also important to check your expectations and make sure you're not setting up situations where an X10 transmission problem has serious consequences. For example, don't rely on X10 to alert you to safety-threatening situations, or use it where the result of a lost or misinterpreted signal would be property damage, or worse. Instead, use it for your own convenience, and if the light over the kitchen sink fails to come on at sunset, just shrug your shoulders and turn it on yourself. That's what you'd have to do if you didn't have a home automation system, and you're no worse for the wear, right? Tomorrow, you can track down what might be causing it, if it really bothers you.

## HACK #87 Avoid Common X10 Problems

**When a problem arises, these handy tips will get you well on the road to resolving it quickly.**

If you're experiencing problems with your X10 modules (lights that won't turn off or on, for example), you always should check the three things addressed in this hack first. If you start with these, you'll solve most problems in a short amount of time.

### Lights That Turn On Unpredictably

When the lights come on randomly, and you're sure your house isn't haunted, it's due to either genuine X10 commands or an interference that masquerades as an X10 command. Let's discuss genuine commands first.

It is possible that one of your neighbors has an X10 system, perhaps without even realizing it [Hack #23]. X10 signals can travel from one home to another when the homes share the same transformer. A single electrical distribution transformer will service from three to eight homes. To avoid conflicts with your neighbor's equipment, change the module's X10 address so that it won't pay attention to the foreign commands.

When you change its address, consider using a house code other than `A`, `B`, or `C`. These are the most common addresses used by newcomers, and are often the default address that modules are set to, so you can avoid extraneous commands by choosing a less-used house code. That, indeed, could be the source of your mysterious `A1 On` command. Similarly, you might choose to use unit addresses in the range of `9` to `16`. This upper half of the address range also is less commonly used, so it will reduce the likelihood of conflicts with your neighbor's devices. Make sure, however, that your transceivers [Hack #5] and minicontrollers [Hack #4] are capable of addressing the upper range of unit addresses; some models work only with units `1` through `8`.

Another source of mysterious commands can be modules that are failing, perhaps due to a weak battery. This causes them to forget the address you programmed and revert to using `A1`.

It's also possible that a *spike* on the power line has triggered the module. When the utility company transfers power around the grid to balance demand, which it often does on a scheduled basis, these microsecond-switching events can wreak havoc on X10, especially older modules. Although this should be a rare occurrence, you might find that installing a whole-house suppressor (*http://www.smarthome.com/4860.HTML*; $190) eliminates the problem.

If lamp and appliance modules turn on when the air conditioner or furnace comes on, it is possible that the *local control* feature is causing the problem. Not all modules work with local control, which senses when you are using the lamp or appliance's switch to turn on the device. Some older modules will interpret small fluctuations in the power line as a request to activate local control and will turn on when these occur. Replacing the module with a newer model usually fixes the problem.

If you've eliminated the possibility of authentic, but spurious X10 commands, the cause might be electrical noise, discussed later in this hack.

## Lights That Turn On Regularly

If you have a computer-based automation system **[Hack #16]**, look in the software's log to see if a scheduled event, or a script that was triggered by another command or event, sent a command to the light. For example, you might have a forgotten event that turns on the Christmas tree lights every night and the module you're experiencing problems with is set to the address **[Hack #1]** you last used for the tree lights.

If you've programmed your controller so that it works even when disconnected from the computer **[Hack #15]**, it might be acting on a schedule you've forgotten or programmed incorrectly. Clearing its memory and reprogramming the controller should fix it.

If you have a mini-timer (*http://www.smarthome.com/1100x.html*; $30) or wall timer (*http://www.smarthome.com/1121.html*; $85), review its programming. You might have a programmed event that you didn't intend to set. You can set both of these devices to trigger each event twice a day, which many people don't notice. To review the schedule of events, slide the switch on the timer to the Review setting, and press each On and Off button *twice*. If you find an event that you don't want, press Clear to delete it.

## Modules That Refuse to Obey

The only thing worse than a dog that won't listen to your commands is a light bulb that refuses to turn off or on no matter how many times you ask politely. This usually is caused by *electrical noise* on your power lines. Electrical noise is any signal on an electrical circuit other than 120 volts AC at 60 Hz. When an X10 transmitter sends a signal to a receiving module, it does so by superimposing an encoded X10 command on the AC circuit. Technically, this means X10 commands are themselves electrical noise, but because the signal is there by our action, it's *good noise*. The problem that you're experiencing is caused by noise that conflicts with the X10 signal by being too similar in frequency, or by drowning it out with a higher amplitude.

One solution to this is to buy more expensive X10 modules that have special circuitry for discerning between X10 commands and noise on the power line. Most high-end light switches, such as those from Leviton (*http://www.smarthome.com/4289.html*), have this capability.

Another approach, and generally a good practice, is to eliminate the most common sources of electrical noise. Here's a list of the most likely suspects:

- Plug-in wireless intercoms
- High intensity discharge (HID) lighting
- Motors (refrigerators, heating systems, pumps)
- Low-voltage lighting that uses solid-state transformers
- Fluorescent-based lighting and ballast transformers
- Failed or failing X10 transmitters, most commonly the TM-751 Transceiver Module

Not all of these devices *will* cause noise, but they *might*. A device that has been noise-free for years might begin to dump excessive noise onto the AC lines as it begins to fail or wear out. Also, most devices emit electrical noise only when they're on, which can add to the difficulty in discovering which device is causing the problems. It can be quite a detective job to find the offending device.

The best way is to isolate the circuits in your house one at a time. Start by turning off all the circuit breakers except for one. Plug in an X10 minicontroller **[Hack #4]** and connect a radio to an appliance module **[Hack #3]**. Turn on the radio, pump up its volume, and plug it into an outlet that's on the circuit you left turned on.

You need to connect both the minicontroller and the radio to the circuit you left on. You can't send X10 commands at all when the power is turned off, as there's no electrical signal on which the commands can piggyback.

Now try to control the radio with the minicontroller. Most likely, it will work. Turn on an additional circuit and try it again. Do you still hear the radio turning on and off? Continue to turn on the circuit breakers one at a time. Each time a new breaker is switched on, send some `On` and `Off` commands to the radio.

You will reach a point when turning on a circuit causes the test to fail. Now you have identified that something connected to this circuit is causing the problem. Turn off all the circuit breakers *except* the one on which you found the problem. Plug the minicontroller and appliance module into that circuit. Then, begin to unplug things on that circuit one at a time.

Don't just turn them off. You must unplug them!

Each time you unplug a device, perform a signal test on the appliance module. Again, you will reach a point when something you unplugged makes the signal test work. The last unit you unplugged is the device that is killing the X10 signals. To fix the problem, install a noise filter between the electrical supply and the offending device, such as a FilterLinc (*http://www.smarthome.com/1626.HTML*; $25). Good job, detective!

For additional methods of tracking down troublesome circuits and devices, see "Identify Trouble Spots" [Hack #98].

—*Smarthome, Inc.*

# HACK #88 Streamline Your AppleScripts

**If you use a Macintosh, you can simplify your scripts by adding your own commands to Indigo and XTension using a little-known AppleScript feature.**

If you've written a snippet of AppleScript that you need to use in several places—such as a routine that converts elapsed seconds to days, hours, and minutes [Hack #97]—you'll appreciate that Indigo and XTension both support *attachability*. Attachability enables you to define new commands (known as *handlers* in AppleScript parlance), and it can greatly simplify your system.

Instead of having to paste a copy of a favorite routine into every script that needs it, you have just one copy that any script can call.

Here's a simple example. Suppose you're using text-to-speech to make announcements. You want to use the voice called Fred to announce regular events and the Zarvox voice to announce anything that's urgent. You can simply hardcode the appropriate voice every time you use it:

```
say "Good Morning. It's time to get up." Using "Fred"
say "Front Door Visitor!" using "Zarvox"
```

This approach works, but if you later decide Ralph is a better voice than Fred, you'll have to find all the places where your computer speaks and change the voice parameter. A better approach is to create a handler for each type of announcement:

```
on sayit_Normal(theString)
  say theString using "Fred"
end sayit_Normal

on sayit_Urgent(theString)
  say theString using "Zarvox"
end sayit_Urgent
```

Now, you can speak phrases by passing the text of what you want announced to the appropriate handler:

```
sayit_Normal("Good Morning. It's time to get up.")
sayit_Urgent ("Front Door Visitor!")
```

Still, these handlers must exist in every script where they're used, unless you move them into an *attachment script*. The attachment script is included automatically when any script is executed, so any handlers that appear in the attachment script become available for use.

## Attachment Scripts and XTension

To edit XTension's attachment script, choose Edit Attachments from the Scripts menu. In the editing window that appears, enter the code for your handlers. You might have other handlers defined already in the attachment script. It doesn't matter where you add new ones (at the top or the bottom), just make sure you don't accidentally alter any that are already there. Also, be sure that each is enclosed by the `On` and `End` statements, as shown in the previous examples. Click Save when you're done, and from now on, you can refer to the handlers without having to copy them into any other location.

In XTension 5.1, some of the commands in the AppleScript Standard Additions might need their own `tell` block when used in an attachment script. In our example script, the say command generates a syntax error when used in an attachment script. To avoid this bug, target the command to the Finder:

```
on sayit_Normal(theString)
  tell application "Finder"
      say theString using "Fred"
  end tell
end sayit_Normal
```

Later versions of XTension might not require this workaround, but it's a handy technique to keep in mind if you ever find yourself unable to compile a script that you know is syntactically correct.

### Attachment Scripts and Indigo

With Indigo, you can have more than one attachment script. Choose Open Scripts Folder from the Scripts menu, and then open the Attachments folder in the Finder window that appears. This is where Indigo will look for attachment scripts to load, so save your compiled script into this folder. Then, either quit and reopen Indigo, or choose Reload Attachments from the Scripts menu. From now on, you can refer to the handlers in your attachment script without having to copy them into any other location. Indigo includes some simple and handy attachment scripts, so it's worth reviewing their code to become familiar with the types of things you can do.

### Final Thoughts

Attachments are powerful, but they can also be somewhat dangerous. When you define your own handlers, be sure the names you use don't conflict with commands already defined by Indigo or XTension. The handler in the attachment script will override the built-in commands. For example, XTension already has a `turnon` command, so don't create your own handler named `turnon`.

## Harness Your Hamster to Power a Night Light

**If your pet hamster drives you crazy with his nocturnal running-wheel marathons, you might as well have him power a night light that helps you find your way to the bathroom during the night.**

The alternative-energy enthusiasts at Otherpower.com (*http://www.otherpower.com*) took up the challenge of advising a schoolgirl on a science project that posed the question, "Can a rodent running on an exercise wheel generate enough electricity to power a light?" The result is a working, hamster-powered night light.

Not only is this a fun project, but also the clever solution embodies the spirit of smart home hacking: think creatively, use existing technology in new and interesting ways, and solve problems that enhance your living space. For the full back story about this project, and for more discussion about how the lessons apply to other alternative energy projects, visit Otherpower.com's web site (*http://www.otherpower.com/hamster.html*).

At first, we thought we could have the hamster's wheel spin a DC hobby motor to generate electricity, which would charge up some NiCad batteries we'd use to run the night light. This didn't prove practical, however, because hobby motors usually are unrated, so you don't know their target revolutions per minute (RPM), which can vary from between 5,000 to 10,000 RPM.

The motor's rated RPM determines at what speed it can generate a certain voltage. None of the DC hobby motors we had available could make even 1.2 volts (the voltage of a single AA NiCad battery) at 40 to 60 RPM. Plus, a diode, which is necessary to prevent the battery from powering the motor instead of vice versa, drops the voltage by at least another 0.7 volts. We could have used gearing or a belt and pulley system to raise the RPM; however, the friction losses in the gearing system would have been too high and the design too complicated. Rodents like to chew, and a rubber belt would be a tempting treat. Besides, it's more fun to build your own alternator than it is to use a premade hobby motor!

To build an alternator that can harness the hamster's energy effectively, you need to know the approximate RPM at which he spins the exercise wheel. The lower the RPM, the more difficult it is to design a good alternator for it. Analise, the eighth grader who asked us to help on this project, arrived at our figures. She took a stopwatch to the local pet store and observed a variety of rodents in action. She recorded how many times the wheel went around in 10-second intervals, then multiplied these figures by 6 to derive the RPM. She found that most rodents achieve between 40 and 60 RPM on the exercise wheel. That gave us a starting point for the alternator design.

So, we decided to make our own low-RPM alternator for the hamster wheel. The first issue we noticed is that the wheel that came with the cage was noisy when Skippy (a Syrian hamster) ran on it—*very noisy*, in fact. Aside from being annoying, this means some of Skippy's energy was being wasted and converted to sound by mechanical vibration. We wanted all of his energy to go into power production instead, so we modified the wheel by adding a new, quiet ball bearing. We simply used a spare, inexpensive brushless DC motor we had on hand, but you can use any free-spinning bearing and attach it to the wheel (a spare bearing salvaged from a roller skate or skateboard, for example). Figure 7-5 shows the bearing we selected.

*Figure 7-5. Brushless DC motor*

To mount the motor to the hamster wheel, center it on the wheel and mark the holes from the top of the hub for drilling. The hub on our motor has six holes: three that are tapped to a weird little SAE 2/56 thread and three that are blank. We used a simple 4/40 tap to thread the blank holes to fit 4/40 machine screws and then used these to mount the hamster wheel to the motor hub. Then we mounted the motor to a thin piece of wood. The entire wheel and motor assembly is freestanding on this wooden bracket, as shown in Figure 7-6, and does not have to mount to the side of the cage.

*Figure 7-6. Wooden motor mount*

The next step is to build an alternator onto the wheel. A *permanent magnet* (PM) is the best approach because brushes are not required for making contact with the spinning part (called an *armature* or *rotor*). On the axis side of the wheel (the flat part), affix a steel ring to which you can affix evenly spaced magnets.

You're creating an *axial-flux* alternator. If you fitted the magnets to the curved surface of the wheel, around the circumference, it would be a *radial-flux* alternator.

Cut a ring of approximately 16-gauge steel to fit the wheel. We used a hole saw to cut the inner circle and a band saw to cut the outer. You could use two coffee can lids stacked together instead, which you can cut using tin snips after you drill some starter holes. Be careful, as the edges will be very sharp! Grind them down so that they are safe for both you and your rodent.

Next, mount magnets on the steel ring. You need to keep in mind that the hamster has a limited amount of power, so choose magnets that are light and powerful. We used a 3/4-inch diameter by 1/8-inch thick Grade N-35 Neodymium-Iron-Boron magnets (*http://www.wondermagnets.com/cgi-bin/edatcat/WMSstore.pl?user_action=detail&catalogno=0030*; $1 each). You can make up for lots of design and mechanical tolerance problems in a homemade alternator by using very strong magnets. If you use weaker magnets, your power output will be reduced significantly.

Compared to the giant, finger-breaking magnets used in 10-foot wind turbines, these little magnets are pretty safe and easy to work with. They are, however, powerful enough to pinch skin and give you a blood blister, so handle them carefully and keep them out of reach of young children. Read more important safety tips at the Wonder Magnets web site (*http://www.wondermagnets.com/disclaimer.html*).

We used 14 magnets because it was the largest *even number* of magnets we could space around the ring, so this number will vary depending on the size of your rodent's wheel. You must use an even number of magnets. We spaced them evenly using a stack of playing cards and trial-and-error until we found the right number of cards to measure the gap between each magnet. It's important to get them spaced evenly, aligned in a perfect circle, and placed so that they alternate between north and south pole orientations.

To begin mounting the magnets, place one on the steel ring and center it between the inner and outer edges of the ring. Tack it down with a small drop of cyanoacrylate glue (i.e., Super Glue) applied around the edges. Use a thin-viscosity glue (*http://www.wondermagnets.com/cgi-bin/edatcat/WMSstore.pl?user_action=detail&catalogno=5501*; $7) so that it will wick underneath the magnets and set up. It also helps to have a tube of glue accelerator (*http://www.wondermagnets.com/cgi-bin/edatcat/WMSstore.pl?user_action=detail&catalogno=5504*; $5) on hand; one spritz of this, and the glue sets up instantly.

You must place each magnet so that its outside-facing polarity is the opposite of its neighbors. To get this right, carefully take each new magnet and hold its face to the previous magnet that's tacked onto the ring. If it repels, tack it to the ring so that the repelling side faces out. Repeat this before you glue each magnet and check it when you are done. The magnets should alternately repel and attract the magnet in your hand as you go around the ring.

Now, make sure the magnets are spaced evenly and are aligned in a perfect circle. Check the inner and outer diameters of the circle of magnets and slide them around to get perfect alignment both ways. When everything is good, and you've checked to be sure the magnets alternate in polarity, glue all of them down by applying glue around their edges.

To mount the ring to the hamster wheel, we simply centered it on the back of the wheel and held it in place with four additional strong magnets stuck to the inside of the wheel! This makes it easy to remove but still holds it tightly. You might need to use glue; remember that the magnets should be away from the metal bars of the cage or other metal pieces. If they're too close, the attraction will slow down the wheel and you'll be wasting lots of hamster power.

Be careful as you experiment with the position of the wheel inside the cage; it will be powerfully attracted to any nearby ferrous metal or other magnets.

The next step is winding the coils. We wound one, tested it, and found we needed more voltage, so we ended up using two coils. Each coil is 400 turns of #30AWG enameled magnet wire (*http://www.wondermagnets.com/cgi-bin/edatcat/WMSstore.pl?user_action=detail&catalogno=930*; $16). We used a simple handheld coil winder, and made up a new center insert to get the coils the right size. The inner hole of the coil should be about the same size as each magnet you are using. You also could wind the coils around a tube that's the right size, but the elliptical coil shape we got by using a winder performs a little better, and the tapered form in the middle makes it easier to get the finished coil off the winder. The important thing is to pack the wire in there as evenly and tightly as possible. The finished coils should be about 1/4-inch thick, with each leg of the coil about 1/4-inch wide and the center hole matching the magnet size. Figure 7-7 shows the winder we built for this project.

Once you get 400 turns on the coil, twist the leads together so that the coil doesn't unwind, and then drip the glue onto the coil. Spray the coil with glue accelerator to speed the drying time and then disassemble the winder and carefully remove the coil. Add more glue if the coil starts to unwind.

Figure 7-7. Coil winder

We coated the winder with crayon wax so that the glue wouldn't stick to it. This made it easier to remove the finished coil.

At this point, you can attach a voltmeter to the ends of the coil and hold it next to the spinning magnetic hamster wheel. Our coil generated about 1.2 volts AC, so we built a second coil which, when hooked up in a series and in phase with the first coil, will double the output.

Let's mount the first coil. Glue it to a small block of wood and glue the block to the mounting frame that you built for the wheel bearing, as shown in Figure 7-8. Get the coil as close to the rotating magnet ring as possible without them touching. We ended up with a 1/8-inch gap here due to the wobbly motion of the wheel, but the big magnets make up for this tolerance flaw.

Figure 7-8. Mounting the coil

Mount the second coil using the same method, but note that its orientation and positioning are critical. Use a voltmeter to make sure you get it right. Connect one lead of the first coil to one lead of the second coil; then, connect the other lead from each coil to your voltmeter. Make sure the meter is set to measure AC power. Hold the second coil away from the alternator and spin the wheel. You'll see the voltage from the first coil. Hold the face of the second coil up to the alternator and spin it again. If you see approximately double the voltage, it is oriented correctly. If you see no voltage, it's backward. Flip the second coil 180 degrees, so the other face points toward the magnets, and measure again. When you know you have the coil oriented correctly, mark the faces; then, mount the coil on a block of wood, as you did before.

Next, you must place the second coil so that it is in phase with the first. The easiest way to do this is to have a helper hold the wheel so that the first coil is exactly centered on one magnet. While your helper keeps the wheel from moving, place the second coil so that it, too, is exactly centered on a magnet and glue the mounting block in place. We placed ours opposite each other, but you can place the second coil anywhere, even right next to the first, as long as both coils are exactly centered on a magnet.

Now, you can test the alternator. Give the wheel a spin and, again, you should see about twice the voltage you saw with one coil. If not, first make sure all the wires are thoroughly stripped—it's trickier than you'd think. It's also possible you got a coil reversed, or that they are out of phase.

Now, you're ready to hook up the night light. As we mentioned before, this simple two-coil design and low-RPM generator don't generate enough energy to charge a battery; rectifying the AC output of this alternator into DC would lose more than half the voltage. Though you could increase the number of coils or the number of turns of wire, the additional resistance would result in more than half the increased output being lost to heat instead of light.

We ran two super-bright, red LEDs directly from the alternator and attached them to a project board, as shown in Figure 7-9.

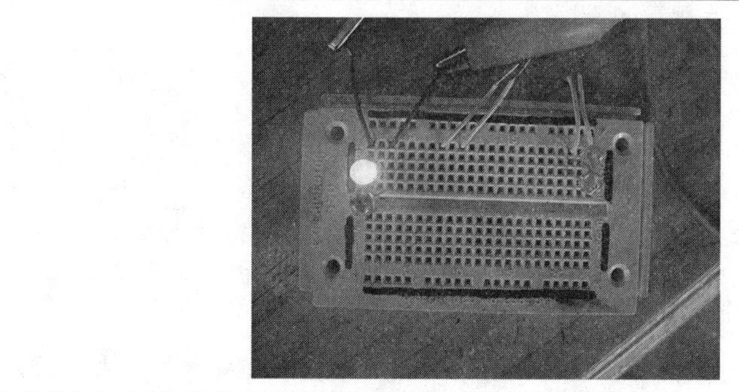

*Figure 7-9. The LED night lights*

We chose red LEDs because they require only about 2 volts to get very bright. Other colors will work, too, but they'll need more power and won't be as bright as the red ones will, at the same speed. The LEDs pass current in only one direction, so connect the two backward from each other. When one is lit, the other is off. The flicker frequency changes as the hamster changes speed. The resulting light is bright enough for finding your way to the bathroom at night, provided that Skippy is running. *Run, Skippy, run!*

## Hacking the Hack

It's just too tempting not to take this project over the top in silliness by gathering data about the hamster's behavior and abilities. So, use an inexpensive bicycle computer, available at any bike store, to monitor the wheel. The computer comes with a sensor and a trigger. The trigger is a cylindrical NdFeB magnet, which you mount to the outside radius of the hamster's wheel, as shown in Figure 7-10.

*Figure 7-10. Wheel sensor mounted in place*

The bike computer's sensor is simply a reed switch. Mount it, or another reed switch that you have handy, to the frame so that the trigger magnet passes within 1/8 to 1/4 inch when the wheel spins by.

Next, you need to tell the computer the circumference of your rodent's wheel, in millimeters. You use a special button to set the wheel size—consult the bike computer owner's manual. Skippy's wheel turned out to be 540 millimeters in circumference. After the computer is set, spinning the wheel should cause the computer to display a miles-per-hour reading. We've found that Skippy regularly achieves 2–3 MPH, and he goes slightly faster when my cat is leering at him through the cage walls.

The bicycle computer gives you a lot of data that you can analyze. It remembers maximum speed, and it keeps track of the total miles your rodent has run. You can use its trip odometer to keep a running total at regular intervals, and its timer mode will show you how many Hamster-Miles (hM) your rodent ran each night, as well as the average speed he maintained.

## Postscript

Skippy (Figure 7-11) passed away recently. One of my cats found a way to open his cage door.

*Figure 7-11. In memoriam*

His data acquisition computer showed a peak speed of almost 7 MPH before his untimely disappearance.

—*Dan Fink*

# HACK #90 Get More Out of Your Motion Detectors

**A computer-based home automation system can derive useful information from even the simplest of X10 modules.**

Instead of simply having motion detectors **[Hack #6]** turn lights on and off, you can use them to determine what state **[Hack #24]** the house is currently in. For example, if it's daytime and a weekday, the motion detector at the front door will create a trigger that puts the house into away mode **[Hack #70]** after 30 minutes. Once the house is in away mode, any activity will trigger a notification **[Hack #73]**. At night, a lack of motion for a long period puts the house into sleep mode **[Hack #48]**.

You implement these functions using an attachment script **[Hack #88]** for Indigo **[Hack #18]**. The script implements a function called `mslogging( )` that, in turn, creates and maintains several Indigo variables to keep track of various states and data:

```
using terms from application "Indigo"
    on mslogging(msunit)
        tell application "Indigo"
```

```
            if not (variable "AwayDelay" exists) then make new variable with
properties {name:"AwayDelay", value:30}
            if not (variable "SleepDelay" exists) then make new variable
with properties {name:"SleepDelay", value:90}
            set AwayDelay to (value of variable "AwayDelay" as real)
            set SleepDelay to (value of variable "SleepDelay" as real)
            set isDaylight to (value of variable "isDaylight" as boolean)
```

The script uses the motion detector's description field to determine an appropriate response to its activation. For example, here it looks for the words `Indoor` or `Main Entry Location` and then sets the variable that tracks whether the home is occupied or resets the sleep mode timer, as appropriate:

```
            if description of device msunit contains "Indoor" then set
(value of variable "HouseMode") to "occupied"
            if description of device msunit contains "main entry location"
then
                if (time/date action "Initiate sleep mode" exists) then
delete time date action ("Initiate sleep mode")
                MakeTrigger("Initiate away mode", (current date) +
(AwayDelay * minutes), "SetHouseAway( )")
                log "Creating trigger for away mode"
            else if description of device msunit does not contain "main
entry location" and description of device msunit contains "Indoor" then
                if (time date action "Initiate away mode" exists) then
delete time date action ("Initiate away mode")
                if (isDaylight is false) and ((calculate sunrise) - (current
date)) / 60 > SleepDelay then MakeTrigger("Initiate sleep mode", (current
date) + (SleepDelay * minutes), "SetHouseSleep( )")
            end if
            set NewUnitName to searchReplace(msunit, " ", "_")
            if not (variable (NewUnitName & "_lastmotion") exists) then make
new variable with properties {name:(NewUnitName & "_lastmotion"), value:
(current date) as string}
```

Keeping track of how many seconds have elapsed since the motion detector was last activated is useful for deciding if the home is unoccupied. You also can use it for turning off the lights after someone has left the room:

```
            set mslastmotion to SubDate((value of variable (NewUnitName & "_
lastmotion")))
            if mslastmotion - (current date) < 1 * hours then
                set msmotioncalc to (current date) - mslastmotion
                if not (variable (NewUnitName & "_motionpermin") exists)
then make new variable with properties {name:(NewUnitName & "_
motionpermin"), value:"600"}
                if (value of variable (NewUnitName & "_motionpermin") as
real) > 0 then set (value of variable (NewUnitName & "_motionpermin")) to
((msmotioncalc + (value of variable (NewUnitName & "_motionpermin") as real)
/ 2) as string)
            end if
            set (value of variable (NewUnitName & "_lastmotion")) to
(current date) as string
```

```
        end tell
    end mslogging
    on searchReplace(theText, SearchString, ReplaceString)
        set OldDelims to AppleScript's text item delimiters
        set AppleScript's text item delimiters to SearchString
        set newText to text items of theText
        set AppleScript's text item delimiters to ReplaceString
        set newText to newText as text
        set AppleScript's text item delimiters to OldDelims
        return newText
    end searchReplace
```

This handler is a nice addition to your attachment script; you can use it to create new triggers by calling it from any script in your system:

```
    on MakeTrigger(TriggerName, TriggerTime, ScriptText)
        if not (time date action TriggerName exists) then
            set newTimeDateAction to make new time date action with
properties {name:TriggerName, absolute trigger time:TriggerTime, date
trigger type:absolute, auto delete:true}
            tell first action step of newTimeDateAction
                set action type to executeScript
                set script code to ScriptText
            end tell
        else
            set absolute trigger time of time date action TriggerName to
TriggerTime
        end if
    end MakeTrigger

     on SubDate(TheString)
        return (date TheString) as date
    end SubDate
end using terms from
```

To install this script, save it as a compiled script using Script Editor and then save it in the *~/Documents/Indigo User Data/Scripts/Attachments* folder. Choose Reload Attachments from Indigo's script menu. To use the script, call it from an Indigo Trigger action with an embedded AppleScript:

```
mslogging("entryway motion detector")
```

The parameter to this script must match the motion detector's device name so that the script can retrieve the information it needs about the device from Indigo.

## Hacking the Hack

Use this technique to enhance other methods for determining what's happening in your home, such as "Which Way Did She Go?" **[Hack #100]** and "Check for an Empty Home" **[Hack #69]**.

—*Greg Smith*

# Track Home Events with iCal

Gain insight into your home's behavior by displaying events in calendar format.

To make your home automation system smarter, you need to have a good understanding of how the system sees your typical, day-to-day activities in the home. That is, discernable patterns derived from your motion detectors, lamp usage, and changes to the home's thermostat can be used to automate these actions for you. For example, by examining your home automation system's logs, you might notice the air conditioning is turned up a few minutes after the garage door sensor is triggered in the late afternoon. This occurs, of course, when you return home from work. It's simple to automate this, but you might not notice it if you weren't looking at the log files.

On the other hand, slogging through log files is tedious work, and they're usually so detailed that finding useful tidbits requires a lot of luck. What would be more useful is a timeline-based graphical view of your home automation data. By combining a little bit of clever scripting with Apple's iCal (*http://www.apple.com/ical/*; free), that's exactly what this hack creates.

This hack works by adding three AppleScript functions via Indigo's support for attachment scripts **[Hack #88]**. The new functions enable you to log device events and status changes to iCal calendars. The hack automatically creates the calendars, groups events, and messages according to how you specify them, as shown in Figure 7-12.

In this example, separate calendars are used to log changes in house mode, Indigo application messages, motion detectors, and various other devices as reflected in the Calendars list on the left side of the iCal window. The main window shows the consolidated view of all the calendars, which is where you can discern patterns and trends in your home. For example, notice that the `Master Bedroom Ceiling Fan` was on from 10:30 a.m. to 11:00 a.m., even though the house was in `Away Mode` at the time. This might be the result of a flaw in your logic for when the house is unoccupied **[Hack #70]**, or it might indicate that the fan came on automatically due to rising temperatures.

It's a neat technique, but it's best to use it in moderation. Just as log files can get overwhelming, overloading your calendar with notes about every event would quickly grow into a mess. Additionally, as nifty as iCal is, it does tend to bog down with hundreds of events. To help you avoid these problems, you have to specifically call the functions provided by this hack's attachment scripts to send data to iCal. This enables you to choose to record only important events.

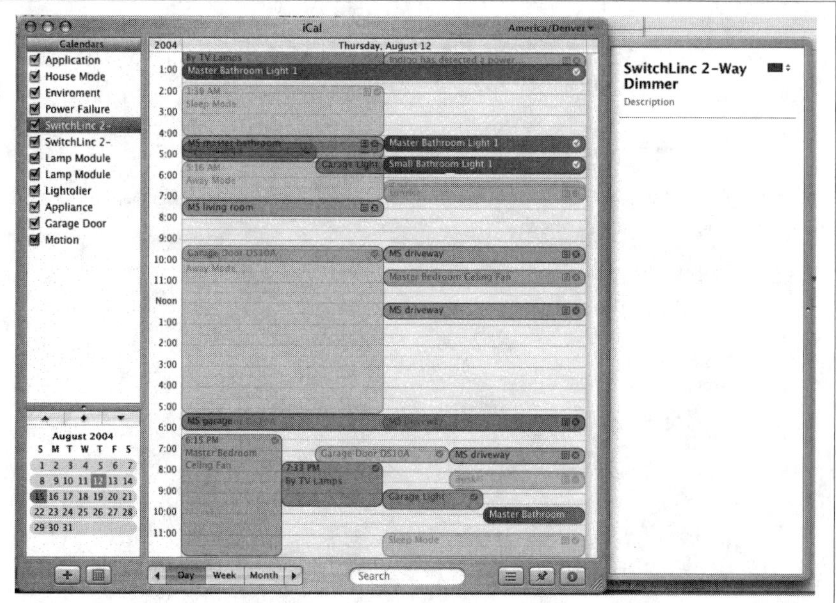


*Figure 7-12. Home data in iCal*

If you want to track the use of a device, such as the ceiling fan mentioned earlier, you define a trigger action that calls the LogDevice( ) function, as shown in Figure 7-13.

The parameter to LogDevice( ) should be the name of the unit (as shown in the action in Figure 7-13). The function uses the name to retrieve information from Indigo about the device and its state. Here's the script that implements the function:

```
using terms from application "Indigo"
    on LogDevice(DeviceName)
        tell application "Indigo"
            log "Log Device: " & DeviceName
            if (name of device DeviceName exists) and (description of device
DeviceName does not contain "future") then
                set eventdate to current date
                set UnitDisc to description of device DeviceName
                set UnitType to type of device DeviceName
                set unitstate to on state of device DeviceName
                set theNote to "" --Optionally put a note here based on
other variables
                set theLocation to UnitType & " " & UnitDisc

                LogIniCal(UnitType, DeviceName, theNote, theLocation,
eventdate, unitstate)
```

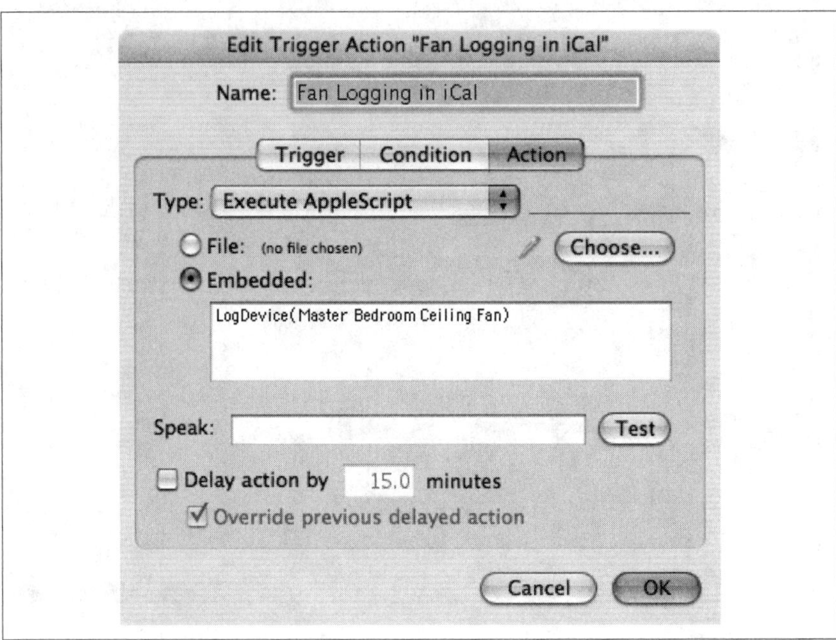


*Figure 7-13. Logging a device action*

```
            end if
        end tell
    end LogDevice
end using terms from
```

The other function you can use to send data to iCal is `LogMessages( )`. You use this function to create iCal events to track things that are not device actions. The house entering `Away Mode` is an example of this type of event. Here's the script:

```
on logmessages(MessageGroup, message, EventKind)
    set MinSummaryLength to 5
    set theLocation to "Message"
    set eventdate to current date
    if MessageGroup contains "Messages" then set EventKind to false

    set OldDelims to AppleScript's text item delimiters
    set AppleScript's text item delimiters to " "
    set theList to the text items of message
    set AppleScript's text item delimiters to OldDelims
    set x to count of items in theList
    if x is greater than MinSummaryLength then
        set i to 0
        repeat with n from 1 to x
            set i to i + 1
            if i is 1 then
```

```
                set LogTemp to item i of theList
            else
                set LogTemp to LogTemp & " " & item i of theList
            end if
            if i is MinSummaryLength then exit repeat
        end repeat
        set LogTemp to LogTemp & "..."
    else
        set LogTemp to message
    end if
    LogIniCal(MessageGroup, LogTemp, message, theLocation, eventdate,
EventKind)
  end logmessages
end using terms from
```

The LogMessages() function takes three parameters: MessageGroup, Message, and EventKind. MessageGroup specifies the calendar to which the event will be added. In Figure 7-12, for example, you might set this parameter to House Mode to add an event to that calendar. The Message parameter is a text string that will become the title of the event when it's added to iCal, such as Sleep Mode. Finally, EventKind can be true, which indicates the event is beginning, or false, which means the event is ending. This determines the duration of the iCal event.

You call LogMessages() using the same method as LogDevice()—using an event trigger with an embedded AppleScript. To log when you're leaving home and the house is entering Away Mode, the action would be:

```
LogMessages("House Mode","Away Mode", true)
```

When Away Mode is turned off, this call ends the iCal event:

```
LogMessages("House Mode","Away Mode", false)
```

Finally, there's one more piece to discuss. Both LogMessages() and LogDevice() rely on another attachment script function, LogIniCal(). This function does all the heavy lifting of communicating with iCal, creating and ending events, adding new calendars, and so on:

```
on LogIniCal(theCalName, theSummary, theNote, theLocation, eventdate,
EventKind)
        tell application "iCal"
            set CalList to (title of every calendar)
            if theCalName is not in the CalList then create calendar with
name theCalName
            set TargetCal to (first calendar whose title is theCalName)

            if EventKind is true then
                --Make new started event
                make event at end of events of TargetCal with properties
{start date:eventdate, summary:theSummary, description:theNote, location:
theLocation, status:tentative}
```

```
            else if EventKind is false then
                try
                    set EventList to (every event whose summary is
theSummary) of TargetCal
                    set TargetEvent to (last item of EventList whose status
is tentative)

                    if status of TargetEvent is tentative then
                        set (end date) of TargetEvent to eventdate
                        set status of TargetEvent to confirmed
                    end if
                on error
                    make event at end of events of TargetCal with properties
{start date:eventdate, summary:theSummary, description:(theNote & " (no
starting event found)"), location:theLocation, status:cancelled}
                end try
            end if
        end tell
end LogIniCal
```

You can combine all three of these scripts into one AppleScript. Save it in compiled format using Script Editor, and install it in Indigo's attachment script folder (*~/Documents/Indigo User Data/Scripts/Attachments*). Then, choose Reload Attachments from Indigo's script menu.

To download these scripts, visit Greg's web site (*http://homepage.mac.com/gregjsmith/icalstats.html*).

## Hacking the Hack

If you want to see what's happening in your home while you're away, set up iCal to automatically publish the calendars you're interested in over the Internet. You will be able to view them in a web browser or subscribe to the remote calendars from another computer's copy of iCal.

For more on using iCal with home automation, see "Remember Important Events" [Hack #25]. If you want to view data about your home automation system using charts and graphs, see "Chart Home Automation Data" [Hack #92].

*—Greg Smith*

# HACK #92 Chart Home Automation Data

**You can mine your home automation system for information about your home, but you need a way to save and view the data you're interested in.**

One of the missing tools in the home automation arena is a means to view simultaneous action and reactions easily and quickly. For example, when the air conditioner comes on, how long does it take before the room temperature drops, and by how many degrees does it fall? If the fan in the greenhouse is set to come on when the inside temperature reaches 78 degrees Fahrenheit, how many times a day does it turn on? You can answer these types of questions by plotting home automation data in graphical form. In my system, I plot weather data, the voltage levels from my solar panel, greenhouse temperatures and cooling fan usage, and the water levels in my fishponds. I can also create a chart showing the number of times, and when, my front gate has been opened, and when there was movement in my house.

To do all this, you need an easy-to-use plotting tool. A look at the applications available finds some heavyweight programs, but the learning curve to adapt them to automation use is daunting. So, I designed my own SuperCard-based program for Mac OS X called *pawsPlot* (shown in Figure 7-14) to solve the problem. It's a free download at my web site (*http://homepage.mac.com/pitchypaw/pocket*).

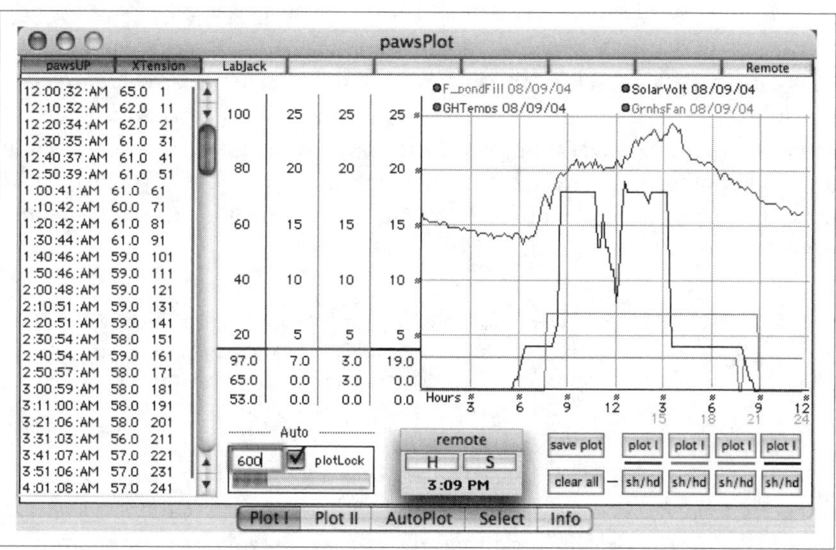


*Figure 7-14. The pawsPlot charting tool*

The pawsPlot application is a *dual-quad* plotter. That is, it can plot two groups of data, with each group containing a maximum of four items. The data is plotted over a 24-hour period. You can save individual or group plots as JPEGs for viewing or for uploading to a web page. pawsPlot does not create the data; it reads the data to plot from files that your home automation system creates. pawsPlot periodically reads the files and creates the graphical plots. You can do all this automatically or trigger it manually if you need to see a graph immediately.

## Getting the Data

pawsPlot can plot data from a file created by any source, but, for our purposes, we'll focus on using an AppleScript from within XTension [Hack #17] to create the datafiles. pawsPlot can handle data values that range from 0 to 125, which is sufficient for storing temperatures, voltages, and other common home automation data. You can even keep track of nonnumerical data by converting it to values that represent each state. For example, track the `On` and `Off` commands received by a unit by saving `10` to indicate `On`, and `0` for `Off`.

pawsPlot has a built-in user manual with directions, sample scripts (shown later in this hack), and other examples. Click the Info tab in the application to view it. You can save a copy of the scripts or manual by selecting what you want and copying it to the clipboard. The help includes extensive discussion of the program's user interface, which is important to review to avoid confusion when setting up your plots.

Most of the data I track is drawn from a serial Weeder board (*http://www.weedtech.com*; $70) from Weeder Technologies. I use the Analog Input Module to read voltages from my solar panels and convert them to digital data. The Weeder board accepts eight inputs per module (you can read more by using multiple modules), which gave me the idea for making pawsPlot emulate an eight-pen plotter.

An alternative to the Weeder board, for data acquisition, is the LabJack U12 (*http://www.labjack.com/labjack_u12.html*; $120). It operates in a similar manner but uses USB instead of serial, which makes it simpler to connect to newer computers.

I use James Sentman's free Weeder Reader application (*http://www.sentman.com/x10/weeder/*) to interrogate the module and store its data in XTension pseudo units [Hack #17]. For example, the current greenhouse temperature is stored in a unit named `grnHTemp`. The Weeder Reader application knows how to read data from the input module, and it can transfer the values directly to XTension. If you want finer-grained control over the board, or you want to send the data to an application other than XTension, Sentman's web site also provides an AppleScript that you can adapt for your own purposes.

## Saving the Data

After the Weeder Reader stores the incoming data in XTension, you need to save it to a file that pawsPlot can read. To work with pawsPlot, the data must be in this format:

```
12:00:32:AM   65.0   1
12:10:32:AM   62.0   11
12:20:34:AM   62.0   21
12:30:35:AM   61.0   31
12:40:37:AM   61.0   41
12:50:39:AM   61.0   51
```

This sample file, showing temperatures from my greenhouse, consists of the time (12-hour format), the temperature, and the minutes since midnight. You must separate each data element by a space and use a short-format date. These are the data points pawsPlot will use to create our plot.

To create the datafiles from within XTension, use a global script that pawsPlot can call to extract the data it needs. The script saves the data that originated from the Weeder board, such as these temperature readings, and any additional data you want to plot, such as motion detector activity. Create a global script named Write_to_Files containing this code:

```
set unitValue to {}
set unitNm to "grnHTemp" --unit name in XTension
set pathFileLoc to "Abacus:WriteFiles:GreenhouseTemps:" & "GHTemps " --path
changes for each unit  to plot
load_Data_pawsPlot(unitValue, unitNm, pathFileLoc) -- do not change
```

Repeat this block of instructions in the script for every value you want to save, changing the parameters for `unitNm` and `pathFileLoc` to reflect the correct XTension unit name and output file location. Here's an example that saves three temperatures and the current fill level of a pond, each to their own datafile:

```
set unitValue to {}--this only needs to be listed once for a group of
entries
---outside temp---
set unitNm to "Outside Temp" --unit name in XTension
set pathFileLoc to "Abacus:WriteFiles:OutsideTemps:" & "OutTemps "
load_Data_pawsPlot(unitValue, unitNm, pathFileLoc)-- do not change
---inside temp---
set unitNm to "Inside Temp"
set pathFileLoc to "Abacus:WriteFiles:InsideTemps:" & "InTemps "
load_Data_pawsPlot(unitValue, unitNm, pathFileLoc)
---greenhouse temp---
set unitNm to "grnHTemp"
set pathFileLoc to "Abacus:WriteFiles:GreenhouseTemps:" & "GHTemps "
load_Data_pawsPlot(unitValue, unitNm, pathFileLoc)
```

```
---front pond---
set unitNm to "F_pondFill"
set pathFileLoc to "Abacus:WriteFiles:FrontPond:" & "F_pondFill "
load_Data_pawsPlot(unitValue, unitNm, pathFileLoc)
```

You can add more blocks to save additional datafiles, up to the maximum of eight supported by pawsPlot.

By naming the global script Write_to_Files, you ensure that pawsPlot can tell XTension to execute it on a regular basis to make sure the datafile contains the latest information.

The previous script calls the handler `load_Data_pawsPlot`, which in turn uses handlers called `ShortDate` and `write_to_file`. You need to add these to XTension's attachment script **[Hack #88]**:

```
on load_Data_pawsPlot(unitValue, unitNm, pathFileLoc)
    set unitValue to value of unitNm
    set theDate to (current date) as string
    set logTime to ""
    set logTime to round ((time of (current date)) / 60)
    set hourWord to word 5 of theDate
    set minWord to word 6 of theDate
    set secWord to word 7 of theDate
    set apWord to word 8 of theDate
    set theTime to hourWord & ":" & minWord & ":" & secWord & ":" & apWord
    set this_data to (theTime & space & space & space & unitValue & space &
space & space & ¬ logTime & return)
    set target_file to (pathFileLoc) & ShortDate( )
    my write_to_file(this_data, target_file, true)
end load_Data_pawsPlot
on ShortDate( ) --example 08/01/04
    set myShortDate to ""
    set myDate to current date
    set theMonth to the (month of myDate) as integer
    set theDay to the (day of myDate) as integer
    set theYear to (text 3 thru 4 of ((year of myDate) as text))
    if theMonth is less than 10 then
        set myShortDate to myShortDate & "0"
    end if
    set myShortDate to myShortDate & theMonth & "/"
    if theDay is less than 10 then
        set myShortDate to myShortDate & "0"
    end if
    set myShortDate to myShortDate & theDay & "/"
    set myShortDate to myShortDate & theYear
end ShortDate
on write_to_file(this_data, target_file, append_data)
    try
        set the target_file to the target_file as text
        if append_data is false then
```

```
            tell application "Finder"
                if (exists of file target_file) then
                    delete file target_file
                end if
            end tell
        end if
        set the open_target_file to open for access file target_file with
write permission
        if append_data is false then set eof of the open_target_file to 0
        write this_data to the open_target_file starting at eof
        close access the open_target_file
        return true
    on error
        close access file target_file
    end try
end write_to_file
```

You might need to change the name of some of the handlers, such as ShortDate, if your attachment script already has handlers that use these names. If that's the case, be sure to change the name everywhere the handlers are called, too.

## Clean-Up Script

To avoid cluttering your hard drive with too many datafiles, add this code to XTension's sunrise script [Hack #20] or schedule it to run once a day. It deletes datafiles that are more than a week old.

```
tell application "Finder" to delete (every item of alias "Abacus:WriteFiles:
FrontPond:" whose modification date< (get current date) - 7 * days) --change
path and days for your setup
```

Duplicate this code once for each datafile you're saving, changing the file path information as appropriate.

## Generating a Plot

To ensure your plots are always up to date, leave pawsPlot running on your home automation computer and set it up to generate plots automatically. To do this, click the Select tab and select the datafiles your script creates, as shown in Figure 7-15.

To set the location where pawsPlot will save the generated plots, click the AutoPlot tab, then choose which plots will be generated automatically, which will be saved, and the folder where the files will be placed, as shown in Figure 7-16.

Select the buttons on the right to specify the data to group together in a plot. If there are eight files to plot, and you select all eight plot buttons, you will get eight separate files saved as *plot1.jpg* through *plot8.jpg*. To specify how

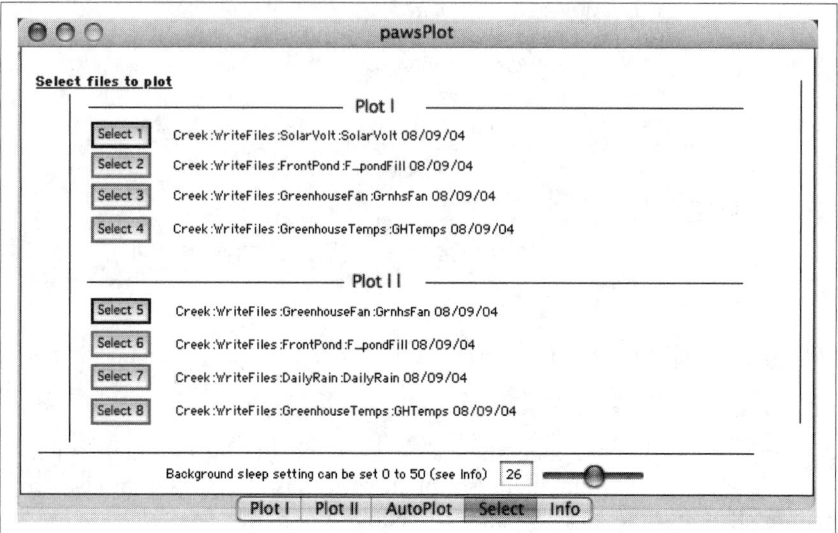


*Figure 7-15. Selecting datafiles to read*

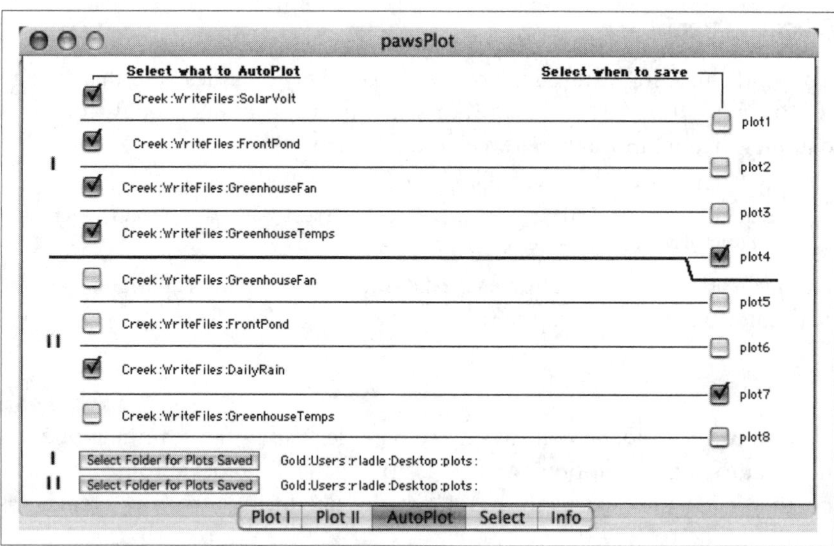


*Figure 7-16. Selecting the plots to save and their location*

often pawsPlot creates the plots, click the Plot I or Plot II tab, then enter, in seconds, how often you want pawsPlot to tell XTension to save new data to disk. It does this by telling XTension to execute the Write_to_Files script you created earlier. Click the XTension button to enable pawsPlot to communicate with XTension. In the example shown previously in Figure 7-14, pawsPlot will tell XTension to execute Write_to_Files every 10 minutes (600 seconds).

Once you have pawsPlot set up, it will generate new plots at the interval you specified, overwriting the old plots each time it updates. Each file is saved in JPEG format so that you can upload it to a web server for viewing with a browser. Each plot has its name in the upper portion of the image, as shown in Figure 7-17.

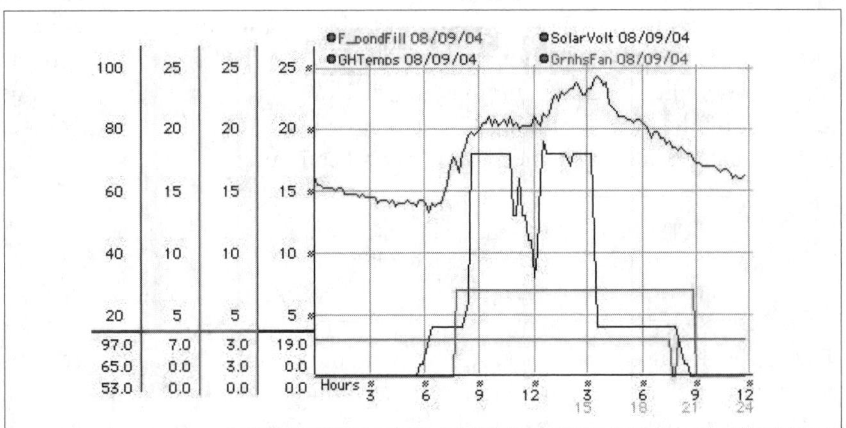


*Figure 7-17. A sample plot*

This example is my final composite plot for a day in my greenhouse. The high, low, and current values (at midnight) are shown below their respective columns on the left side of the image. The time of day is shown across the bottom of the image. The plot of `SolarVolt` shows a dip in the solar voltage that is the result of fir trees blocking direct sunlight at noon. The `GrnHsFan` plot shows when XTension turned on the greenhouse fan and for how long it was on. By analyzing this plot, I can tell the fan did inhibit the rate of temperature increase (`GHTemps`) but lost hold about 1:00 p.m. This makes me consider that I need a second fan to keep the temperature in the greenhouse lower than 90 degrees. The dip in the bottom plot at about 8:00 p.m. shows that my front fishpond required additional water and was refilled automatically shortly after. By watching this plot over several days, I can tell if more water is needed due to evaporation or if the pond has sprung a leak.

Happy plotting!

*—Robert Ladle*

# Share Your Home Automation Mac with Other Users

Use Mac OS X's Fast User Switching to safely run your home on a computer that's also used for other tasks.

Traditionally, Macintosh users have had to dedicate a Mac to running their home. It was necessary to ensure that the computer would always be on and running the necessary home automation software. It was simply too easy for a user doing a little word processing, or playing a game, to quit or otherwise interfere with the system. But then Mac OS X Panther came out, with its superior stability and preemptive multitasking architecture, and I was intrigued by the idea of running XTension as a background task on my main office Mac, eliminating the need for a second computer.

The technique used in this hack will also work for Indigo and many other Mac OS X applications.

The secret is to use Panther's Fast User Switching feature. While logged in as an administrator, create a separate user account for running XTension, with its own login name and password. To do this, open the Accounts pane of System Preferences and click the plus (+) button at the bottom of the Accounts list. Fill in the fields to create a new account. Be sure to select the "Allow user to administer this computer" option in the Security pane, as shown in Figure 7-18. XTension requires this setting.

Next, click the Login Options button. Check the "Enable fast user switching" checkbox. This allows multiple users to stay logged on to the computer, which is key for this hack to work, as discussed later. Check the "Automatically log in as:" checkbox and select the XTension user account from the pop-up menu, as shown in Figure 7-19.

The next step is to set up XTension so that it starts automatically when this user logs in. To do this, you need to log in using the account you just created. Choose Log Out from the Apple menu. Log back in using the XTension account, and then open the Accounts pane of System Preferences. Click the Startup Items tab and add XTension to the startup list.

You also might add other applications that you use for home automation, such as WeatherManX **[Hack #64]** and X2Web **[Hack #99]**.

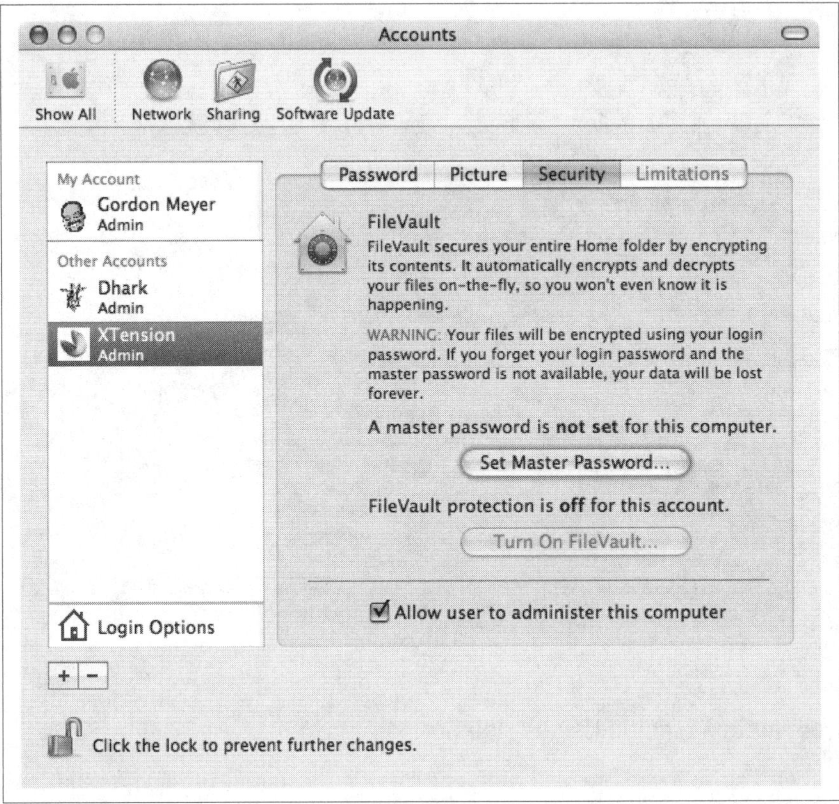


*Figure 7-18. Users with admin authority*

Now, you have an account dedicated to running XTension. When the computer is turned on, the XTension user will log in automatically and XTension (and all your other home automation applications) will launch. When you want to use your computer for other tasks, there's no need to quit XTension or log out. Simply use Fast User Switching to change to another account. Click the account name in the upper-right corner of the screen and select a different account to log into. You'll see Panther's sexy rotating-cube graphic effect, and XTension will be switched into the background—still running, but invisible! Whenever you want to go back to XTension, just switch back to its user account.

XTension generally consumes very little CPU power, so, on a fairly new Mac, you probably won't even notice it's running while you do other work. If you want to pass files back and forth between the XTension user account and another account, you can drop them into the */Users/Shared* folder or the *Public Drop Box* folder for the account you want to pass them to.

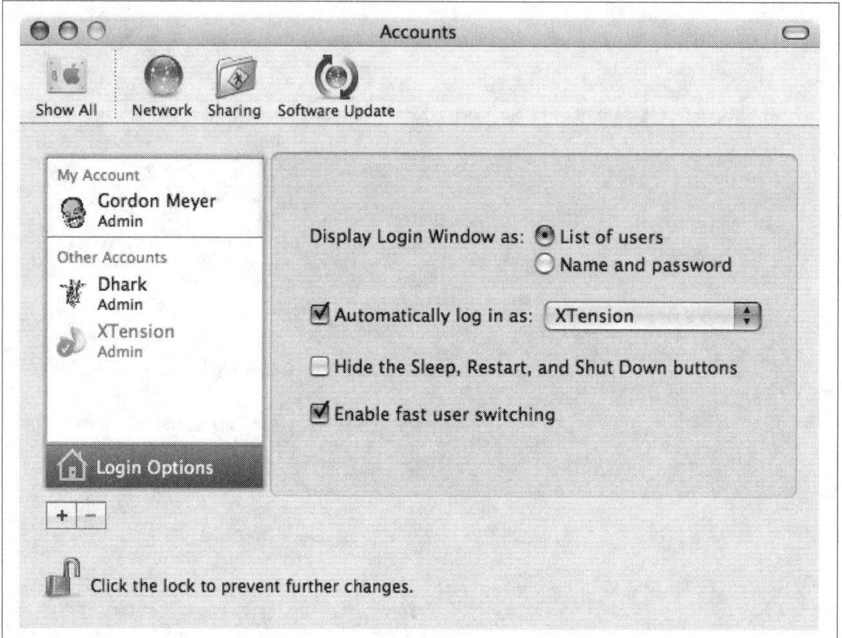


*Figure 7-19. Auto login options*

## Talking to XTension Using AppleScript

When you're logged in as a different user, you can still run scripts that talk to XTension, provided that your scripts use a special form of the `tell` statement. This is a handy way to change a setting without having to switch back to the XTension account. For example, I have a script that will turn on the audio system in my office:

```
tell application "XTension" of machine "eppc://192.168.1.100" to turnon
"Office Sound System"
```

You use the `machine` parameter to specify the IP address of the computer that is running XTension—in this case, the same computer executing the script. Alternatively, you can use the computer's Rendezvous name, which looks like `MyMac.local`. To find your IP address or Rendezvous name, look in the Sharing pane of System Preferences.

This script can control XTension running on another computer on your network, but to make it work, you must turn on the Remote Apple Events option in the Sharing pane of System Preferences on the computer you want to control. When you run the script, you'll be asked for a name and password for the remote computer.

### Hacking the Hack

James Sentman (*http://www.sentman.com*) provides this useful AppleScript to enhance this hack:

```
do shell script "/System/Library/CoreServices/Menu\\ Extras/User.menu/
Contents/Resources/CGSession -suspend"
```

Save the script as an application, using Script Editor, and then set it as one of the login items for the XTension user account. Drag it to the bottom of the Startup Items list so that it's the last application to launch.

This script uses the CGSession application to suspend the current user's session. That is, when it runs, the XTension user will be switched out and the computer will return to the login screen. XTension is still running, but now other users won't have to remember to switch to their own accounts before using the computer, and they can't log the XTension user out accidentally, which would quit the home automation system and all the other programs you need running.

—*Rob Lewis*

## HACK #94 Remap X10 Addresses

**If you've run out of house codes for your modules, or if you just need to rearrange a few addresses without redoing your entire system, this handy script will save you a lot of work.**

*Address remapping* is an advanced technique for having your home automation software change command recipients on the fly. For example, you might want your system to change every command for house code `P` to the equivalent device on house code `L`. That is, when `P10 On` is received, your system sends `L10 On` to the power line.

You might occasionally need to remap X10 addresses for a few reasons, the most common of which is to allow a Palm Pad **[Hack #5]** that is set to one house code to control devices that are set to a different house code. In other words, when you press the `P10 On` button on the Palm Pad in the upstairs bedroom, you want the light in the downstairs kitchen (device `L10`) to turn on in response.

You certainly can accomplish this without having to resort to remapping. Simply set the Palm Pad's house code to `L`, and it will send `L10 On` directly. That's the simplest, and often the best, solution. However, as your home automation system grows more complex, you are likely to find that your addressing scheme **[Hack #1]** and the things you want to accomplish don't always match up so cleanly. For example, if you set the Palm Pad to house code L so that you can control the kitchen light, this means every other button on the Palm Pad also sends commands for L. If you're using L11, L12, and L13 for devices you don't want to control directly, such as motion detectors, you're in a bit of a bind. Not only will those buttons on the Palm Pad be wasted because you'll never want to use them, but also, pressing them accidentally might set off a confusing chain of events. For example, if L12 is the motion detector on your front porch, an inadvertent L12 On might result in turning on your porch light and outside lights, and activating a camera to see who's at your front door. Oops!

Although you might avoid some of these problems by carefully planning your system, it's almost inevitable that at some point you'll run into an address allocation conflict. When this conflict does arise, you could try to resolve it by changing device addresses to accommodate the new need—reprogramming the motion detectors to use house code K, for example—but this is tedious work, and it often disrupts other parts of your system.

House code remapping is so commonly used for wireless devices—Palm Pads and motion detectors in particular—that Indigo and HomeSeer provide ways of automatically remapping signals that are received using direct-to-computer wireless X10 receivers. Figure 7-20 shows the settings that remap commands for house code P to house code L before sending the command to the power line.

XTension solves the problem differently, by keeping wireless house codes distinct from power-line house codes, but the result is similar. If you have a wireless receiver, see "Improve the Response Time of Motion Detectors" [Hack #83] for information about these techniques.

If you're using transceivers [Hack #5] to receive wireless commands, you'll need to do some scripting to accomplish the remapping. HomeSeer user Frank Perricone accomplishes this with one cleverly constructed command:

```
&hs.ExecX10 "L" & mid(hs.StringItem(hs.LastX10( ),2,";"),2), hs.
StringItem(";on;off;dim;bright",hs.StringItem(hs.LastX10( ),3,";"),";"),6
```

Although you can avoid any scripting by setting up individual device events that retransmit your commands, that's a tedious approach. For example, you'd need to add an event for P10 On whose only action is to send L10 On. For P11 On, you'd add another event, and so on. This certainly will do in a pinch, but a programmatic approach is more efficient for remapping an entire house code.

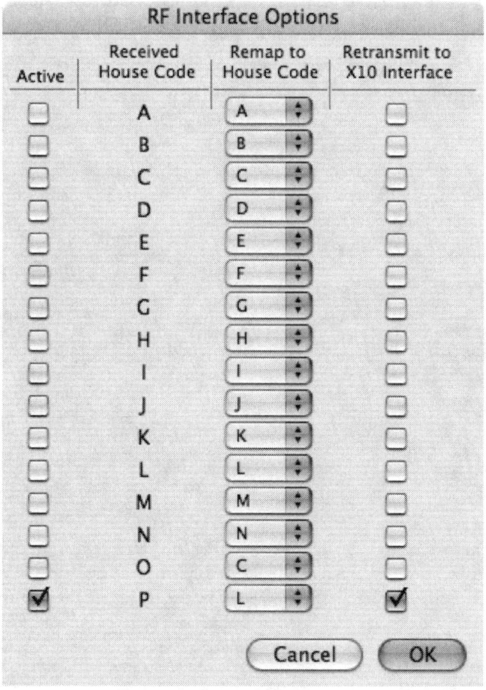

*Figure 7-20. Remapping wireless commands in Indigo*

This script maps X10 commands to house code `L`. You determine which house code the command is remapped *from* by setting up an event that executes this script for all commands received on the originating house code. Figure 7-21 shows an event that is triggered by all commands received on house code `P`. Then the preceding script is specified as the Scripts/Speech action for this event.

The key to understanding the formula used by this script is to work from the inside out, not from left to right. Let's start with the second half because it's a little harder:

```
hs.StringItem(";on;off;dim;bright",hs.StringItem(hs.LastX10( ),3,";"),";"),6
```

The `hs.LastX10( )` function returns the X10 command that triggered the event. If the command was `P11 On`, the output looks like this: `P;P11;2;`. The third parameter can be one of many values, but the important ones are `2` (on), `3` (off), `4` (dim), and `5` (bright). See HomeSeer's command reference in the onscreen help for full details.

This expression uses `hs.StringItem` to pull off the third item returned by `hs.LastX10`. In our example, the result is `2`. The `hs.StringItem` function parses

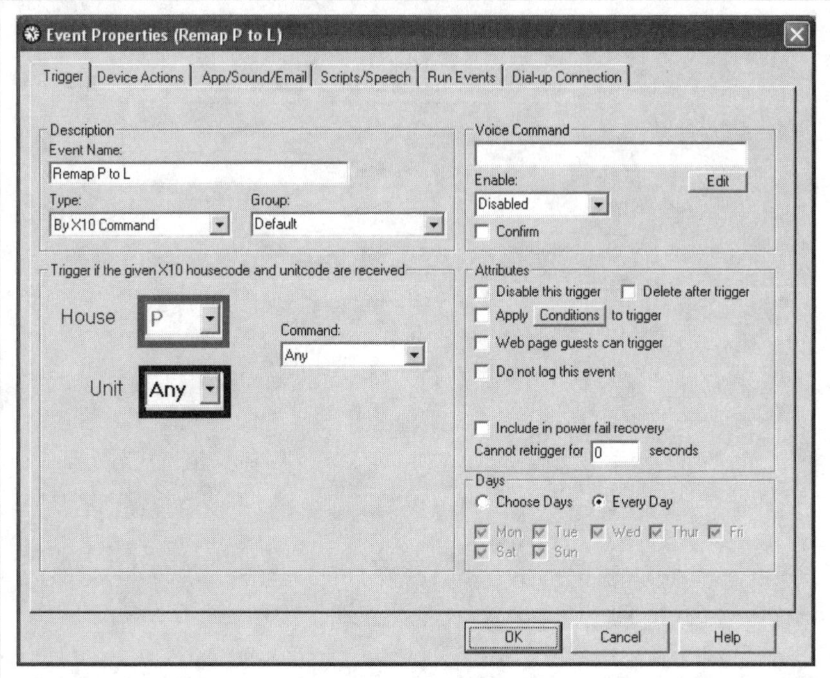


*Figure 7-21. An event for all commands in a house code*

the string using the separator (;) which was specified when the function was called:

```
hs.StringItem(hs.LastX10( ),3,";")
```

Next, `hs.StringItem` is called again to translate the command digit (`2` in our example) to a command string:

```
hs.StringItem(";on;off;dim;bright",hs.StringItem(hs.LastX10( ),3,";"),";"),6
```

In this example, the result is `on`, the second item in the parameter list supplied. If you need to support additional commands before the four shown (`on`, `off`, `dim`, and `bright`), simply add them to the list.

We've finished the second half of this one-line script. Now let's look at the first half:

```
"L" & mid(hs.StringItem(hs.LastX10( ),2,";"),2)
```

As before, `hs.LastX10` and `hs.StringItem` are used to extract a portion of the incoming X10 command. This time, the third item (the device address, `P11` in our example) is extracted:

```
mid(hs.StringItem(hs.LastX10( ),2,";"),2)
```

The mid function in this case takes all of the string starting from position 2 to the end of the string. In other words, this strips off the first character of the command (i.e., the P from P11). Now, look at the whole expression:

```
"L" & mid(hs.StringItem(hs.LastX10( ),2,";"),2)
```

This appends an L to the remaining portion of the address, resulting in L11.

Now, assemble the two pieces of the new command using the concatenation operator (&):

```
&hs.ExecX10 "L" & mid(hs.StringItem(hs.LastX10( ),2,";"),2), hs.
StringItem(";on;off;dim;bright",hs.StringItem(hs.LastX10( ),3,";"),";"),6
```

The L11 we derived from the first half of the equation and the on we derived from the second half become L11 on. This result is passed to hs.ExecX10, which causes HomeSeer to send the command to the power line. *Voilà*! Just one line of code to map all commands from one house code to another.

It might seem weird to write inside-out formulas such as this, but in some programming languages most things are done this way. My favorite example is the extremely obscure SML, but even LISP tends to look like this.

### Hacking the Hack

You can use an attachment script with Indigo to accomplish the remapping. Add a handler named receive X10 event, and Indigo will pass the incoming command to this handler before taking any other action, which enables you to change it to another address, if necessary, using a technique similar to the one described here.

—*Frank Perricone and Gordon Meyer*

HACK #95

## Control Lights in a Group

**Most home automation software enables you to define groups that contain many units so that you can control multiple devices at once. This is a handy feature, but sometimes you might want to control the entire group except for one or two members.**

Groups offer a convenient way to control a lot of devices all at once. For example, in the script that awakens me in the morning, I turn on several lights on the first floor so that I can see where I'm going when I fetch my morning cup of coffee. To do that, the script simply commands a group of lights on the first floor to turn on. Each member of the group is sent an On command, in the order in which they're listed in the group definition. Figure 7-22 shows the group I used, as defined in XTension [Hack #17].

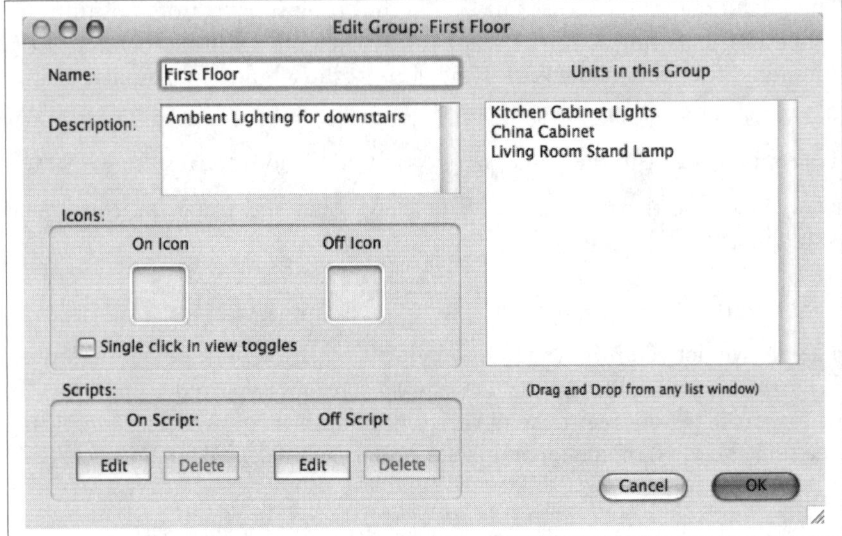


*Figure 7-22. A group of lights*

You can define as many groups as you need, and groups can contain other groups, so with a little planning, you can subdivide your devices in very useful ways. For example, you might define groups for each part of your yard:

- Front yard lights
- Back yard lights
- Porch lights

You can control any of these groups individually, but for cases when you want to turn on all outdoor lights, create a group called `Outdoor Lights` that contains each of the other groups, as shown in Figure 7-23. Sending an `On` command to this group will turn on each device in all the groups.

Sometimes you might want to control only some of the members of a group. If you're interested in just one or two of the units, it's probably best to command them individually—for example, `turnon "Front Porch Light"`. However, if the group is large and you want to command more units than you want to omit, consider blocking a member of the group temporarily.

In XTension, blocking a unit prevents the computer from sending any commands to it. Any command that a script or event tries to send will be blocked and a note will be made in the log.

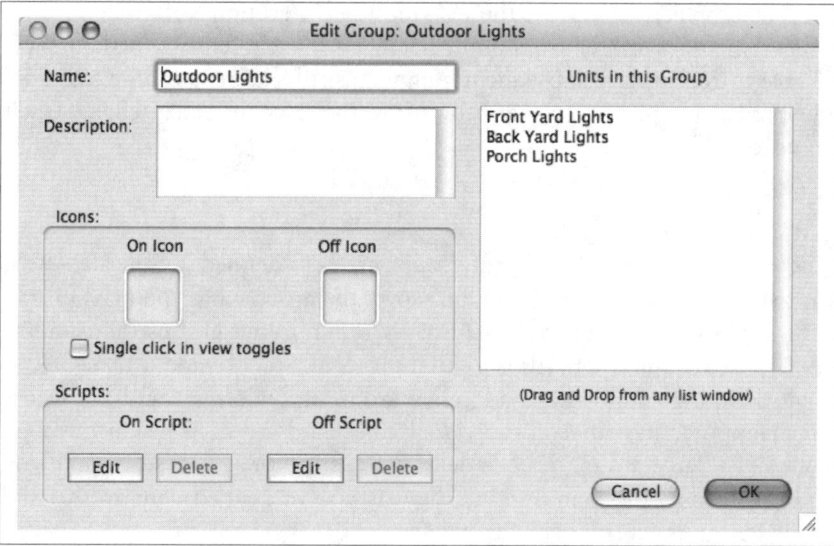


*Figure 7-23. A group containing other groups*

Blocking a unit in XTension does not prevent other controllers, such as a motion detector or Palm Pad, from sending commands to the unit. It simply prevents XTension from initiating any commands for that unit. In other words, it doesn't disable the X10 device itself; it will still be listening to the power line and will respond to commands that originate from other sources.

The following script illustrates this technique:

```
-- stop sending commands to some lights
block unit "Office Light"
block unit "GBR Lamp 1"
-- tell the whole group to turn on
turnon "Second Flr Lights"
-- reactive the two blocked lights
unblock unit "GBR Lamp 1"
unblock unit "Office Light"

-- turn on all outdoor lights except the rope light
block unit "Outside Rope"
turnon "Outside Lights"
unblock unit "Outside Rope"
```

This script temporarily blocks two units that are members of the group `Second Floor Lights`, then commands the group to turn `On`. Because the two units are blocked, they'll remain off. This process is repeated for the `Outside`

Rope light, which is part of the Outside Lights group. Note that in both cases, the blocks are removed immediately after use to ensure the lights will be able to receive any subsequent commands. Blocks remain in effect until you programmatically, or manually using the user interface, change their status to unblocked.

### Final Thoughts

Most home automators find that groups are best defined organically—that is, as you grow your system, you'll discover the groups that you need to simplify your scripts and processes. Also, try to limit your groups to just a few devices. Remember that X10 is a serial protocol, so although it takes only a single scripting command to address a group of devices, behind the scenes your computer has to send individual commands to each group member. With a very large group, this can delay your script for several seconds. If you find this to be a problem, rethink the way you've defined your groups and see if you can organize them into smaller lists.

## HACK #96 Block Units for Easier Scripting

**XTension's ability to temporarily block commands for a unit provides an easy way to refine group handling.**

One of the best things about being able to define groups of units [Hack #17] is that you can control multiple devices with a single command. For example, you might assign all your outside lights—front porch, back porch, patio, and sidewalk lights—to a group called Outdoor Lights. When you want all these lights to turn on, a single command does the trick:

```
Turnon "Outdoor Lights"
```

However, there might be times when you want to address most, but not all, of a group—for example, if the Outdoor Lights group were to also include the Christmas Lights that you have on the Blue Spruce in the front yard. During the holidays, you'd want the tree to be lit, but the rest of the time the holiday lights are packed up for storage. Or, perhaps they're still in place but you don't want them to come on in June and broadcast to the neighborhood that you haven't gotten around to taking them down.

One way to address this problem is to edit the Outdoor Lights group and remove the holiday lights from the list. But that's not very hackish. A better approach is to change the unit's status in XTension to *blocked*.

To block a unit, select it in the Master List window, then choose Control Panel from the Window menu. In the Control Panel, click the stop-sign button located to the left of the New Unit button, as shown in Figure 7-24. In the Master List, you'll see a filled-in circle in the first position in the Flags column, indicating the unit is blocked.

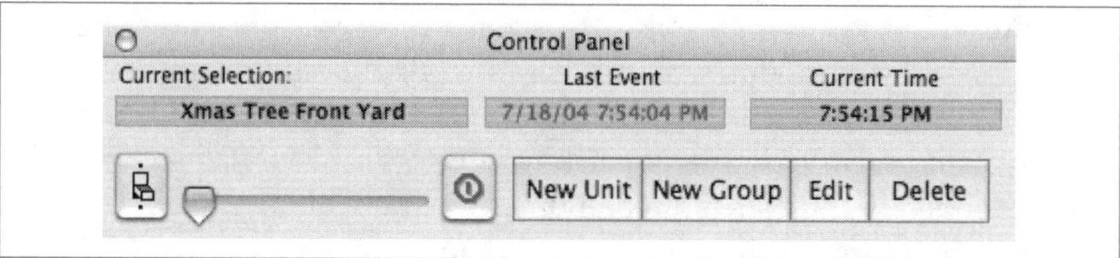


*Figure 7-24. Blocking a unit*

Once a unit is blocked, it cannot be turned on, turned off, or dimmed by any script that XTension runs or by using the XTension user interface. Commands sent to the unit are logged, but they are not sent over the power line. Note, however, that only XTension knows about the unit's blocked status, and it can't prevent other X10 devices from sending commands to that address. For example, if the holiday lights are set to address `K3`, you can still turn them on by sending `K3 On` from a minicontroller. There's nothing XTension can do to prevent, or stop, a command once it's already on the power line.

You can also block and unblock units using a script. This is handy for when you want to alter a group of units temporarily. For example, perhaps you're concerned about the safety of the holiday lights, so you want them to come on only when you or someone else is at home **[Hack #70]**.

To do this, set up a repeating event **[Hack #17]** that runs at 8:00 p.m. every night, executing this script:

```
If (status of "nobody home") is false then
 Turnon "Outdoor Lights"
Else
 Block unit "Xmas Tree Front Yard"
 Turnon "Outdoor Lights"
 Unblock unit "Xmas Tree Front Yard"
End if
```

If the house is unoccupied, the holiday lights are blocked before turning on the `Outdoor Lights` group. This will turn on every member of the group except `Xmas Tree Front Yard`. Then, the unit is unblocked so that it's back to regular status. This is good housekeeping practice, as it returns the unit to a ready state for other scripts or events that might want to control it.

# Calculate Elapsed Time

**If you want to know how much time has passed between events, such as how long it has been since the rain sensor was activated, or how many hours it has been since the motion detector in the library last sent a command, this technique will give you an answer in human-readable format.**

Computers tend to count the passing of time in seconds, not bothering with anything as silly as minutes, hours, days, weeks, months, and years. Ask your computer how long ago it was that you last turned on a particular light and it will dutifully report that the light was last touched 6,026 seconds ago, which is about 10 minutes in human time. We puny humans tend to have a hard time keeping track of anything more than around 59 seconds.

In XTension [Hack #17], the `time delta` property of a unit tells you how many seconds ago the status of the unit was last changed. If you want to announce or log this information in a more human-friendly fashion, you'll have to do a little conversion first.

## The Code

This script, inspired by a similar one from Jeffrey Laughter, provides a handy subroutine for converting from seconds to seconds, minutes, and days:

```
-- IntervalToString(numseconds)
-- Returns a text string decribing an interval in calendar format.
-- i.e. 791766 seconds returns " 9 days 3 hours 56 minutes"
-- left over seconds are dropped as being insignificant

on IntervalToString(numseconds)

  set IntervalString to ""
  set daysnum to numseconds div 86400 -- number of seconds in a day
  set numseconds to numseconds mod 86400 -- save the remainder
  if daysnum = 1 then set IntervalString to "1 day"
  if daysnum > 1 then set IntervalString to " " & daysnum & " days"

  set hoursnum to numseconds div 3600 -- seconds in an hour
  set numseconds to numseconds mod 3600 -- save the remainder
  if hoursnum = 1 then set IntervalString to IntervalString & " 1 hour"
  if hoursnum > 1 then set IntervalString to IntervalString & " " & hoursnum
& " hours"

  set minutesnum to numseconds div 60 -- left over minutes
  if minutesnum = 1 then set IntervalString to IntervalString & " 1 minute"
  if minutesnum > 1 then set IntervalString to IntervalString & " " &
minutesnum & " minutes"
```

```
  if IntervalString = "" then set IntervalString to " less than a minute" --
occurs because we dropped the seconds remainder
  return IntervalString

end IntervalToString
```

In XTension, it's best to add to your attachment script [Hack #88]. This enables you to call it as a function from any other script:

```
set goneTime to IntervalToString((time delta of "Nobody Home"))
say "Welcome Back. The house was empty  " & goneTime
```

In this example, the elapsed time since the last time someone was home [Hack #70] is announced so that the person returning home knows how long the pets have been without supervision or a bathroom break.

Although it's used in XTension here, this script should work with any AppleScript-enabled home automation software. Plus, it's simple enough to convert to just about any other language you need.

## HACK #98 Identify Trouble Spots

**If you're serious about making X10 as reliable as possible, these tools and techniques will help you find problem areas.**

It happens to everyone, eventually: you're faced with an X10 module that just won't turn off, or turn on, or dim properly when you send it commands. You know your system is working—other devices respond as they should—but for some reason, a lamp or two just refuse to cooperate. What you have, my friend, is a transmission trouble spot. What you need is a way to identify its source so that you can eliminate the problem and get back to work on more interesting projects.

No single method is available for finding and eliminating the source of X10 problems. There are tradeoffs between budget, convenience, nature of the gremlins, how technical you are, and how persistent you care to be in resolving the problem.

First off, look for the likely suspects: appliances that are *signal suckers* and prevent all X10 commands from propagating through the house, *noise inducers* that interfere with some commands but not others, and *poor coupling* that results in unpredictable and sporadic problems. You can readily troubleshoot problems caused by poor coupling and signal suckers with an X10 signal meter, discussed further in this hack. You can find some noise inducers, such as televisions and refrigerators, by unplugging everything in the problem area and then plugging items back in, one at a time, until the problem occurs again.

Too bad every problem isn't so easy to solve. The nastiest gremlins are things such as loose connections and flaky modules; the problems they cause are intermittent. When you run into one of these, there are some techniques to help you cope.

For instance, here's an inexpensive way of getting an overall idea of whether X10 signals are reaching all areas of the house. Set a chime module **[Hack #9]** to an unused address, and then set up a repeating event in XTension **[Hack #17]** that sends an `On` command to the chime, as shown in Figure 7-25.

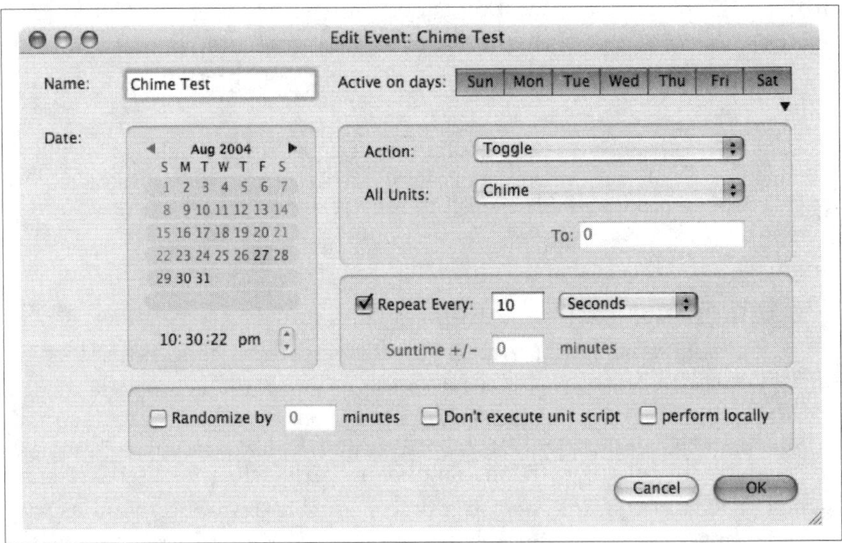


*Figure 7-25. Repeating chime event*

Walk around the house, plug the chime module into various outlets, and see if it sounds off. If you find an area where the signals aren't getting through, turn off the circuit breakers for other areas to narrow down where the source of interference lies, or begin unplugging everything in the area and see if the chime suddenly starts working. If it does, you've just found a signal sucker.

For a more precise approach, buy an ELK EL-ESM1 signal meter (*http://www.hometech.com/tools/signal.html#EL-ESM1*; $65). The EL-ESM1 has an LED bar graph that displays X10 signal strength and flashes a green LED when any X10 traffic is received. It doesn't decode the signal, like the more expensive Monterey Power Line Analyzer (*http://www.hometech.com/tools/signal.html#MI-PLSA*; $320), but it will give you an idea of how signals are traveling in your home.

You can set XTension to transmit continuously (as in Figure 7-25) and walk around the house measuring the signal strength. You also can activate two-way modules and measure how strong their signals are at the outlet where the XTension system is connected. Finally, keep an eye on the green LED and the XTension log file to identify signal noise that appears to be X10 commands.

Speaking of two-way modules, if you're planning to use any, it's a good idea to get one and survey your house for how well they'll work in different areas. Place the module in an outlet and have XTension ask the module for its status. If you find that the status report is not received, you can begin looking for the source of interferences. Alternatively, if you have a two-way wall module that you can't move around the house easily, install XTension on a laptop and walk around with it, plugging it into various outlets and making sure you can query the module.

If it seems you have a noise problem, most often indicated by sporadic interference, an oscilloscope is the best way to find it. The Oscilloscope Dual Trace PCC Adapter (*http://www.act-solutions.com/PCCSpecFrame.htm*) displays the X10 signal on one channel, while you examine the power line on a second channel.

If you find that you need to generate a specific X10 signal to help with your testing, the best way to do it is using XTension. But if that's not practical, the ACT AT004 Multi-Tester (*http://www.smarthomeusa.com/Shop/Lighting/X10-Test-Equipment/Item/AT004/*; $370) can generate specific X10 signals and analyze received signals in detail.

*—Bill Fernandez*

## HACK #99 Control Your Home from a Web Browser

**If your home automation computer is tucked away in a closet or the garage, it can be inconvenient when you need to access it directly or add a new event. Set up web-based access, however, and you can take control from any computer in the house.**

Software that enables you to control your computer remotely is invaluable for home automation. Let's say you want to add a new scheduled event, or you just want to brighten a nearby lamp that's not already set up for access from a minicontroller. Simply open your laptop, connect it to the home automation server, and make the change. You can do all of this without leaving the comfort of your easy chair. Could life possibly get any better?

If you want to control your system via the Internet from a location outside of your home network, you'll need to set up your computer for outside access and take the appropriate security precautions. These steps vary depending on your operating system, your Internet service provider, and the type of network connection you have, and they are beyond the scope of this hack.

If your Unix-based home automation system is built using command-line-based programs, such as Heyu and Xtend **[Hack #16]**, all you need for remote control is a terminal and an SSH connection to the server. If you're using one of these applications, this approach probably has occurred to you already. But even the most hardcore command-line-interface guru can use a break from remembering arcane syntax, and that's exactly what an application called BlueLava can provide.

BlueLava (*http://www.sgtwilko.f9.co.uk/bluelava/*; free) is a Perl CGI script that works in conjunction with the Apache web server and Heyu (or other programs) to provide access to your home automation system in a web browser. It presents a sparse, but very usable, text-only interface and even supports WAP access so that you can control your system using a cell phone.

If you're using MisterHouse **[Hack #16]** to run your home, you know that its primary interface is a web browser, but perhaps it never occurred to you that you can access it using any computer on your network, without any extra configuration.

If you're using HomeSeer for Windows **[Hack #19]**, remote access to your system via a web browser is just a couple of clicks away. HomeSeer's built-in web server enables you to control virtually every aspect of the system. To turn on the Web Control feature, select Options from the Tools menu and then click the Web Server tab in the Options dialog, as shown in Figure 7-26.

The choices available here give you control over every aspect of your home's automation web site. If you plan on accessing your system while you're away from home using your broadband Internet connection, the first thing you want to do is set up appropriate user settings. Click Define Users to change the password for the default account, which has full administrative access to your system, or delete the default and add a new one you can use for the same purpose. The other settings you can set for the web server also are worth examining. See the Web Server section in the Options section of the HomeSeer Help for all the details.

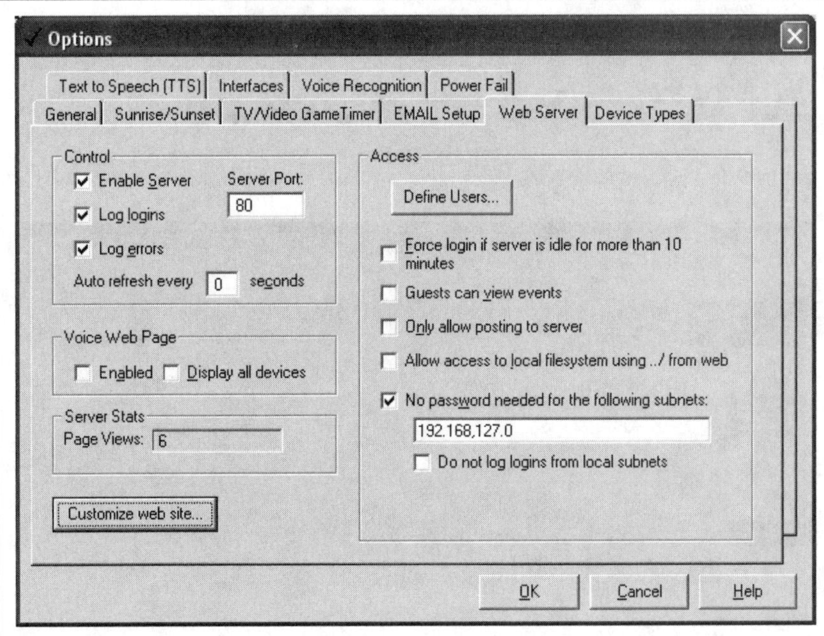


*Figure 7-26. Setting Web Server options for HomeSeer*

To access your HomeSeer system using a web browser, simply go to the IP address of your home automation computer, as shown in Figure 7-27. If you've configured HomeSeer to allow Guest access, what you'll be able to do will be limited, unless you log on using an account with more power.

HomeSeer provides a functional, if staid, default design for its web pages. If you want to tweak the web pages' appearance, click the Customize Web Site button in the Web Server Options dialog. You'll be able to change the style sheets that control how the site appears, as well as select which functions and data appear on the pages.

You can get even fancier, if you're so inclined. For example, when you define a custom status message for a device **[Hack #19]**, you can embed HTML formatting within the status message to further format how the device's status appears in the web view. (Finally, a good use for the `blink` tag!) Furthermore, you can have HomeSeer call a scripted function every time a page is loaded, which provides you with the hook to get even fancier with your programming.

Additionally, HomeSeer provides a built-in TouchPad view, designed for use with touchscreen computers, which works especially well for home-based web access to your system. See Figure 7-28 for an example.

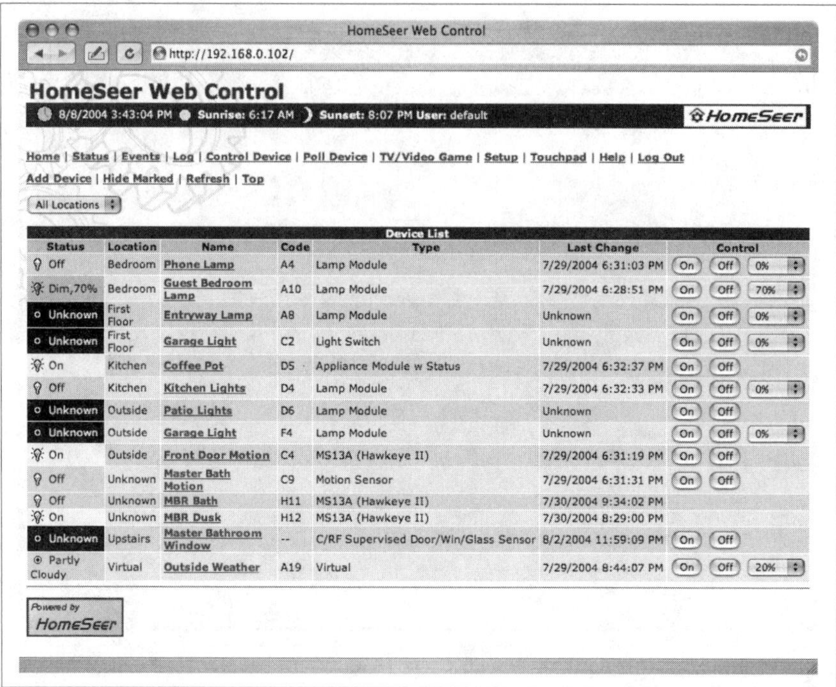


*Figure 7-27. HomeSeer's web control*

If you're an Indigo for Macintosh user **[Hack #18]**, providing web access to your home automation system requires a bit more effort to set up, but it works well once you get it going. Indigo doesn't have a built-in method for this, so instead you have to configure Apache (the web server built into Mac OS X) so that it can execute AppleScript programs. The AppleScripts, in turn, control Indigo. Apache doesn't know how to run AppleScript-based CGI programs, so, to make it work, you'll also need James Sentman's acgi dispatcher application (*http://www.sentman.com/acgi/index.php*; $15). A web page at Perceptive Automation (*http://www.perceptiveautomation.com/indigo/web_ui_info.html*) has all the information you need to get started. Figure 7-29 shows the results.

If you use XTension for Macintosh **[Hack #17]**, James Sentman (who also co-authored XTension for Mac OS X) has created an add-on package that works nicely. The basic idea is the same as what Indigo uses—it executes AppleScripts via Apache—but Sentman's X2Web package (*http://www.sentman.com/x2web/*) takes care of all the details for you and adds additional capabilities. You simply install the package in the root directory of your web server and then access the special pages, whose names end in *.x10*, to gain access to your system. Figure 7-30 shows a sample of what it looks like.

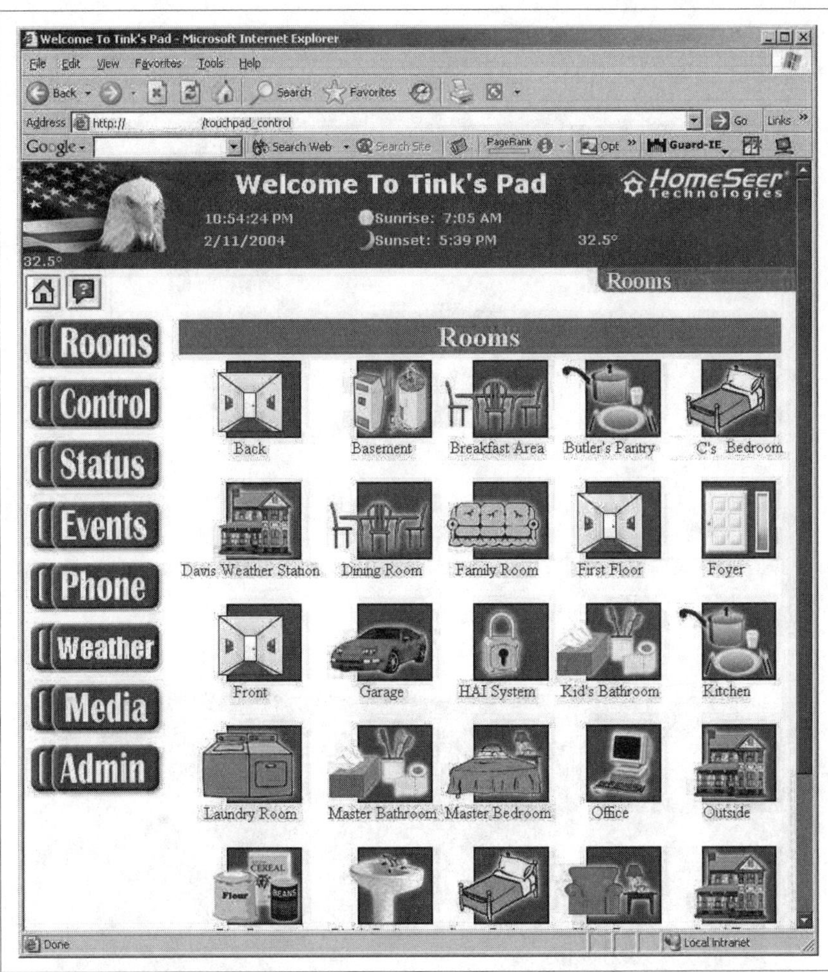


*Figure 7-28. HomeSeer's TouchPad interface*

If you'd rather not use a web browser, another approach is to use a program that enables you to view, and control, a computer across the network. Programs such as Windows Remote Desktop (*http://www.microsoft.com/windowsxp/using/mobility/getstarted/remoteintro.mspx*; included with Windows XP Pro), Apple Remote Desktop (*http://www.apple.com/remotedesktop/*; $300), and Timbuktu (*http://www.netopia.com/software/products/tb2/*; $180) offer this capability, and more, but for a price.

Regardless of the method you use, making your home automation system accessible remotely, even if just within the confines of your own home, can be very useful. In fact, having easy access to all your devices can actually

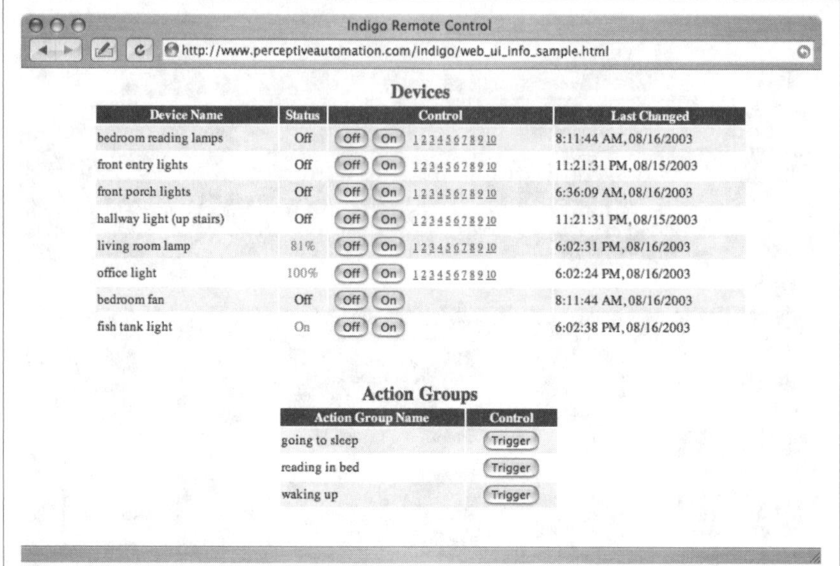


*Figure 7-29. A sample web page for controlling Indigo*

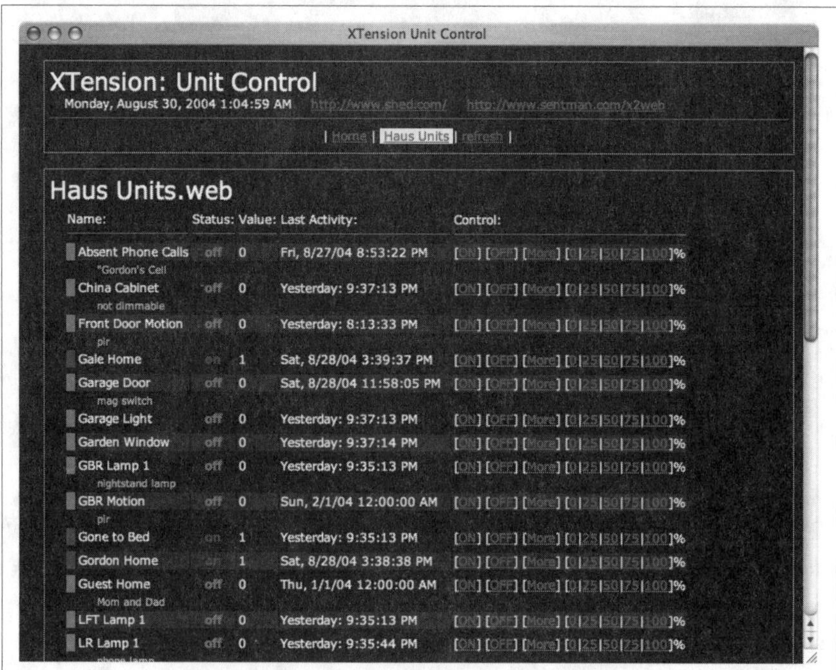


*Figure 7-30. A sample web page for controlling XTension*

decrease the amount of equipment you need, particularly if you have a laptop and WiFi network at home. For example, instead of having to reach for a Palm Pad when you want to turn on a light or execute a script, simply connect to your system and make the change with your mouse. Additionally, having remote access to the system's log files, which each system we've discussed can provide, can be a big help when troubleshooting a problem.

## HACK 100 Which Way Did She Go?

**Combine a few motion detectors and some simple logic to teach your home how to anticipate what you want to happen next.**

A hallmark of a good home automation system is the ability to predict what people are about to need and take care of it for them, such as a light coming on in a dark room right before a person enters and then turning it off shortly after she departs. An essential ingredient to this type of automation intelligence is detecting in which direction she is going. Is she:

- Walking up the sidewalk toward the house, or away from it?
- Climbing up the stairs, or down them?
- Walking down a hall toward the bedroom, or up the hall away from the bedroom?

This might sound like a complicated problem that would require video cameras and sophisticated visualization software to solve, but in reality, you can determine all of these things fairly accurately using two inexpensive X10 motion detectors [Hack #6].

The trick is to place one X10 motion detector at one endpoint (end of hall, sidewalk, stairs) and the other detector at the other end. Based on which motion detector triggers first, the home automation system can make a good guess about the direction of travel.

For example, consider a hallway leading to a bedroom, as shown in Figure 7-31.

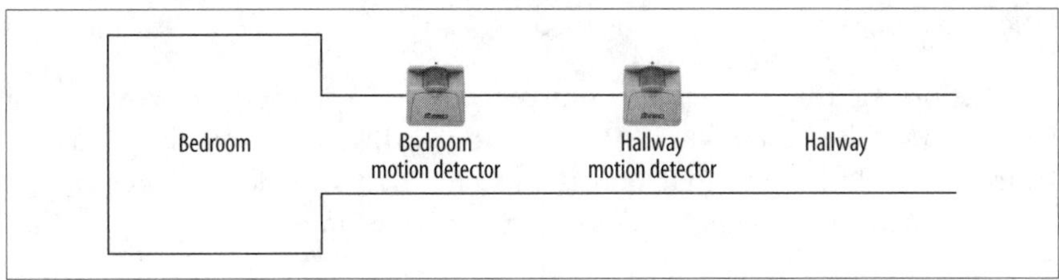


*Figure 7-31. Knowing the landscape*

If a person walks down the hall toward the bedroom, the Hallway Motion Detector will trigger first. At this point, we can predict that someone likely is going into the bedroom, and we can turn on the bedroom lamp in anticipation of her arrival. Likewise, when she leaves the bedroom, the Bedroom Motion Detector will trigger first and the system knows to turn off the lamp.

A person walking in the hallway will trigger both detectors eventually, so the key here is tracking which detector triggers first. Otherwise, the light would blink on and off as she approaches the bedroom, and off and on as she travels away from the bedroom. Let's look at how to do this using Indigo **[Hack #16]** for Mac OS X, beginning with turning on the light as a person approaches the bedroom.

After you've added the motion detectors to Indigo's device list **[Hack #18]**, create a trigger action that occurs when Indigo receives an X10 On command from Hallway Motion Detector, as shown in Figure 7-32.

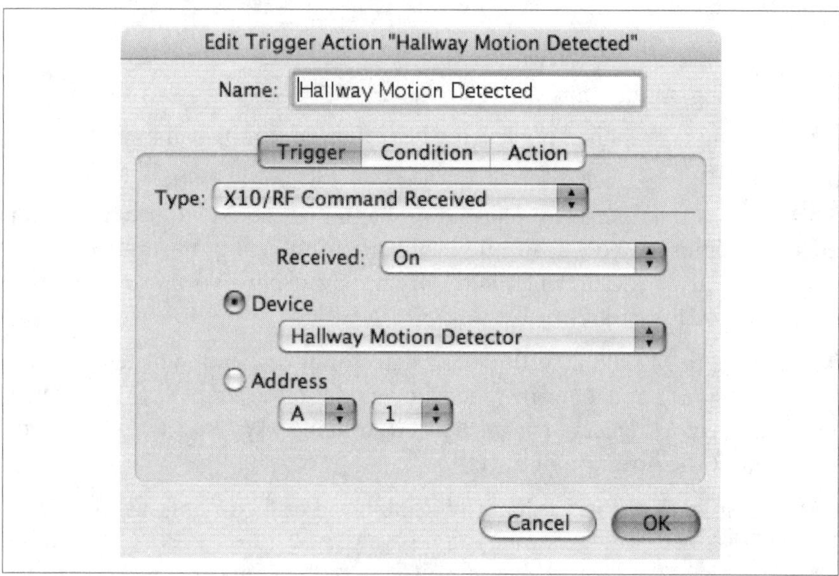


*Figure 7-32. Hallway traffic prediction trigger*

The action for this trigger is a simple two-line AppleScript, as shown in Figure 7-33. The second line is straightforward; it turns on the bedroom lamp. It's the first line that makes this smart. It disables another trigger action (which we'll discuss in a moment) for 10 seconds.

This gives the person who triggered the motion detector just enough time to scoot down the hallway and pass the other motion detector, Bedroom Motion Detector. That detector will still send a command to Indigo, but there will be no *enabled* trigger actions for the command, so no action is taken.

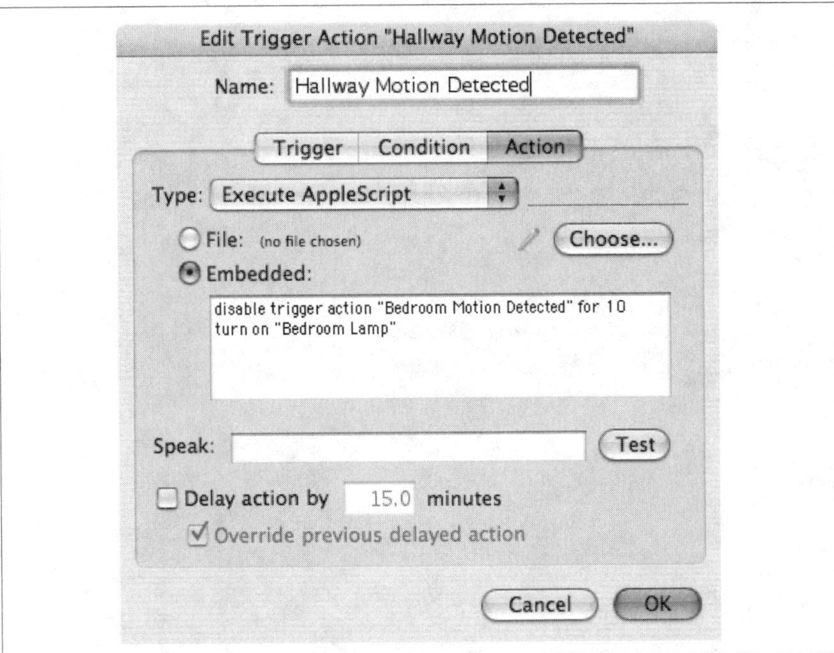


*Figure 7-33. Hallway traffic prediction action*

You could replace the AppleScript commands with a two-step action group that accomplishes the same thing.

Now, let's set up the trigger action for the `Bedroom Motion Detector`. As in the previous example, its trigger is the receipt of an X10 `On` command, as shown in Figure 7-34.

The action for this trigger turns off the bedroom lamp and then disables the action trigger for the `Hallway Motion Detector` we defined earlier, as shown in Figure 7-35.

Similar to the technique described earlier, this disables the action so that it does not occur when the person triggers the second motion detector. As long as she makes it past the detector within 10 seconds, the bedroom light will not turn back on.

When you implement this hack on your system, make sure you specify the correct names of the trigger you're disabling.

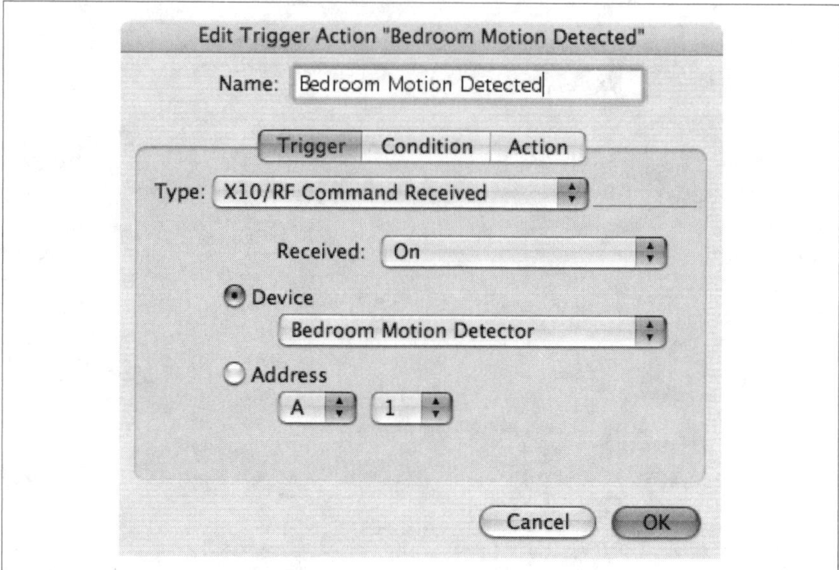


*Figure 7-34. Bedroom traffic prediction trigger*

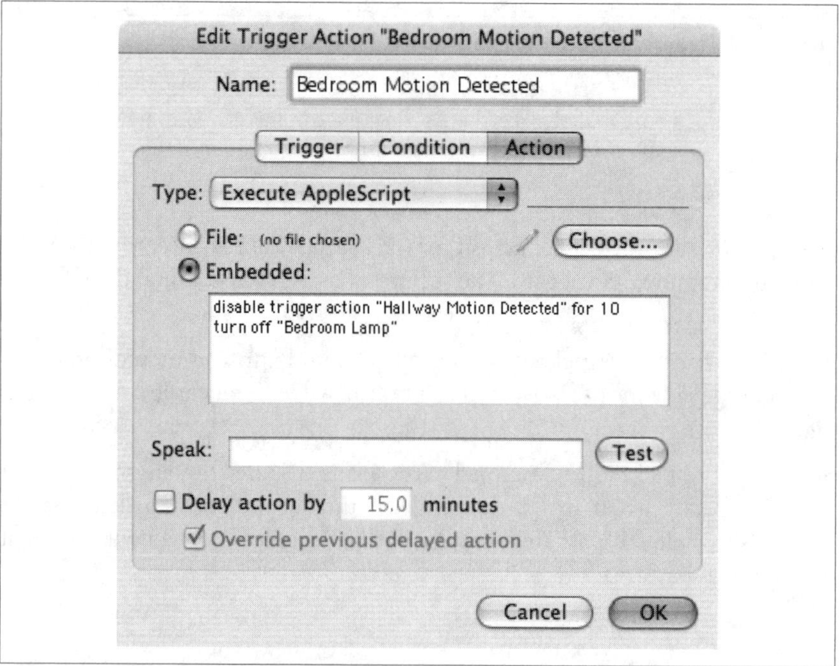


*Figure 7-35. Bedroom traffic prediction action*

You might need to tweak the 10-second delay so that it works better in your home. It needs to be long enough for the person to pass from the view of one detector and into the view of the other. Don't make the duration too long, though, or you'll miss a valid trigger if she reverses direction quickly, such as running back into the bedroom to grab a forgotten item.

You'll need to carefully consider the placement of the motion detectors **[Hack #85]**. For example, if the `Bedroom Motion Detector` was actually in the bedroom and not in the hallway, we would end up turning off the lights as the person walked around the bedroom. You can place the two motion detectors fairly close to one another; they just need to be far enough apart that the closest one will reliably trigger first in both directions.

Finally, as smart as this hack is, it is not perfect. In our example, if two people enter the bedroom, the lamp will turn off when the first person leaves. This is not necessarily a flaw in our scheme for predicting a person's direction of travel; it's caused by the fact that this system does not keep track of how many people have entered a room.

You could, indeed, extend the logic to increment a counter as people enter the room and decrement it as they leave—keeping the lights on until the counter reaches 0. However, you will likely run into problems with the counter getting out of sync if two or more people walk down the hallway together. Two people passing each other between the detectors will also confuse our simple logic. But despite these flaws, this hack demonstrates that it is possible to create a somewhat intelligent system with a little careful analysis.

*—Matt Bendiksen*

# Index

## A



We'd like to hear your suggestions for improving our indexes. Send email to *index@oreilly.com*.

## B



## C

## N



## O



## P

## R

## S

## T



## U

## V



## W



## X



## Y



## Z

## Colophon

Our look is the result of reader comments, our own experimentation, and feedback from distribution channels. Distinctive covers complement our distinctive approach to technical topics, breathing personality and life into potentially dry subjects.

The tool on the cover of *Smart Home Hacks* is a key ring of skeleton keys. A skeleton key is an old-fashioned key used in warded locks. Warded locks, first developed by the ancient Romans, consisted of concentric plates protruding outwards to block the rotation of the inner mechanism. When the correct skeleton key was inserted into the maze of wards, with slots to correspond to the protrusion in the locks, the key rotated freely in the lock, causing it to press against the latch or bolt and open what was locked. When warded locks and skeleton keys were in vogue, a well-designed skeleton key opened a wide variety of locks. Based on that fact, many believed a specially cut skeleton key existed that could open any lock, but it proved to be a myth.

Today, skeleton keys are a popular collectable, and when worn around the neck or carried as an amulet, skeleton keys are believed to open the doors of opportunity and success.

Marlowe Shaeffer was the production editor and proofreader for *Smart Home Hacks*, and Audrey Doyle was the copyeditor. Matt Hutchinson and Mary Anne Weeks Mayo provided quality control. Johnna Dinse wrote the index.

Hanna Dyer designed the cover of this book, based on a series design by Edie Freedman. The cover image is a photograph from photos.com. Clay Fernald produced the cover layout with QuarkXPress 4.1 using Adobe's Helvetica Neue and ITC Garamond fonts.

Melanie Wang designed the interior layout, based on a series design by David Futato. This book was converted by Julie Hawks to FrameMaker 5.5.6 with a format conversion tool created by Erik Ray, Jason McIntosh, Neil Walls, and Mike Sierra that uses Perl and XML technologies. The text font is Linotype Birka; the heading font is Adobe Helvetica Neue Condensed; and the code font is LucasFont's TheSans Mono Condensed. The illustrations that appear in the book were produced by Robert Romano and Jessamyn Read using Macromedia FreeHand 9 and Adobe Photoshop 6. This colophon was written by Reg Aubry.